Advances in
the Human Side of
Service Engineering

Advances in Human Factors and Ergonomics Series

Series Editors

Gavriel Salvendy

Professor Emeritus
School of Industrial Engineering
Purdue University

Chair Professor & Head
Dept. of Industrial Engineering
Tsinghua Univ., P.R. China

Waldemar Karwowski

Professor & Chair
Industrial Engineering and
Management Systems
University of Central Florida
Orlando, Florida, U.S.A.

3rd International Conference on Applied Human Factors and Ergonomics (AHFE) 2010

Advances in Applied Digital Human Modeling
Vincent G. Duffy

Advances in Cognitive Ergonomics
David Kaber and Guy Boy

Advances in Cross-Cultural Decision Making
Dylan D. Schmorrow and Denise M. Nicholson

Advances in Ergonomics Modeling and Usability Evaluation
Halimahtun Khalid, Alan Hedge, and Tareq Z. Ahram

Advances in Human Factors and Ergonomics in Healthcare
Vincent G. Duffy

Advances in Human Factors, Ergonomics, and Safety in Manufacturing and Service Industries
Waldemar Karwowski and Gavriel Salvendy

Advances in Occupational, Social, and Organizational Ergonomics
Peter Vink and Jussi Kantola

Advances in Understanding Human Performance: Neuroergonomics, Human Factors Design, and Special Populations
Tadeusz Marek, Waldemar Karwowski, and Valerie Rice

4th International Conference on Applied Human Factors and Ergonomics (AHFE) 2012

Advances in Affective and Pleasurable Design
Yong Gu Ji

Advances in Applied Human Modeling and Simulation
Vincent G. Duffy

Advances in Cognitive Engineering and Neuroergonomics
Kay M. Stanney and Kelly S. Hale

Advances in Design for Cross-Cultural Activities Part I
Dylan D. Schmorrow and Denise M. Nicholson

Advances in Design for Cross-Cultural Activities Part II
Denise M. Nicholson and Dylan D. Schmorrow

Advances in Ergonomics in Manufacturing
Stefan Trzcielinski and Waldemar Karwowski

Advances in Human Aspects of Aviation
Steven J. Landry

Advances in Human Aspects of Healthcare
Vincent G. Duffy

Advances in Human Aspects of Road and Rail Transportation
Neville A. Stanton

Advances in Human Factors and Ergonomics, 2012-14 Volume Set:
Proceedings of the 4th AHFE Conference 21-25 July 2012
Gavriel Salvendy and Waldemar Karwowski

Advances in the Human Side of Service Engineering
James C. Spohrer and Louis E. Freund

Advances in Physical Ergonomics and Safety
Tareq Z. Ahram and Waldemar Karwowski

Advances in Social and Organizational Factors
Peter Vink

Advances in Usability Evaluation Part I
Marcelo M. Soares and Francesco Rebelo

Advances in Usability Evaluation Part II
Francesco Rebelo and Marcelo M. Soares

Advances in the Human Side of Service Engineering

Edited by
James C. Spohrer
and
Louis E. Freund

CRC Press
Taylor & Francis Group
Boca Raton London New York

CRC Press is an imprint of the
Taylor & Francis Group, an **informa** business

CRC Press
Taylor & Francis Group
6000 Broken Sound Parkway NW, Suite 300
Boca Raton, FL 33487-2742

First issued in paperback 2019

© 2013 by Taylor & Francis Group, LLC
CRC Press is an imprint of Taylor & Francis Group, an Informa business

No claim to original U.S. Government works

ISBN-13: 978-1-4398-7026-6 (hbk)
ISBN-13: 978-0-367-38111-0 (pbk)

Visit the Taylor & Francis Web site at
http://www.taylorandfrancis.com

and the CRC Press Web site at
http://www.crcpress.com

Table of Contents

Section V. Service Design

Section VI.   Organizations and Change

Preface

This book is concerned with an emerging field we refer to as "The Human-Side of Service Engineering." If there is any one element to the engineering of service systems that is unique, it is the extent to which the suitability of the system for human use, human service, and excellent human experience has been and must always be considered. Contributors to this book explore the wide range of ways in which Human Factors Engineering, Ergonomics, Human Computer Interaction (HCI), Usability Testing, Attitude and Opinion Assessment, Servicescape Designs and Evaluations, Cognitive Engineering, Psychometrics, Training for Service Delivery, Co-Creation and Co-Production, Service Levels and Cost Effectiveness, Call Center Engineering, Customer Support Engineering, and many other areas relate to and impact the human-side of engineering service systems.

The book is organized into seven sections that focus on the following subject matters:

I: Health Service
II: Service Innovation
III: Societal Factors
IV: Service System Frameworks
V: Service Design
VI: Organizations & Change
VII: Value Co-Creation & Customers

Forty-six papers include a wide range of topics: self-service comes to healthcare, human voice and service system design for health information systems, visualizing service processes for nurses, social media and nursing service systems, engaging patients in service systems re-design, intelligent service systems for the elderly, mobile technology and redesigning medical practice service systems, mobile-banking service system design, higher education as a service system, service design in museums to evoke enthusiasm, computer-supported quality control in restaurants, value co-creation in public transit service systems, smart home service system design, telepresence in future service systems, predicting budgetary risks caused by human factors in B2B ecosystems, models of compliments and complaints, increasing productivity in product-support service value chains, approaches to value co-creation in business and society, disciplinary bridges from user experience to customer experience, comparing employee and customer perceptions of service, leading organizational change by playing games, augmenting human performance as service design, importing product innovation to service innovation, from user interface design to service-oriented-architectures (SOA) design, lean techniques for service operations, lean techniques for service system safety, value orchestration platforms, R&D as a service to accelerate value co-creation, multi-agent approaches to service system development, organizational change and service system engineering, governing service systems to promote human capital development, communications models in service systems as viable systems, internationalization: when a local service system goes global, smart governance in service systems, distributed cognition in service systems, distributed cognition in virtual environments as service system design, creativity and learning in service innovation, interactive design methods for service systems, service systems as B2B platforms, human-centric approaches to value proposition design, computational models of society as service systems, service and cooperative behavior, human resources in service systems, social networking and creativity in outsourced service systems.

Each of the chapters of this book were either reviewed or contributed by the members of Editorial Board. For this, our sincerest thanks and appreciation goes to the Board members listed below:

This book will be of special value to a large variety of professionals, researchers and students interested in the human-side of service engineering from multiple perspectives ranging from industry sectors (healthcare) to tools and methods (service innovation, service design, organization & change) to broader issues (societal factors, service system frameworks, value co-creation). We hope this book will excite curiosity in many discipline areas and lay a foundation to attract others to contribute to this emerging area of research and practice. We want to thank the contributors, and encourage the readers to get involved - there is much work to do, and great opportunities to make lasting contributions!

April 2012

Louis E. Freund
San José State University
San Jose, CA USA

James C. Spohrer
IBM Almaden Research Center
San Jose, CA USA

Editors

Section I

Health Services

Service Process Visualization in Nursing-care Service Using State Transition Model

Hiroyasu Miwa, Tomohiro Fukuhara, Takuichi Nishimura

Center for Service Research,
National Institute of Advanced Industrial Science and Technology (AIST)
Tokyo, JAPAN
h.miwa@aist.go.jp

ABSTRACT

Employees at nursing-care facilities co-operate with each other und provide nursing-care service according to care plans designed to meet residents' physical condition and needs. Traditionally, improvement of service productivity has been based on experience and inspiration of managers and employees. Engineering methods to achieve improvement is a relatively new and necessary approach. In this study, we developed task classification for nursing-care service and a description of nursing-care service process using state transition model in order to visualize the nursing-care service process and to support the improvement of service productivity. We then visualized the nursing-care service process of 20 employees at three nursing-care facilities to evaluate the proposed methods.

Keywords: service process, nursing-care service, task classification, state transition, visualization

1   INTRODUCTION

Nursing-care services are categorized into facility care services and home care services in Japan. There are about 15,000 facilities for facility care services such as health institute on long-term care for the aged, welfare institute on long-term care for the aged and specialized care house in 2009. These types of facilities are increasing every year (MHLW, 2011). Facilities' employees such as care workers, nurses and care managers provide nursing-care service according to care plans designed to meet residents' physical condition and needs.

On the other hand, employees are required to improve service productivity while maintaining service quality and workload to increase profitability. Generally, measurement, visually understandable tool and evaluation of service are effective for improvement of service productivity However, traditionally these measures have not been based on engineering methods but on the inspiration of managers or employees involved in nursing-care services.

The authors defined service process as flow of humans, objects, information, tools, time and money in the service field, and defined nursing-care service process as the service process at nursing-care facilities. Time & motion studies, which measure an employees' behavior with self-observation of observation by a third person, are commonly used to measure nursing-care service processes. It is possible to compare facilities or employees using a task classification code.

Task classification has been studied in various fields. International Classification of Functioning (ICF) is a famous behavior classification method (WHO, 2008). However, it is insufficient to describe employees' behaviors in nursing-care service because it mainly describes human behavior and physical condition. For task classification of care workers, a care code was proposed by the Ministry of Health, Labour and Welfare (MHLW), Japan (MHLW, 2007). That code described typical assistance provided by care workers using three-layer structure. However, there are many functions provided during nursing-care service in a nursing-care facility. The care code cannot cover the multiple tasks or duties that are performed. Numasaki and colleagues proposed a task classification code for nurses in hospital wards (Numasaki, Harauchi and Ohno, et. al., 2007). Their code also classified nursing acitivities with a three-layer structure, and has sufficient capacity to describe nursing service. Zheng proposed STAMP (Suggested Time And Motion Procedures) to improve the methodology and results by reporting consistency of time and motion research (Zheng, Guo and Hanauer, 2011).

In the field of mass production, active research has been conducted to design optimal product-service systems. Flow chart, Gantt chart and Business Process Model Notation (BPMN) have been used to describe employees' behaviors and work in chronological order (White, 2009) (Aalst, 1999). Shimomura proposed service engineering as a modeling technique of service (Shimomura, 2005). Hara applied these methods for mass production in order to design and visualize the service process (Hara, 2009). These methods work well in highly automated service or similar production service systems.

Nursing-care service has different characteristics as below to proactive service.

(1) Many irregular tasks: employees frequently receive interruptions from both customers and other employees.

(2) Wide variety of customers' needs and service methods: nursing-care service process varies according to the customers' condition, and there are many ways to provide the same service.

(3) Concurrency of tasks: an employee can work multiple tasks in parallel.

Previous methods used to describe employees' behaviors in chronological order are difficult to apply nursing-care service. We, therefore, developed a new visualization method of nursing-care service process, which would be effective to improvement of service productivity. In this paper, we describe a new method to visualize the nursing-care service process.

2   TASK CLASSIFICATION FOR NURSING-CARE SERVICE

For service process measurement, it is necessary to properly classify and code the employees' behaviors and work. Domestic laws, environment and culture in human centered service are important consideration. Information within the service field is also considered very important. In order to clarify the employees' behaviors and work in nursing-care service, we visually observed and recorded work of care workers, nurses, occupational therapists, a care manager and a registered dietician, and interviewed employees at three nursing-care facilities; Wakouen (health institute on long-term care for the aged located in Nanao, Japan), Super court Hirano (specialized care house located in Osaka, Japan) and Super court Minami-hanayashiki (elderly housing located in Kawanishi, Japan).

As a result, the employees' work included many operations such as position change, ambulation assistance and excretion assistance. We then extracted 108 fundamental operations by our discussions.   In addition, we classified these operations into a three-layer structure consisting of "service type", "situation" and "operation" using references to previous studies (MHLW, 2007) (Numasaki, Harauchi and Ohno, et. al., 2007). Figure 1 shows the structure of task classification for nursing-care service.

The service type layer, which is the upper layer, included seven factors; "care", "nursing", "rehabilitation", "care management", "indirect work", "off duty" and "measurement error". These factors were determined by the job category of each operation. Care, nursing, rehabilitation and care management were tasks related to direct service for residents. Indirect work included task related business operations such as meetings and employee's transfers.

Then, we extracted the situations associated with each operation, and found 12 situations in care service, five situations in nursing service, three situations in rehabilitation service, one situation in care management, three situations in indirect work, one situation in off-duty and one situation in measurement error. We assigned these to the situation layer. After that, we assigned all operations obtained by behavior observation and interviews to the operation layer, which is the lower layer.

In addition, we found that several behaviors were common to all service types and situations. We added four situations, which were "information sharing", "data recording", "confirmation" and "interaction with residents' family", to each service type, and two operations, which were "preparation" and "hand washing", to each situation. Also, we added "other service" to each situation in order to cover unexpected behaviors. Finally, we gave all operations unique codes to develop task classification for nursing-care service.

6

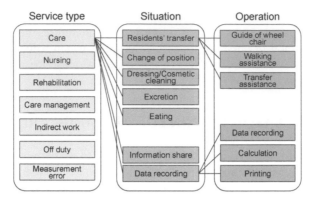

Figure 1 Three-layer structure of task classification for nursing-care service

3   DESCRIPTION OF NURSING-CARE SERVICE PROCESS WITH STATE TRANSITION MODEL

Nursing-care service process has three characteristics; "many irregular tasks", "wide variety of customers' needs and service methods" and "concurrency of tasks". The authors selected the state transition model to describe these characteristics because employee and customer condition varies during the nursing-care service process. We then regarded employees' behaviors or tasks as a state, and observed employees at nursing-care facilities to record employees' work style. As a result, six patterns of state transition were found in the work behaviors shown in Figure 2. Figure 2 shows the state transition and transition order. In most cases, employees' state transition from one state to another is represented in Figure 2 (a). We named this "Serial process". We also found "Parallel process" where the employees' transition from one state to two or more states (see Figure 2 (b)-(e)) and "Interruption process" where the employees' state is suspended and resumed to transition to another state (see Figure 2 (f)). These six state transition patterns combined in the actual nursing-care service process.

Then, we described the nursing-care service process with state transition. To describe the employees' state, we regarded each task as the employees' state, and added seven properties to each task; "id", "start time", "end time", "task", "place", "target resident" and "detailed information". The task property was selected from the task classification for nursing-care service. The nursing-care service process was described as transition of employees' state with seven properties. Even when the transition condition was independent from chronological order such as during an interruption, the proposed method expressed the nursing-care service process. "Many irregular tasks", one of the characteristics of nursing-care service process, could be described with the proposed method and transition from one state to multiple states could be incorporated. "Concurrency of tasks" was also described. Moreover, when employees' state transitioned from one state to other, we could choose any intermediate state. This transition route included how to provide the service, and a variety of routes were acceptable. "Wide variety of customers' needs and service methods" could also be described. Thus, the proposed method could describe three characteristics of the nursing-care service process.

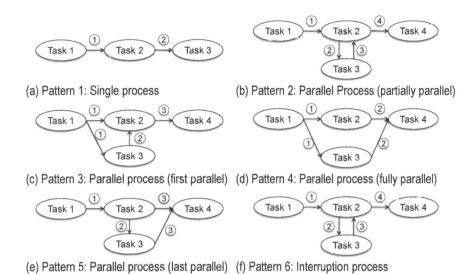

(a) Pattern 1: Single process   (b) Pattern 2: Parallel Process (partially parallel)

(c) Pattern 3: Parallel process (first parallel) (d) Pattern 4: Parallel process (fully parallel)

(e) Pattern 5: Parallel process (last parallel) (f) Pattern 6: Interruption process

Figure 2 State transition pattern in the nursing care process

4   VISUALIZATION OF NURSING-CARE SERVICE PROCESS

4.1 Measurement of nursing-care service

In order to evaluate the proposed method, we displayed the nursing-care service process of 11 care workers, four nurses, three occupational therapists, a care manager and a registered dietician at Wakouen, Super court Hirano and Super court Minami-hanayashiki. The authors visually observed each employee for half or one day and made observation notes in order to measure their behaviors. We were careful not to disturb the employees or the work flow. Before the measurements, we explained the purpose and procedure of the experiment to all employees and obtained orally consent. After the measurement, we constructed their nursing-care service process with the proposed method.

4.2 Visualization of nursing-care service process

Easy-to-understand visualization of service process is important to improve service productivity. We visualized the nursing-care service process with three modes; "time-line mode", "statistics mode" and "state transition probability map".

The time-line mode expressed the transition of employees' states in chronological order. It arranged time on a horizontal axis and a property arbitrarily selected from the seven properties of the employees' state on a vertical axis, and was marked according to the employees' state. When there were multiple marks at one time, the employee was concurrently conducting multiple tasks. Figure 3 shows the state transition of a task property of a care worker at Wakouen with the time-line mode. We could change the resolution of the task property as shown in Figure 3 (a)-(c) because the task property is described using a three-layer structure. If the vertical axis changed to other properties such as place and target resident, we could understand the employee's movement and service history.

The statistics mode expressed results of statistical processing similar to the time study. This mode allows for the number of states to be counted and calculates the sum, average, and rate of time spent for each property for any period with statistical methods. The results can be displayed graphically. Figure 4 shows the nursing-care service process similar to Figure 3 with the statistical mode. The care worker spent more than half of her work time in sharing information with other employees, communication with residents and her own transfers.

The state transition probability map expressed how often state transition occurred in the nursing-care service process. It extracted the anteroposterior relation of each state and calculated the probability. Figure 5 shows the nursing-care service process similar to Figure 3 with the state transition probability map. The care worker shared her information after most tasks. We considered that the state transition probability map would standardize the state transition or standardize the work task.

4.3 Comparison of nursing-care service process

Wakouen changed their nursing-care service process. Before the change, all data was input into an electric health record system by employees. After the change, the data entry work was farmed out to experts in a group organization. We compared the nursing-care service process doing the same work before and after the change using statistical mode shown in Figure 6. The recording task was reduced by 37.6 points. And, employee's transfer, excretion and communication with residents increased 9.93 points, 7.90 points and 3.69 points respectively. In addition, the rate of direct service time increased 22.8 points, from 27.6 % to 50.4 %. Thus, these results could visually and quantitatively show the improvement that occurred because of the change. We considered that the proposed method was effective for evaluation of improvement in service.

5   DISCUSSIONS

We interviewed directors and employees at Wakouen, Super court Hirano and Super court Minami-hanayashiki with predetermined questions to evaluate visualization of nursing-care service process. The interviews were recorded by IC

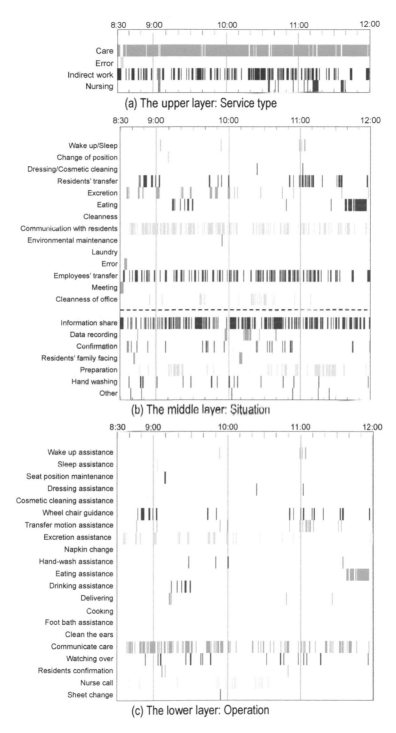

(a) The upper layer: Service type

(b) The middle layer: Situation

(c) The lower layer: Operation

Figure3 Nursing-care service process of a care worker at Wakouen by time-line mode

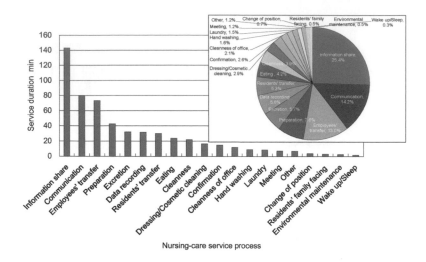

Figure 4 Nursing-care service process of a care worker at Wakouen by statistical mode

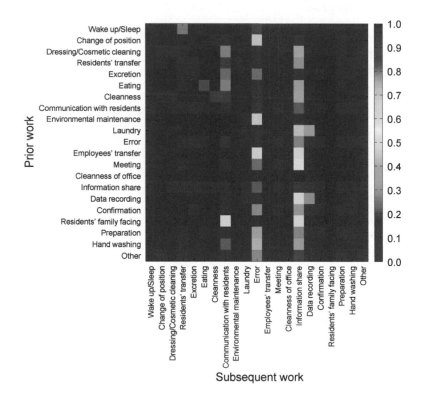

Figure 5 Nursing-care service process of a care worker at Wakouen by state transition probability map

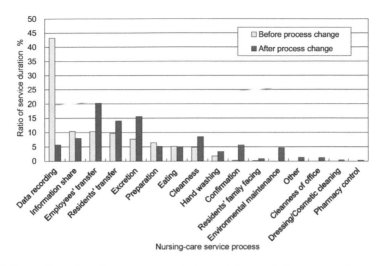

Figure 6 Comparison of nursing-care process between before and after process change

recorder. As a result, most employees thought that it was effective for review and design of their nursing-care service and employee training. The director at Wakouen also reported that it was effective to evaluate improvement of service because it could visually and quantitatively show the increase of customer contact time caused by a change of nursing-care service process. In this experiment, we visualized nursing-care service with three methods. The time-line mode and statistical mode were well suited for the purpose of improving business and employee training. These modes were considered especially useful and effective. Because the improving business and employee training generally achieve to improve service productivity, we considered that the proposed method was effective for improving service productivity.

However, a few employees thought that the method would not help to improve business and train employees. We discovered that some employees could not understand the state transition probability map. We hypothesized that alternative visualizations and evaluation indicators should be tailored to different types of job and purposes. In this study, we could not fulfill everyone's needs or the gaps between those needs. Therefore, the proposed visualization methods were insufficient to address every concern for every job. We recommend that employees' needs should be assessed and understood, in order to improve their ability to give quality service. Alternative methods of visualization should be developed to help various work groups.

6   CONCLUSIONS AND FUTURE WORK

In this study, we developed the task classification for nursing-care service and a description of nursing-care service process using the state transition model in order to visualize the nursing-care service process and to support improvement of service productivity. The task classification for nursing-care service included three-layer structure. And, the nursing-care service process was described as transition of employees' state with seven properties. We then visualized the nursing-care service process of 20 employees at three nursing-care facilities using three modes. We confirmed that the proposed methods could visualize the nursing-care service process, and they were effective for improving service productivity. However, visualization methods were insufficient to cover all employees' needs. Employees' needs must be assessed and then additional visualization methods can be developed.

In the future, the authors hope to implement the proposed system to an actual service filed. We developed methods to evaluate nursing-care service process in this study. Automatic measurement employees' behavior, more visualization methods and service process simulation are still necessary for the loop. We hope to develop the linkage to these technologies.

ACKNOWLEDGMENT

This study was supported by the Project of Service Engineering Research in 2011 from Japanese Ministry of Economy, Trade and Industry (METI). And, the authors would like to express their thanks to Tousenkai Healthcare System, City Estate Co., Wakouen, Super court Hirano and Super court Minami-Hanayasiki.

REFERENCES

Aalst, W. M. P. van der. 1999. Formalization and verifica-tion of event-driven process chains. Information and Software Technology, Vol. 41, Issue 10: 639-650.
Hara, T., T. Arai and Y. Shimomura. 2009. A CAD system for service innovation: integrated representation of function, service activity, and product behaviour. Journal of Engineering Design, Special issue on PSS, Vol. 20, No. 4: 367-388.
Ministry of Health, Labour and Welfare (MHLW). 2007. 3rd investigative commission of certification of long-term care need (in Japanese).
Ministry of Health, Labour and Welfare (MHLW). 2011. Survey results of nursing-care service facilities and companies in 2009 (in Japanese).
Numasaki, H., H. Harauchi and Y. Ohno, et. al. 2007. New classification of medical staff clinical services for optimal reconstruction of job workflow in a surgical ward: Application of spectrum analysis and sequence relational analysis. Computational Statistics & Data Analysis, 51: 5708-5717
Shimomura, Y. and T. Tomiyama. 2005. Service Modeling for Service Engineering. IFIP International Federation for Information Processing, 167: pp. 31-38.
White, S. A. 2009. Introduction to BPMN. http://www.bpmn.org/Documents/Introduction_to_BPMN.pdf.
World Health Organization (WHO). 2008. International Classification of Functioning, Disability and Health.
Zheng, K., M. H. Guo and D. A. Hanauer. 2011. Using the time and motion method to study clinical work processes and workflow: methodological inconsistencies and a call for standardized research. Journal of American Medical Informatics Association, 18:704-710

CHAPTER 2

Self-service for Personal Health Monitoring and Decisions

Zbigniew J. Pasek, Gheorhe M. Bacioiu

University of Windsor
Windsor, ON, Canada
zjpasek@uwindsor.ca

ABSTRACT

The exploding demand for health services and escalating costs are a growing concern for the health care systems, which may not be able to handle growing volume and complexity of cases. Considering that obesity is responsible for over 60% of all leading death causes in developed societies and world-wide spread of that preventable condition, it is critical to understand the mechanisms that may lead to its control. The obesity model presented here was developed using system dynamics and is based on three fractional models that were used to create energy balance equation. The results were verified on weight loss clinic data. Tools for quantified self-tracking may provide an automated source for data collection needed to refine the model and estimated individual characteristics needs as model parameters.

Keywords: obesity, modeling, self-service

1   GLOBAL HEALTH CARE TRENDS

In the period from 2000 to 2008, the Canadian population grew less than 5% (from under 31 million to over 33 million). At the same time, the total personal expenditures on medical care and health services increased continuously in the last 8 years by more than 70% (Statistics Canada, 2010). Considering the projection of population growth by 11.7% in 2026, it can be expected that demand for the healthcare services will substantially increase relative to the Canadian population growth (Lassard 2006). If the health care costs will continue to grow at that rate, the healthcare spending will soon overwhelm provincial and federal budgets. This predicament is common to many countries, independent of their economic development level. Proper working solutions, however, despite vigorous national discussions, have not been found yet. To prevent this from happening, many governments are implementing multiple measures of cost control, which, despite being called otherwise, are various forms of health care rationing, either by access to the health care system (in the US approx. 17% of population has no health insurance) or selective access to procedures (like in UK or Canada, which set budget limits for annual health expenditures, that in turn creates waiting lines for elective procedures). The root causes of continuous cost increases in health care are many: demographic trends, new technologies, insurance bureaucracy overhead, etc.

In October 2011 the size of world population has reached 7 billion. While over the past century the longevity and quality of life of the global population have increased (although to a varying degree), increasing fraction of that population is adversely affected by a variety of industrial society ailments. Out of top 10 leading death causes in Table 1, lifestyle-related account for almost 79%. While heart disease can be fully attributed to lifestyle, #2, 3, 6, 9 and 10 are perhaps impacted in at least 50 percent.

Table 1 Leading death causes in the industrialized societies

#	Cause	Prevalence [%]
1	Heart disease	29.0
2	Malignant neoplasms (cancer)	22.9
3	Cerebrovascular disease (stroke)	6.8
4	Chronic lower respiratory disease	5.1
5	Accidents	4.2
6	Diabetes	3.0
7	Pneumonia and influenza	2.6
8	Alzheimer's disease	2.2
9	Kidney disease	1.6
10	Septicema	1.3

The lifestyle expresses itself most obviously in the body weight: out of 7 billion people, approximately 1.5 billion are overweight, and over 400 million are suffering from obesity (IASO 2012). While the US population has taken the lead in the obesity epidemic (in some states over the past 30 years, obese fraction of the population has tripled), this is a global trend: in the European Union population of Germany is leading that trend, but other countries like China also see it as a side effect of urbanization; even in some African countries obesity is spreading as well.

Obesity is commonly recognized as a condition leading to many other medical conditions, such as hypertension, diabetes, and joint failures, to name a few. It is responsible (directly or indirectly) for over 60% of all leading death causes in developed societies, not to mention significant contribution to the health care costs (Bhattacharya 2011).

While there are many reasons for growing demand for health services, it is exacerbated by the corresponding escalating costs, reasonable solutions are unfortunately not yet apparent. One potential source of relief for health care systems is to shift some (if not majority – but in the long term) of the responsibilities to patients themselves. To do so effectively, however, new health care delivery paradigms need to be developed on one hand, and on the other - personal health care education has to be ramped up along with availability of personal health monitoring and management tools.

2 UNDERSTANDING OBESITY

While there exists a multitude of research related to obesity, the efforts are generally not coordinated and focused, and they typically lead to only fractional models addressing some narrow aspect of the phenomena. Even though the global spread of obesity is being acknowledged, no successful efforts to fight it have been developed. As (Lustig 2006) pointedly described it: "The Centers for Disease Control and Prevention says obesity results from an energy imbalance, by eating too many calories and not getting enough physical activity. Big Food says it's a lack of activity, the TV industry says it's the diet. The Atkins people say it's too much carbohydrate, the Ornish people say it's too much fat. The juice people say it's the soda, the soda people say it's the juice. The schools say it's the parents, the parents say it's the schools. How are we going to fi x this, when no one will accept responsibility?"

One of the key reasons for that is lack of full understanding of that phenomena. Most of the contemporary obesity mechanism models are based either on glucostatic theory (Mayer 1953) or lipostatic theory (Kennedy 1953). Based on either of those theories the origin of obesity is perceived as a disorder in one of the two feedback systems (e.g., signaling equilibrium of either fat deposition or glucose blood levels). In practice, however, neither of those two approachs could satisfactorily explain the obesity phenomenon.

Recent multidisciplinary theory of the "selfish brain" (Peters 2004, Peters 2011) combines these two theories and considers blood glucose and fat feedback control systems as a joint, complex system. While this theory is still undergoing extensive experimental validation, it offers some important insight into complexity of human body operation. The essential elements of this theory is a control system model with multiple feedback loops and the centrality of the brain as a main body decision center.

3 OBESITY MODEL USING SYSTEM DYNAMICS

The body weight model presented in this paper is based on the principle of energy balance (see Fig. 1). The two major elements associated with the energy intake are Food and Drink consumption. On the other side, the main two elements associated with energy expenditure are Physical Activity and Basal Metabolic rate.

However, an additional element is introduced in the model based on relatively recent research in the field. That is the energy consumed to transform food in either fat or lean tissue. While not as impactful as the other two elements, it can be up to 20% of the overall contribution.

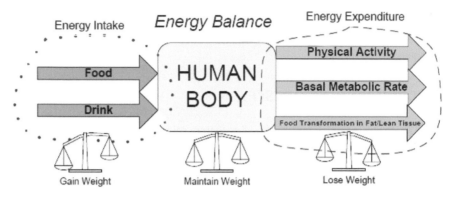

Figure 1 Human body energy balance principle

Using available equations (derived through published experimental research) which relate energy intakes and expenditures, key model elements were formulated. These included:

- According to (Christiansen, Garby, & Sørensen, 2004), an obesity condition typically develops over time and is reflected in an energy imbalance, leading to a body mass increase too small to be measured or controlled. A mathematical model has been defined for describing the relation between the change in body mass, values of the energy intake (EI) and the energy expenditure (EE), controlled by the physical activity factor (PAF).
- (Butte, Christiansen, & Sørensen, 2007) developed an energy balance model based on empirical data and human energetics to predict the total energy cost of weight gain and obligatory increase in energy intake and/or decrease in physical activity level associated with weight gain in children and adolescents.
- (Lazzer, et al., 2009) created a simple energy balance model based on Body Weight, Age and Gender which is as accurate as more complex models based on body composition.

These key building blocks were then connected by feedback loops and complemented by variety of factors considered in conjunction to the obesity condition as a systemic phenomenon. Overall arrangement of all the components is shown in Fig. 2. It is worth noting that this layout is aimed at monitoring the body weight and generating targeted adjustments (to gain or to lose it) given certain amount of control over energy intake (food) and expenditure (physical activity).

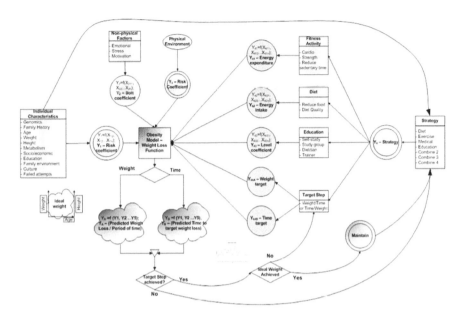

Figure 2 Predictive analytical structural model for self-healthcare

The first layer is represented by the yellow boxes that are actually the obesity subsystem high level components, which include, for example individual characteristics, fitness activity, diet, physical environment, etc.. Each high level component contains the model elements for which transfer functions need to be identified. The transfers are symbolically represented by the purple circles.

The integration of all these functions is carried out by the main "brain" of the model, the blue colored box. From this point on the user follows the decision structure by building strategies based on his/her own preferences (Bacioiu 2011).

Since the key equations have a form of differential equations, they could be integrated into a comprehensive model using System Dynamics approach (Sterman 2000). System Dynamics is a modeling approach to understanding the behavior of complex systems over time. It deals with internal feedback loops and time delays that affect the behavior of the entire system. What makes using system dynamics different from other approaches to studying complex systems is the use of feedback loops and stocks and flows (for quantitative analysis). These elements help describe how even seemingly simple systems may display complex nonlinearities. The final system dynamics form of the model is shown in Fig. 3.

To understand the usefulness of the model, it was tested and validated using actual field data provided by a clinic specializing in weight loss. The patients went through a very particular type of diet based on Human Chorionic Gonadotropin, which allows patients to not experience hunger, which in turn allows them to stay in a restricted diet of only 500 calories/day. That is complemented by a particular choice of foods, administered under the supervision of the trained personnel. The patients keep records of calories consumed and types of food in their daily diet.

18

The graph in Fig. 4 illustrates how close (no significant statistical difference and R^2 around 90%) the predictive model approximates the actual field data. To reinforce the conclusions related to the accuracy of the dynamic model, non-parametric comparisons between the original data and simulated data have been performed. Particularly, the results of the Mann-Whitney non-parametric procedures are incorporated in the graphs they represent. All the non-parametric tests had p-values well over 0.1. That shows confidence in the fact that there is no statistical difference between the model and the field data.

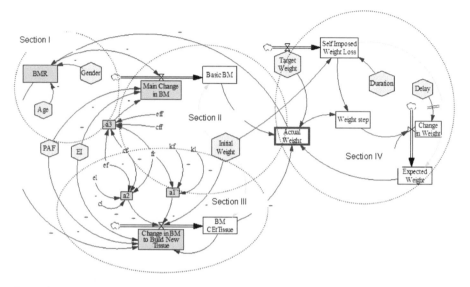

Figure 3 Weight loss/gain model using system dynamics

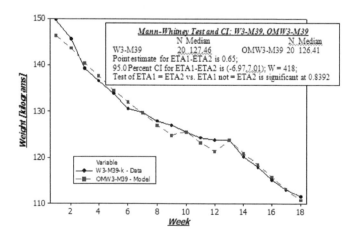

Figure 4 Weight loss clinic patient data compared with simulation result

Based on the data, sensitivity analysis was performed, as well as a variety of scenario analyses. Assuming that certain conditions are met, the model developed through this work can predict weight changes using energy balance information (energy intake and expenditure), as well as an individual's characteristics with an accuracy greater than 90%. The user can personalize the strategy for weight loss based on individual preferences and specific constraints. While the model can be used at a very personal level, it has the potential of becoming a cornerstone for personal health status management.

4   EMERGING HEALTH CARE DELIVERY PARADIGMS

As mentioned before in section 1, only of the potential solutions (albeit a long-term one) for the health care cost/capacity conundrum is creation of the new care delivery paradigms. It is happens, that with progress of technology (in particular IT) and spread of collaborative technologies such as social networks, new health care models started to emerge in recent years. These include (Agus 2011; Swan 2009; Topol 2012):

- Health social networks
- Consumer personalized medicine
- Quantified self-tracking.

In particular the last form is of interest in furthering the development of more accurate health-related models. Quantified self-tracking involves easy-to use online tools for condition, symptom, treatment and other biological information. The information can then be displayed visually or further processed or aggregated with group/population. Self-tracking gains on popularity (Anderson 2008; Wolf 2010; Singer 2011), and even though it currently focuses on collection on relatively simple personal data, it may offer a lot of potential when through aggregation is can enable tracking of group/population health (Dubberly 2010). Particular promise can perhaps be achieved when the data collected can be used to estimate individual parameters need to custom-fit or personalize models like described in section 3. Currently many of the model parameter values were only rough estimates. With repeated data collection and also possibly experimentation, the accuracy of these estimates can be improved.

5   CONCLUSIONS

Increasing pressure on the health care systems in terms of cost management and capacity requires search for new care delivery methods. One of the potential avenues is to empower patients themselves by allowing them to collect and track their own health data and monitor their health status. Considering that obesity is one of the wide spread health conditions, control of which can bring down the overall care delivery costs and also improve individual's health, having thorough understanding of this phenomenon is essential. Developing causal obesity models and using them for simulation to conduct exponentially many experiments, will possibly expand ability to predict desirable outcomes of controlling it.

REFERENCES

Agus, D. 2012. A Doctor in Your Pocket; What does the future of medicine hold? Tiny health monitors, tailored therapies – and the end of illness. *Wall Street Journal* Jan 14

Anderson, C. 2008. The End of Theory: The Data Deluge Makes the Scientific Method Obsolete. WIRED Magazine

Bhattacharya, J. and Sood, N. 2011, Who Pays for Obesity? *Journal of Economic Perspectives* 25 – 1: 139–158

Bacioiu, G. M. 2012. System Design of an Analytical Model for Health Self-Care Based on System Dynamics: Implementation and Case Study in Obesity. *Doctoral dissertation.* University of Windsor, Windsor, Canada

Butte, N. F., Christiansen, E., and Sørensen, T. I. 2007. Energy Imbalance Underlying the Development of Childhood Obesity. *Obesity Journal* , 3056-3066

Christiansen, E., Garby, L., and Sørensen, T. I. 2004. Quantitative Analysis of the Energy Requirements for Development of Obesity. *Journal of Theoretical Biology* , 99-106

Dubberly, H., Mehta, R., Evenson, S., and Pangaro, P. 2010. Reframing Health to Embrace Design of our Own Well-being, *Interactions*, May-June, 56-63

Kennedy, G. 1953. The role of depot fat in the hypothalamic control of food intake in the rat. *Proc R Soc Lond., B, Biol Sci.* 140 -901: 578–596.

Lassard, R. 2006. *Improving the Health of Canadians: Promoting Healthy Weights*. Canadian Institute for Health Information, Ottawa

Lazzer, S., Bedogni, G., Lafortuna, C. L., Marazzi, N., Busti, C., Galli, R., et al. 2009. Relationship Between Basal Metabolic Rate, Gender, Age, and Body Composition in 8,780 White Obese Subjects. *Obesity Journal* 71-78

Lustig, R. 2006. The Skinny on Childhood Obesity: How Our Western Environment Starves Kids Brains. Pediatric Annals 35-12: 899-907.

Mayer, J. 1953. Glucostatic mechanism of regulation of food intake. *New England Journal of Medicine* 249-1: 13–16.

Peters, A. 2011. Das egoistische Gehirn, Ullstein Buchverlage GmbH, Berlin

Peters, A; Schweiger, U; Pellerin, L; Hubold, C; Oltmanns, KM; Conrad, M; Schultes, B; Born, J et. Al. 2004). The selfish brain: competition for energy resources. *Neuroscience & Biobehavioral Reviews* 28-2: 143–180.

Singer, E. 2011. The Measured Life. *MIT Technology Review* 114-4: 38–45.

Sterman, J. D. 2000. *Business Dynamics - Systems Thinking and Modeling for a Complex World*. Mc-Graw Hill Higher Education

Swan, M., 2009. Emerging Patient-Driven Health Care Models: An examination of Health Social Networks, Consumer Personalized Medicine and Quantified Self-Tracking, *Int. J. Environ. Res. Public Health* 6: 492–525.

Topol, E. 2012. *The Creative Destruction of Medicine: How the Digital Revolution Will Create Better Health Care*. Basic Books

Wolf, G. 2010. The Data Driven Life. *New York Times*, Apr. 26

World Map of Obesity. 2012. International Association for the Study of Obesity. http://www.iaso.org/publications/world-map-obesity/ Accessed March 10, 2012

A Study of the Productivity Enhancement in Medical Practice by the Introduction of a Handheld Tablet Computer

Yamada, KC[1,2], Ishikawa, S[1], Sakamoto, Y[1], Motomura,Y[2], Nishimura, T[2], Sugioka, T[1]

1) Saga University, Saga, Japan
2) National Institute of Advanced Industrial Science and Technology (AIST), Tokyo, Japan
yamadakc@saga-u.ac.jp

ABSTRACT

This study examined the effects of the introduction of a handheld tablet computer to the productivity enhancement in medical practice. When the medical staff refers to patient information in medical practice, they usually access the information system such as electronic health record by use of fixed special terminal. For the medical staff who provides medical care in unfixed space, to reach a fixed special terminal decrease their productivity in order to get patient information each time. Therefore, we developed the system that the medical staff is able to refer to patient information everywhere by use of a handheld tablet computer. This system use the environment of wireless local area network (WLAN) in the hospital, and it can get information on the actual location of the handheld tablet computer. To investigate the effects of the introduction of this system, the estimated behavior of the medical staff from their location was compared with the behavior before the introduction of this system.

Keywords: medical service, productivity enhancement, tablet computer

1 ELECTRONIC MEDICAL RECORD (EMR) IN JAPANESE MEDICAL SERVICE

The introduction of electronic medical record (EMR) in Japanese medical service has begun as electronic order-entry system and administration system since 1970s (Figure 1). Some major information technology (IT) companies, which called vendor, sent their system programmers into medical settings such as hospitals to develop EMR system. In 1999, The Ministry of Health, Labour and Welfare approved that the medical records are saved as electronic media. After that, almost all vendors have devoted to develop their system. Nowadays, over 800 major hospitals (with at least 400 beds) have the most advanced hospital information systems (Seed Planning, Inc., 2012).

Given that history, separated organization have established by medical institutions and vendors. In other words, the current EMR systems have developed both the user and the developer. The EMR system has two problems mainly as a result. First, the medical staff became increasingly dependent on the vendor to develop, operate, and maintain each EMR. Second, the IT engineers have some lack of understanding the workflow and tasks of the medical staff and their needs.

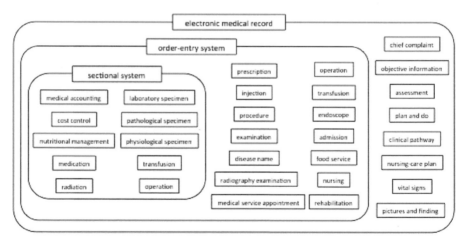

Figure 1 The component of Japanese electronic medical record (EMR) (Tahara, 2011, modified by the author).

2 UTILIZING INFORMATION TECHNOLOGY IN MEDICAL SETTINGS

With the spread of the EMR system in the past decades, medical staff commonly deals with digital information and works on computers. Such as digitized pictures that scanned with computed tomography (CT) or magnetic resonance imaging

(MRI) and digitized clinical laboratory test values are routinely used. Vast amounts of data are included in the EMR system, such as texts, numerical values, and pictures.

Recently, there are various activities with tablet computers such as iPod and iPad in medical settings over the world and Japan is no exception (Takao et al., 2012). Behind the scenes is not only that the tablet computers is available at low cost but that the workflow and tasks that medical staff perform everyday has changed. A cross-sectional work of medical care have increasingly needed in addition to the traditional one because modern medical care provides a holistic perspective against a patient. For example, patients with cancer are in different department and elderly patients have several disease. The cross-sectional medical staff must go to different places in hospital to care for these patients. It is hard, however, to work with the current EMR system because that system has a limit to its reliability and safety. Namely, the number of available desktop/laptop computer terminal for the EMR system is restricted, and no restricted laptop terminal can carry because the battery of these laptop terminal runs down during work hours.

3 INTRODUCTION OF A HANDHELD TABLET COMPUTER INTO MEDICAL PRACTICE

We conducted job analyses and interviews with several cross-sectional medical staff in a hospital such as nurses worked for pressure ulcer care, stoma outpatient, infection control, and so force. The job analysis consisted of a questionnaire and an investigation of amount of activity for the day. The questionnaire was used to examine the tasks during a day. The amount of activity was measured by the physical activity monitor (Active style Pro: Omron Healthcare, Inc., Tokyo, Japan).

As a result, following is a brief description of their typical workflow. First, the cross-sectional medical staff is supposed to search her/his patient from the schedule in the EMR system. Next, they seek or hold a room to treat their patients and available terminal to use the EMR system from the schedule or by phone. Then, they carry out actual treatment or operation. In this case, they have never access to the EMR system if they want when they move away from the terminal, such as explain to the patient about their own illnesses and moving from room to room. Figure 2 and 3 shows the amount of activity of a nurse of normal workday.

A hypothesis derived from our preliminary investigation of the cross-sectional medical staff is that they take much time to gather information before going into hospital ward. The problems (limited number of terminal and non-portability) could prevent the medical staff from executing the work effectively. Therefore, we experimentally constructed a thin client system to avoid the influence of other users and leak of information. This experimental thin client system is available for iPad (Apple, Inc.) and displays the EMR system of oneself (Figure 4). This system use the environment of wireless local area network (WLAN) in the hospital, and it can get information on the actual location of the handheld tablet computer.

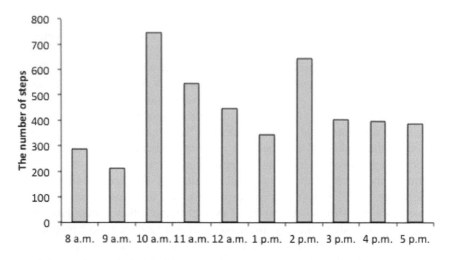

Figure 2 The mean number of steps of one week during work time.

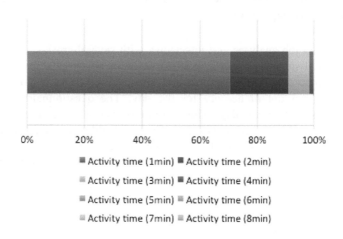

Figure 3 The percentage of the activity time for the mean of each duration of task.

It is predicted that our thin client system could minimize the wasted time for the cross-sectional medical staff because they access the EMR system with the handheld tablet computer to gather and confirm the information anytime and anywhere in hospital if they want to. However, further study is needed to demonstrate the hypothesis.

Section II

Service Innovation

CHAPTER 4

Human-centric Approach of Value Proposition in New Generation Digital Business

Toshiro Kita

Doshisha University
Kyoto, Japan
tokita@mail.doshisha.ac.jp

ABSTRACT

"The Times They Are a-Changin" as Bob Dylan sang. In the electronics industry, many changes have been observed such as the shift from "integral to modular" and from "closed to open". As a result of these changes, Japanese electronics firms (e-firms) have lost not only their profitability but also their competitiveness. Now, they are no more than the followers who are struggling to catch up with the winners of the "modular and open" way, e.g. Microsoft, Intel, Google, HP, Dell or Samsung.

But "The Times They Are a-Changin" again! Several phenomena occurring in the smart-phone/tablet and cloud computing markets are signs of new era in the electronics industry. It is not a simple recurrence of the shift to "modular and open" or "integral and closed". Indeed, it is another type of change, namely "product-centric to human-centric".

Keywords: product architecture, modular, integral, open innovation

1 HARD TO LAUGH STORY

I would like to start this paper from my personal experience. Buying a brand-new iPad2 just after its launch last spring and using it for about three months, I was impressed very much with its usability, and I gave it to my wife who had hated using PCs due to their complicated operations. As expected, she was able to use it without my assistance, and has not only enjoyed web-browsing but also downloaded/used many applications freely. But then the problems began for me. As I had lost my iPad2 to my wife, I started to look for my own tablet in late summer, and I ordered a brand-new tablet released in September from an unnamed company, S, a major Japanese e-firm. I trusted S in terms of not only its product performance and quality but also its pioneer spirit. I hoped for it to be superior to iPad2 because S had readied a bow and stand at ready according to its news release.

One week later a package was delivered to my office. I opened it leaving everything else and found a tablet, several accessories and a start-up guide sheet instead of thick manuals. Relieved that the packaging was same as iPad, I switched on and started to use it, but...... Setting-up the web browser and reading e-mail presented no problems. After downloading several free applications and deleting uninteresting ones, however, I was deadlocked. In iPad, it was easy to delete apps by tapping and holding down the icon of the app that I wanted to delete for a few seconds. For S's tablet, such an operation was invalid. I tried different kinds of operations, and finally I found the complicated app deletion procedure in the on-line manual. Eventually, I solved this problem by downloading a free application for app deletion. It was a sort of petty or funny story, but it did not finish there. There is another incredible story concerning this tablet. Two weeks later, I recounted this story to one of my old friends who was using S's smart-phone. He had several alternatives including Korean smart-phones, but being a patriot he finally bought one from S. He used S's smart-phone for three months but felt that it had several inconveniences. In the morning when we met he was nonplussed that he was completely unable to stop the alarm clock function of his smart-phone and had to resort to pushing it under his pillow! It was his "hard to laugh" experience. This story continued moreover. We exchanged my tablet and his smart-phone which were both based on Android. Surprisingly, both operation procedures were different. We looked at each other in open-mouthed surprise.

S has been admired as a company with advanced technologies and innovative products which have changed our very lifestyles. Why did the company develop such disasters? This is my motivation to write this article. I have considered the following:

- What are the problems of company S?
- Is it only in S? Or are there problems in other Japanese e-firms which have been struggling in economic downturn for more than 20 years?
- If so, can we find out the solutions for this problem?

The answer to these questions obtained from my case studies is, actually, very simple and may be defined in abstract terms as a lack of "warm heart" which Japanese e-firms used to have. My friend and I did not perceive the warm heart from my tablet and my friend's smart-phone. By using Google's Android and the display panel produced by rivals, is it possible to put a warm heart into the product?

In this article I analyze what is changing and discuss what is necessary to recover our warm heart.

2 ANALOGUE TO DIGITAL, INTEGRAL TO MODULAR AND CLOSED TO OPEN

Pablo Picasso, one of the greatest artists in the 20th century, quotes "Bad artists copy. Good artists steal. " T. S. Eliot, an important English-language poet of the 20th century, said exactly same thing; "Good Poets Borrow, Great Poets Steal." In 1996 PBS documentary "Triumph of the Nerds", Steve Jobs quotes Picasso's "good artists copy, great artists steal" and adds about Apple: "We have always been shameless about stealing great ideas." The intention of these three great genius is that a third rate copy cannot be superior to the original, but that a first class copy can absorb everything and exceed the original. In this context we can understand that Japanese companies between the 1950s and the 1980s were the real embodiment of "Copy and Steal". They inundated global markets with their Made-in-Japan high cost/performance products. But their prosperity collapsed in the beginning of 1990. There is no doubt that the crash of bubble economy was one of the reasons. Here, however, it is pointed out that a big change in technology from analogue to digital disrupted many Japanese e-firms [Ogawa, 2006; Yoshimoto, 2009]. This had a dramatic impact on the electronics industry which, along with the automotive sector, was one of the most important breadwinners for the Japanese economy. Figure 1 shows the comparison of profit margins between seven Japanese major electronics firms and five foreign firms. In 1980s, their performance was almost the same, but the downturn of Japanese e-firms after 1990s is quite obvious. This figure shows clearly the reality of Japanese e-firms that had left behind the European and American firms which they once had caught up now overtaken by firms from emerging economies which they had not regarded as their competitors.

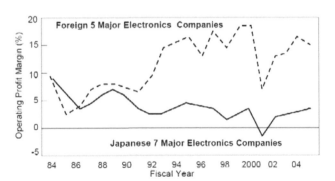

Japan: Hitachi, Toshiba, Mitsubishi, Panasonic, SONY, Fujitsu, NEC
Foreign: Intel, TI, Motorola, Nokia, Samsung

Figure 1. Comparison of Profit Margin

The paradigm shift in technology, e.g. "analogue to digital" was behind the decline of Japanese e-firms. Furthermore, the drastic change from "closed to open" in the value chain combining R&D, design, production, sales and support, etc. also contributed to their downfall. In the analogue era companies who had the design skills like artisans and the technical know-how of integration could deliver their excellent products into market. But the spread of digital technology which enabled any company to develop the excellent products by purchasing micro-processors and writing code, accelerated the decline of Japanese e-firms.. High-tech products such as flat-panel TVs, DVD and Blu-ray recorders, which Japanese e-firms had created before rivals, became commoditized in an extremely short time period and Japanese e-firms consequently lost their market share.

One of the typical products in the digital era is the personal computer (PC). In the era of the main-frame computer only Japanese companies such as NEC, Fujitsu and Hitachi, were recognized as contenders against the giant, IBM. The conditions of competition, however, changed completely by the quick spread of the PC which was developed in 1980s. In the 2011 PC market share[1] ,no Japanese e-firms are listed within No.5. In the PC market dominated by two giants; Microsoft and Intel, PC assembly manufacturers only can purchase the *de facto* components, assemble them into PCs, sell them to users and support them. In an industry such as PCs, which is highly modularized, the differentiations by assembly manufacturers are only found in customer support and the design for miniaturizing at most. The important action in such an industry is the control of supply-chain-management (SCM) and the intimate relationship with Electronics Manufacturing Service (EMS). Japanese e-firms have fallen behind in establishing the open processes required for collaboration with partners.

Another major industry beside electronics in Japan is the automotive sector. While the electronics industry has been beaten by the paradigm shift as mentioned above, the automotive industry still maintains its advantage in the market in spite of the terrible East Japan earthquake in March 11th, 2011. Fujimoto [2007] explained the difference between the automotive and electronics industries using his concept of 'product architecture'. Japanese companies have excellent skills to realize highly integrated products based on the closed relationship with their inner divisions or partners. Regarding automobile with "closed-integral" architecture, Japanese companies, Toyota and Honda *et al* still maintain competitive advantages.

3   NEW MOVEMENT IN DIGITAL

In computer and consumer electronics the transformations from analogue to digital, integral to modular, and closed to open, have been observed. Even in the automotive industry a change from closed-integral to open-modular by the entry of electric cars, which implement the motor and battery in spite of the internal-combustion engine, is expected.

Are such changes of architecture inevitable in the history of industries? If so, Japan has no future. However, most Japanese e-firms seem to believe that they must change their architecture from closed-integral to open-modular. To compete

---

[1] http://www.gartner.com/it/page.jsp?id=1893523

not only with big Korean firms such as Samsung and LG and companies in emerging economies including China and India, they are struggling to absorb the capabilities of the open-modular architecture by the purchasing the core-components and the utilization of EMS. One of the typical examples is my tablet from S; just a disaster for me.

But now we have an alternative to "open-modular" in Apple's iPhone/iPad. As is well-known, Apple's Macintosh was beaten by Wintel's open-modular platform leadership strategy [Gawer, 2002]. Macintosh's product architecture of was closed-integral based on Steve Jobs' strong conviction which was to have "end-to-end control of the customer experience". In the iPhone/iPad, however, Apple introduced a new type of architecture which enables not only this end-to-end control but also utilizes the best key-components and manufacturing. Fujimoto claimed that open and integral are contradictory in concept. Contrarily to his theory, however, Apple's new product architecture seems "open-integral". Consequently, Apple has reborn by the iPhone and iPad based on open-integral. iPhone had 3.9% market share but 56% profit share in the cellular phone market of 2011 Q3[2] , and iPad dominates the tablet market with a share of 68% [3]. These are surprising business results. In the smart-phone and tablet markets products with non open-modular architecture are overwhelming ones with open-modular architecture.

Another recurrence from open-modular is observed in the corporate computing service industry. Until the 1970s corporate computing services were provided using the closed-integral mainframe computer systems. By the entry of Unix-based computing system in 1980s the major stream of corporate computer systems was changed to the so-called "open system". Recently, "cloud computing" is widely noticed. Cloud computing is a model for enabling ubiquitous, convenient, on-demand network access to a shared pool of configurable computing resources (e.g., networks, servers, storage, applications, and services) that can be rapidly provisioned and released with minimal management effort or service provider interaction [Mell, 2011], and its architecture seems closed-integral because it is a completely invisible "black box" from outside.

4   COOL HEAD, SPIKY PASSION OR WARM HEART?

The clear difference of business situation between PC and smart-phone/tablet may provide us an implication for Japanese e-firms. Is there an answer here for the puzzle of Japanese e-firms?

Apple is a company of finespun distinctions as a German architect. Mies van der Rohe noted " God is in the detail". The mirror-like finish and delicate round beveling of the first iPod touch back-plate are Apple's typical obsessions. In the touch panel of iPhone/iPad, Apple's requirements to suppliers are detailed not only

---

[2] http://www.asymco.com/2011/11/04/the-end-of-the-independent-phone-brand/

[3] http://9to5mac.com/2011/09/14/idc-ipad-market-share-at-68-lead-growing-into-holidays/

technically but also aesthetically whereas it is said that other Android smart-phone/tablet manufactures send them only the technical specifications of touch panel. It came from Steve Jobs' personal passion. His spiky passion was ineffective in PCs delivered into the business-use market where "cool headed" rationality counted, but it did attract customers who buy iPhone/iPad for their personal use.

What do Japanese e-firms have? Cool head like Bill Gates or spiky passion like Steve Jobs? No, they have "warm heart". Warm heart is the fundamental of the Japanese virtues, e.g. politeness, hospitality and patience. Like appealing to rationality with cool head and penetrating with spiky passion, Japanese e-firms can wrap up the customer gently with their warm hearts shown in Figure 2. It seems, however, that most Japanese e-firms including unnamed company S hope to have "cool head" for open-modular smart-phone/tablet and compete with rivals who are much better at handling the open-modular approach than Japanese e-firms. Japanese e-firms have to abandon such a "product-centric" approach based on cool head and change to a "human-centric" approach based on a warm heart. The meaning of human-centric is not "user-oriented". It involves "innovative and unmet user oriented", "encouraging workers" and "co-petition [Brandenburger, 1997] with partners". This is my revival proposal for Japanese e-firms in this new digital generation.

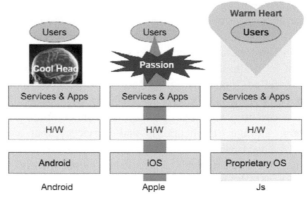

Figure 2 P Cool Head, Spiky Passion and Warm Heart.

5   REQUIRED CAPABILITIES TO REVIVE

Formerly, Japanese e-firms could put their warm hearts into their products and dominated the world market by doing so. But their glory days have passed. Now Japanese e-firms are left behind by the very competitors whom they once outstripped and are copied/stolen by Apple. They are just the pathetic followers of the open-modular concept. It is not enough for Japanese e-firms to look back to the past. They have to go back to the future. To be new warm hearted and human-centric companies, Japanese e-firms must not only remember the past successful Japanese ways but also develop three new capabilities for competing with their strong rivals except for product-centric management skills. These capabilities are reflect "Value", "System" and "Culture".

5.1   Capability on Value

The first capability which Japanese e-firms have to have is to redefine the value proposition. The value which they have to create is no more than the product value, e.g. performance, functions or quality. The product value is at its highest when released to the market and declines over time. Companies have to take countermeasures against this value decline, such as the release of new a product with improved value, to sustain their business. But new products are confronted with the same value decline, too. This death-march of the development of new products and the value decline has been impoverishing Japanese e-firms. To escape this death-march they have to redefine the value proposition. One aspect of redefinition is "Love" and another is "Experience".

Most of workers in automotive companies, especially designers and engineers, are car enthusiasts. They design and produce their cars with love. They are very happy people whose passions are identified with their jobs. Workers in other industries are not in such a happy work situation. There are very few workers in electronics companies who fall in love with a refrigerator and press their cheeks with it! However, they can design and produce an excellent product or service with their love if they want to have or use it. It can be easily imagined that the engineers of unnamed company S have no motivation to develop the tablet by using commodities, e.g. Android OS, ARM CPU, rival's LCD panel etc. It may be a fatal mistake for S to follow the open-modular type production and abandon development from the scratch. This means that the company should provide the motivations for their workers to enable them to put their love into their products.

Regarding "Experience", there are good practices in not only by Apple but also by many internet start-ups, e.g. Facebook, Dropbox, Evernote etc. Utilization of experience for value-capture is one of the most ignored activities in high-tech Japanese e-firms. In the non high-tech business arena, however, some Japanese e-firms set strategy based on experience for their success. One of the examples is MYFARM [4], a Kyoto-based start-up in the agriculture business. This company offers many agricultural services, e.g. personal farm rental, agriculture academy, party in farm for singles, etc. MYFARM's services are designed to increase the experience of not only customers but also the partners including land owners. . Although the customer-side experience has been paid attention to, Japanese high-tech companies have to notice the importance of partners to increase the experience.

---

[4] http://www.myfarm.co.jp/index.html

5.2 Capability on System

Japanese companies have produced high quality and high performance products based on the closed-integral product architecture and recognized that the closed vertical integrated organization was the most effective to operate their business. However, they lost confidence in their Japanese way, especially in the electronics industry. They adopted the open-modular architecture and let their organization move from closed to open. It is not always a bad choice for them to turn the rudder from closed innovation to open innovation. Open innovation is one of the best choices for companies in the globalization of business. But it must be fatal for them to try to implement open innovation without right understanding. The open-modular architecture does not mean open innovation, and the closed-integral does not mean closed innovation either. Apple, recognized as the flag-bearer of closed-integral, is one of the best experts of open innovation. While Apple establishes the very intimate relationship with the key-components suppliers and EMS in the production of hardware, Apple has created an almost closed and well-controlled relationship with application developers and contents suppliers. Japanese e-firms must learn Apple to construct the most suitable business system for them. Table 1 shows the examples of right and bad business architecture in smart-phone/tablet.

Table 1 Business Architecture of Smart-phone/Tablet

	Android		Apple
	Samsung, HP, Sony etc	Kindle Fire & Sharp Galapagos	
Application /Contents Services	Open	Closed	Closed
Product	Open-Modular	Open-Modular	Open-Integral

There are two types of business architecture in Android. One is a main stream, open/open-modular which is adopted by most manufacturers, e.g. Samsung, HP, Sony etc. Among them, only Samsung goes well. Other companies with no economies of scale but also with no competitive core-device like the LCD panel put up a poor fight. Only two companies adopt the closed/open-modular model based on Android. Amazon's strategy is to set a low price tag and to encourage e-books sales through its strong web site. Sharp, which is a consumer electronics company and has few experiences to sell the applications and content, also selected the closed/open-modular model. Unlike Amazon, Sharp set a high price tag for its product Galapagos because Sharp was confident of its core device LCD panel as the key differentiator. It was as clear as day that the sales of Kindle Fire got a smooth start while Galapagos sales were unimpressive. No doubt, Sharp put its warm heart into Galapagos. It was the fatal, however, that Sharp had no capability to select the most suitable architecture for its organization, assets and culture. Japanese e-firms should establish their business architecture carefully to maximize the performance of the human-centric approach based on their warm heart.

5.3 Capability on Culture

It is very difficult to obtain new capabilities. The biggest obstacle is "culture". At a leadership conference celebrating IBM's centennial on September 20, 2011 , Sony CEO Howard Stringer told that change is never easy for companies that have successful pasts, including his own company. And he added "all great companies have a rowboat mentality," and "you row hard, but you're still always looking behind you to the past." His cited rowboat mentality must be shed. It is not enough, however, for Japanese companies to shed only this mentality. They have to overcome more dangerous way; the " boat race syndrome". It is an implication of the boat race like "Oxford vs. Cambridge". Crews (workers in company) row the boat hard with their back toward the goal. Only cox (CEO in company) seems to look forward. But the reality is that he/she merely looks forward. His/her main jobs are to look around to check the position of rivals and to offer his/her staff " Good luck!". It should be necessary for them not only to look behind or around but also to look ahead like rowing a sea kayak. In other words, features of both rowing and kayaking are necessary. Japanese firms must develop such ambidexterity in culture.

Regarding management logic, the ambidexterity between causal and effectual [Read, 2010; Saravathy, 2001], must be owned by them. The causal logic to minimize the false-positive error is powerful when the concrete business goal is set, and the effectual logic to minimize the false-negative error is very effective at the dynamic and uncertain business situation.

By having these types of ambidexterity for the functioning of a human-centric approach based on warm heart, Japanese e-firms will be able to be "Japan as Number One [Vogel,1999] " again.

6 THE TIMES THEY ARE A-CHANGIN, AGAIN!

In the mid 1980s, Japanese e-firms dominated the world market with their high cost performance products. It was a dreamy time. But they were rudely awakened from their reverie and the long nightmarish downturn started. The decision of Japanese e-firms to change the product architecture from the familiar closed-integral to the global standard open-modular is rational in terms of re-catching up with rivals. However, they seem to lose their most important possession during this struggle. That is, they lost their heart not only to users but also to workers and partners.

I have interviewed several companies which provide key-components, manufacturing equipment or measuring instruments. Most of them said "they had very intimate relationship with Japanese e-firms to produce high cost performance products and were able to create new technological seeds from the discussion with them. But now we have such relationships with foreign companies." In order to collect commodity components and to produce the open-modular type products effectively, strong commitments with their partners are not necessary. The relationship with partners must be very helpful to obtain deep insights and clear vision for future invisible markets and the warm heart must be distilled from this mixture of discussions. Their testimony shows clearly the Japanese e-firms' present status and mistakes.

The Tohoku earthquake and tsunami on March 11, 2011 gave a hair-raising shock all over the world. At the same time, however, the patience and warmth of the victims left a deep impression with people all over the world. Japanese companies have to remember this warm heart which they used to value in their business. Here, I proposed that Japanese firms be "human-centric based on a warm heart" as a means for those firms to overcome their challenges. . The meaning of human-centric is not "user-oriented" as mentioned above. It involves not only "user-oriented" but also "encouraging workers" and "co-petition with partners". By establishing three capabilities around value, system and culture, Japanese companies will be able to put their warm hearts into t products which embrace users gently in contrast to the way in which Steve Jobs used his spiky passion to penetrate users.

ACKNOWLEDGMENTS

The author would like to acknowledge the support by the Grants-in-Aid for Scientific Research (C) by Japan Society of the Promotion of Science, the MEXT Supporting Project for Forming a Strategic Research Base of Private Universities, and the Telecommunications Advancement Foundation Research Grant.

REFERENCES

Brandenburger, A. J. and Nalebuff, B. J.(1997) Co-opetition, Doubleday, New York, 1997
Clark, K. B., and T. Fujimoto. Product Development Performance: Strategy, Organization, and Management in the World Auto Industry. Boston: Harvard Business School Press, 1991.
Clark, K. B., and T. Fujimoto. "The Power of Product Integrity." Harvard Business Review 68, no. 6 (November-December 1990): 107-118.
Fujimoto, T., Evol. Inst. Econ. Rev. 4(1): 55–112 (2007) JEL: F10,
Gawer, G. and Cusumano, M. A., Platform Leadership: How Intel, Microsoft, and Cisco Drive Industry Innovation, Harvard Business Review Press; 1st edition (April 29, 2002)
Ogawa, K., Shintaku, J. and Yoshimoto, T., Architecture-based Analysis of Competitive Advantage Between Japanese and Catch-up Country's Firm and Introduction of New Global Alliance, Annals of Business Administrative Science, 4(3), pp. 21-38. (2005)
Mell, P. and Grance, T. The NIST definition of cloud computing, Special Publication 800-145, National Institute of Standard and Technology (2011)
Read, S., Sarasvathy, S., Dew, N. , Wiltbank, R. & Ohlsson, A., Effectual Entrepreneurship, Abingdon/New York: Routledge (2010≫
Sarasvathy, S. D., Causation and effectuation: Toward a theoretical shift from economic inevitability to entrepreneurial contingency, Academy of Management. The Academy of Management Review. Vol. 26, Iss. 2; pp. 243-263 (2001)
Vogel, E., Japan As Number One: Lessons for America, iUniverse (1999)
Yoshimoto, T., "Modularization, design optimization, and design rationalization: A case study of electronics products" , Annals of Business Administrative Science pp.8 (2009)

CHAPTER 5

Value Co-Creation in R&D

Yuriko Sawatani[1], Tateo Arimoto[2]

[1]S3FIRE, RISTEX, JST, [2]Director-General, RISTEX, JST
yurikosw@gmail.com

ABSTRACT

The structural change in society of the shift to a service-based economy is advancing. This affects R&D as well. In this paper, we propose service research model positioning value co-creation in R&D and apply it to Research Institute of Science and Technology for Society (RISTEX) R&D program. One of characteristics of RISTEX research program is the existence of a management team. The management team interacts with project teams for the project success.

It would be not easy to meet both research impacts as well as social impacts. Our current assessment shows that the research maturity affects the research project selection as well as the portfolio management. When the research maturity is low, then a research program tends to focus on social issues. As the result, the research project portfolio is managed from problem-solving viewpoints. Even though the research maturity is low, activities to link research projects have a good potential to influence positively to create strategic research themes from the bottom-up research projects. In the future research, we study the management system at RISTEX may work positively for the research theme creation and social impacts.

Keywords: Service Science, R&D Management, Social Technology, Multidisciplinary Research

1   INTRODUCTION

The structural change in society of the shift to a service-based economy is advancing. This phenomenon, common to the economies and societies of both developed and developing countries, results from the growth of the share of the economy accounted for by the service sector, spurred by rapid growth in service industries against a background of increasing social sophistication and diversification. The service industry's share of Japanese GNP has grown to 60.7% (the percentage of nominal GDP in fiscal 2009 not including the agriculture, forestry, fisheries, mining, and manufacturing industries), and according to a study by the Organization for Economic Co-operation and Development (OECD) the shift to a service-based economy has advanced steadily in other countries too. However, productivity in service industries is lower than in manufacturing industries, and the need to achieve innovation and productivity improvements has become an important issue (Ministry of Economy, Trade and Industry of Japan 2007).

Regarding this issue, the report Innovate America: Thriving in a World of Challenge and Change submitted to the Bush Administration by the U.S. Council on Competitiveness in December 2004 (commonly referred to as the "Palmisano Report") offered the view that there was a need to create an interdisciplinary field of service science to resolve issues originating in the shift to a service-based economy (IfM and IBM 2007, Chesbrough and Spohrer 2006). A new movement has appeared toward the development of such an interdisciplinary field of service science as science and engineering researchers join the domain of service research that until now has advanced chiefly in the fields of social science, service marketing, and service management. In Japan, addressing emerging and interdisciplinary domains over the years 2006 through 2010 was planned in March 2006 under the Third Science and Technology Basic Plan. In the Fourth Science and Technology Basic Plan, the focal point has shifted further from field-specific to issue-driven innovation in science and technology, with research and development activities in interdisciplinary domains such as service science positioned as important topics and serving as forerunners of the issue-driven approach. The June 2006 outline of the Economic Growth Strategy from the Ministry of Economy, Trade and Industry discussed innovation in service industries, and a movement to create a field of service science has begun in Japan too. In May 2007 Service Productivity & Innovation for Growth (SPRING) was established, as was the Center for Service Research (at the National Institute of Advanced Industrial Science and Technology, or AIST) in April 2008, in April 2007 the Ministry of Education, Culture, Sports, Science and Technology began the Service Innovation Human Resource Development Program, and in April 2010 the Japan Science and Technology Agency's Research Institute of Science and Technology for Society (JST-RISTEX) began seeking R&D projects under its Service Science, Solutions and Foundation Integrated Research Program.

At the same time, the definition of services has been reconsidered as the manufacturing industry, together with the shift to a service-based economy, has seen the appearance of firms that fuse product development with the provision of services, and thus do not fit within the framework of traditional industry categories. In everyday use in the Japanese language, the word "service" frequently is used to express the nuance of "freebie." In industry categories, the service industry refers to what is left over after grouping into the agriculture, forestry, fisheries, and manufacturing industries. In service marketing, an attempt was made to separate products and services and define services using characteristics differing from those of products. In recent years, service-dominant (S-D) logic has been proposed, seeing the essence of services as the co-creation of value and identifying service as a fundamental principle of exchange, and research based on this concept is in the process of spreading not just in service marketing but to other fields as well. In the following section, service R&D and the positioning of value co-creation in R&D are discussed.

2   SERVICE R&D

R&D definition has been changed influenced by economic changes and servitization. The definition of R&D has been worked in National Experts on Science and Technology Indicators, NESTI since 1960, and updated in Frascati Manual version 6 (OECD 2002). Originally R&D focused on natural science and engineering, however, the wider R&D definition including social sciences, humanities was considered by Djellal, et al. (2003) when Frascati Manual version 6 was created.

"Research and experimental development (R&D)
Research and experimental development (R&D) comprise creative work undertaken on a systematic basis in order to increase the stock of knowledge, including knowledge of man, culture and society, and the use of this stock of knowledge to devise new applications." (OECD 2002 p.20)

The R&D definition includes "knowledge of man, culture and society", and becomes wider from technological outputs to social sciences based knowledge creation. The R&D in social sciences and humanities is described as the following.

"The social sciences and humanities are covered in the Manual by including in the definition of R&D "knowledge of man, culture and society" (see Chapter 2, Section 2.1). For the social sciences and humanities, an appreciable element of novelty or a resolution of scientific/technological uncertainty is again a useful criterion for defining the boundary between R&D and related (routine) scientific activities. This element may be related to the conceptual, methodological or empirical part of the project concerned. Related activities of a routine nature can only be included in R&D if they are undertaken as an integral part of a specific research project or undertaken for the benefit of a specific research project. Therefore, projects of a routine nature, in which social scientists bring established methodologies, principles and models of the social sciences to bear on a particular problem, cannot be classified as research." (OECD 2002 p.48)

"Defining the boundaries of R&D in service activities is difficult, for two main reasons: first, it is difficult to identify projects involving R&D; and, second, the line between R&D and other innovative activities which are not R&D is a tenuous one. .... Identifying R&D is more difficult in service activities than in manufacturing because it is not necessarily "specialised". It covers several areas: technology-related R&D, R&D in the social sciences and humanities, including R&D relating to the knowledge of behaviour and organisations. ....

Also, in service companies, R&D is not always organised as formally as in manufacturing companies (i.e. with a dedicated R&D department, researchers or research engineers identified as such in the establishment's personnel list, etc.). The concept of R&D in services is still less specific and sometimes goes unrecognised by the enterprises involved. As more experience becomes available on surveying R&D in services, the criteria for identifying R&D and examples of service-related R&D may require further development. ....

"The following are among the criteria that can help to identify the presence of R&D in service activities:

– Links with public research laboratories.

– The involvement of staff with PhDs, or PhD students.

– The publication of research findings in scientific journals, organisation of scientific conferences or involvement in scientific reviews.

– The construction of prototypes or pilot plants (subject to the reservations noted in Section 2.3.4)." (OECD 2002 p.48-49)

Even though some guidelines to determine whether routine works or Service R&D are suggested, those are not enough. The guidelines depend on the establishment of people, organization or academia, which do not exist in immature research areas, such as service science.

3   POSITIONING VALUE CO-CREATION IN R&D

The S-D logic has been recognized as the theoretical foundation of service research (Vargo and Lusch, 2004a, 2004b). It expects to integrate the traditional G-D logic view of innovation model based on goods vs. services dichotomy to a unified innovation model from the value co-creation point of views. The S-D logic is appropriate for studying service innovation since it removes the limitation of goods vs. intangible goods (services) dichotomy approaches, and synthesizes customers and service pro-viders (Drejer, 2004).

Moeller (Moeller, 2008) shows service processes based on collaboration between service providers and customers influenced by S-D logic. The service processes are divided by Facilities, Transformation, and Usage. The stage of facilities exhibits only when potential value for customers is created by company resources. The next stage, transformation is divided into two parts, Company-induced transformation or Customer-induced transformation. At the company-induced transformation, the transformation includes only company resources for the potential value for customers. On the other hand, at the customer-induced transformation, companies and customers co-create value. At the last stage, the usage, customers act as the prime resource integrator to receive benefits from the transformation. The facilities and the company-induced transformation are both activities of service providers. So that the Moeller's model has three types of activities, such as activities done by service providers alone (facilities and company-induced transformation), value co-creation activities by both (customer-induced transformation), and activities of customers (usage).

We introduce Service Research Model for Value Co-Creation extending the Moeller's model. The model has three spheres, such as R&D, Value co-creation, and Site. "R&D activities" and "New research theme creation" are activities contained in a R&D sphere (Sawatani and Niwa, 2008). There are bidirectional links between a R&D and a Value co-creation sphere. The first link from "R&D activities" to "Value co-creation" implies that a R&D sphere provides technologies and knowledge to a Value co-creation sphere. The service innovation success depends on technologies and knowledge created in a R&D sphere integrated by the design methods created in a Value co-creation sphere (Sawatani and Niwa, 2009).

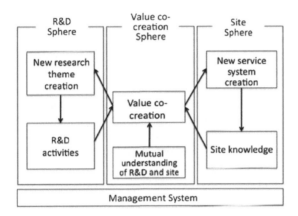

Figure 1. Service research model for value co-creation

The second link illustrates that a R&D sphere gains research value thorough the value co-creation interaction with customers and members in a service organization, not only providing their technologies and knowledge to them. The knowledge created through the value co-creation interaction includes technologies, integrated design methods and service domain knowledge (Sawatani and Niwa, 2009). Adding to the knowledge creation, new research themes are discovered when researchers are practicing activities in a value co-creation sphere (Sawatani and Niwa, 2008). The value co-creation interaction is beneficial not only to a Site sphere for customers, but also to a R&D sphere (Vargo, et al. 2010).

In this model, researchers are not only providing technologies and knowledge into a service sphere, but also receiving new research ideas by proactively joining to the value co-creation interaction (Sawatani and Niwa, 2008, 2009). The S-D logic states that "value created through exchange is based on the mutually beneficial relationships among service systems", and "all parties are simultaneously both producers and customers of value" (Vargo, et al., 2010). That is, the value co-creation interaction is beneficial to both of a R&D and a Site sphere. The key element of the service research model is a Value co-creation sphere where requires the collaboration of researchers and customers/service members.

In addition to these R&D and site activities, Management system is added to cross these three spheres. This works for a supplementary system to execute R&D projects, but will provide vital functions. We will see these functions in the case study.

4   Comparison with the related concepts on R&D

4.1   Service innovation research

We look into service innovation research. The macro level innovation surveys are conducted including service industries. Miles pointed out that there are two issues on service innovation surveys (Miles, 2002, 2007). One is the survey design which is biased to the technological innovation. The current survey questionnaires depend on the innovation studies based on goods innovation (Miles, 2007, Drejer, 2004), and could not capture the wider scope of service innovation. The other issue is the immature understanding of service innovation by service industries in particular (Miles, 2007, Sundbo, 1997). Service industries do not have a specialized innovation organization such as R&D in the most cases (Miles, 2007, Sundbo, 1997), so that it is difficult for them to recognize activities and knowledge contributing to the service innovation.

On the other hand, case studies of service innovation identified non-technological innovation, such as process and organizational innovation adding to the technological innovation (Miles, 2002, Sundbo, 1997). The empirical findings of the service innovation show that the characteristics of the service innovation processes are dynamic and ad-hoc (Sundbo, 1997, Mamede, 2002, Edvardsson and Olsson, 1996). Despite the growing studies on service innovation, the literature from marketing and innovation research continues to improve our understanding of service innovation, however, these service innovation studies show the tendency to emphasize the service distinctive features (Gallouj and Weistein, 1998, Gallouj, 2002, Sundbo, 1997, de Vries, 2006, van der Aa and Elfring, 2002) or the assimilation of goods and services (Gallouj, 1998, Vargo and Lusch, 2004a, 2004b, Drejer, 2004). Both of approaches are based on the separation of goods or services, and provides incomplete view of service innovation (Drejer, 2004). The modern service theory is being formed based on value co-creation by customers and service providers (Vargo and Lusch, 2004a, 2004b, Vargo, et al., 2008, 2010).

4.2   Customer focused product innovation research

So far seeds based product innovation was done driven by R&D. The needs based product innovation happens by the collaboration with R&D and the marketing organization, which is the information gathering functions of customers for the new product development. These innovations are the product innovation sponsored by companies. Because of that, the decision making about the new product is done by a company although it makes use of collected customer information by the marketing.

On the other hand, there are innovations where R&D is not involved in. In the case of user innovation (von Hippel, 1986, 1994, 1998, 2001), customers innovate a product by improving the existing products or developing a new product based on tool-kits. R&D organization doesn't usually exist in the service industry, such as Book-Off, which started the service innovation of the secondhand bookstores. A traditional craft by the craftsman is given as an example of the product innovation which R&D organization don't exist in. However, both of the cases assume that company and customers are separated.

4.3   Issue based research

A new knowledge production concept focusing on issue based research was proposed as a mode 2 (Gibbons and et al. 1994). Mode 2 knowledge productions in an application context, which includes the experiential elements as well as theoretical elements, have the similar characteristics as the service R&D activities from the viewpoints of the participation of various stakeholders and the quality control by them. The presented framework is confined to the general idea which describes the characteristics of the knowledge creation with transdisciplinary, however, the positive research for the elucidation of the R&D behavior isn't being done fully.

In addition, the relations between mode 1 and mode 2 were not focused enough. As for the service research model for value co-creation, mode 2 is similar to activities in Value co-creation sphere, and traditional R&D activities are equal to the knowledge production of mode 1. Mode 1 and mode 2 knowledge production are related dynamically in the proposed model.

5   CASE STUDIES: SOCIAL TECHNOLOGY R&D PROGRAM IN RISTEX

Next we look into Research Institute of Science and Technology for Society (RISTEX)'s research programs. The RISTEX fosters innovation addressing social challenges though science and technology, and aims to create science and technology for society by solving problems in society. Here, science and technology for society means that "technology for the purpose of building new social systems that integrate the knowledge from multiple areas in the natural sciences, humanities, and social sciences". RISTEX has the following five activities for research programs; 1. Identifying social issues, 2. Establishing R&D focus programs, 3. Promoting R&D, 4. Presenting prototypes for the return of R&D results to society, 5. Assisting the application of R&D results to wider areas.

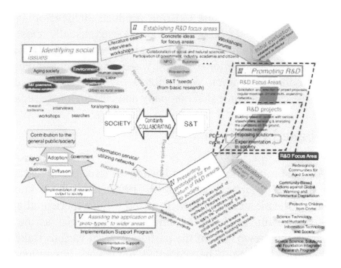

Figure 2. RISTEX activities cycle

RISTEX R&D has characteristics, such as high intensity of interaction with society and low maturity of research area (Figure 3.) comparing with traditional applied research (Kobayashi and et al. 2012). The research maturity, i.e. high if academic societies exist, affects the research project selection as well as the portfolio management. When the research maturity is low, then a research program tends to focus on social issues. As the result, the research project portfolio is managed from problem-solving viewpoints. Even though the research maturity is low, activities to link research projects have a good potential to influence positively to create strategic research themes from the bottom-up research projects.

For a research management, inputs (maturity of research area, social linkage, management staffs), process (program, project), and impacts (research, social) are important attributes to consider. One of characteristics of RISTEX research program is its management system. In order to promote R&D programs, RISTEX calls for proposals that are in line with R&D goals. The management team of RISTEX's research program, which consists from a program director, a program officer and program advisors, selects appropriate proposals. Then the management activities, such as Research program meetings, Lodging together meeting (Once a year), Interaction (advice, QA) on research plans, research reports, Site-visits (visiting research project meetings), Outreach activities are provided by interacting with selected research projects. Figure 4 shows a logic model of RISTEX's research program.

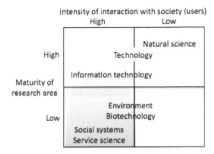

Figure 3. RISTEX's R&D focus (shaded part)

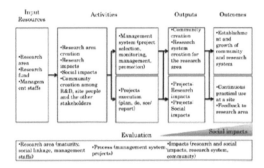

Figure 4. Logic model of RISTEX research program

Service Science, Solutions and Foundation Integrated Research (S3FIRE) Program case

6   FUTURE RESEARCH

As we see RISTEX's research programs, it would be not easy to meet research impacts as well as social impacts. The research maturity and social linkage guide how to manage the research program. However, management staffs, organization of management team, leadership collaborating with various stakeholders, interaction methods with a project team and a management team, also give huge impacts to the program success. We continue to study these attributes and the evolvement of R&D transformation toward a service-based economy through the execution of S3FIRE program.

REFERENCES

Chesbrough, H., and Spohrer, J., "A Research Manifesto for Services Science", Communications of the ACM, Vol. 49, No. 7, pp. 35-40, 2006.
de Vries, E.J., "Innovation in services in networks of organizations and in the distribution of services", Research Policy, Vol. 35, No. 7, pp. 1037–1051, 2006.

Djellal F., Francoz D., Gallouj C., Gallouj F., and Jacquin Y., "Revising the definition of research and development in the light of the specificities of services", Science and Public Policy, Vol. 30, No. 6, pp. 415-429, 2003.

Drejer, I., "Identifying innovation in surveys of services: A Schumpeterian perspective", Research Policy, Vol. 33, No. 3, pp. 551–562, 2004.

Edvardsson, B., and Olsson, J., "Key concepts for new service development", The Service Industries Journal, Vol. 16, No. 2, pp. 140–164, 1996.

Gallouj, F., "Innovating in reverse: services and the reverse product cycle", European Journal of Innovation Management, Vol. 1, No. 3, pp. 123-138, 1998.

Gallouj, F., Knowledge-intensive business services: processing knowledge and producing innovation, Edward Elgar, 2002.

Gallouj, F., and Weistein, O., "Innovation in services", Research Policy, Vol. 26, pp. 537-556, 1997.

Gibbons, M., Limoges, C., Nowotny, H., Schwartzman, S., Scott, P., and Trow, M., et al, The New Production of Knowledge, London: Sage, 1994.

IfM and IBM, "Succeeding through Service Innovation: A Service Perspective for Education, Research, Business and Government", Cambridge, United Kingdom: University of Cambridge Institute for Manufacturing, 2007.

Kobayashi, N., Akamatsu, M., Okaji, M., Togashi, K., Harada, A., Yumoto, N., "Analysis of synthetic approaches described in papers of the journal Synthesiology", Synthesiology, Vol. 5, No. 1, pp.36-52, 2012.

Mamede, R., "Does Innovation (Really) Matter for Success? The Case of an IT Consultancy Firm", DRUID Conference on Industrial Dynamics of the New and Old Economy, Elsinore, 6–8 June 2002.

METI, http://www.meti.go.jp/report/data/g70502aj.html, 2007.

Miles, I., "Service innovation: towards a tertiarization of innovation studies", In Gadrey, J. and Gallouj, F. (Eds.), Productivity, Innovation and Knowledge in Services, Edward Elgar, Cheltenham, UK, 2002.

Miles, I., "Research and development (R&D) beyond manufacturing: the strange case of services R&D", R&D Management, Vol. 37, pp. 249-268, March 2007.

Moeller, S., "Customer Integration—A Key to an Implementation Perspective of Service Provision", Journal of Service Research, Vol. 11, No. 2, pp. 197-210, 2008.

OECD, Oslo Manual: OECD Proposed Guidelines for Collecting and Interpreting Innovation Data, 1st edition. OECD, Paris, 1992.

OECD, Frascati Manual—Proposed Standard Practice for Surveys on Research and Experimental Development. OECD, Paris, 2002.

OECD, Oslo Manual: Guidelines for Collecting and Interpreting Innovation Data, 3rd edition, OECD, Statistical Office of the European Communities, Luxembourg, 2005a.

OECD, Enhancing the Performance of the Services Sector, OECD, Paris, 2005b.

RISTEX (Research Institute of Science and Technology for Society) http://www.ristex.jp/EN/aboutus/principle.html

Sawatani, Y., "Research Program Assessment from Research and Social Impacts", PICMET 2012 (submitting), 2012.

Sawatani, Y. and Niwa, K., "Services Research Model for Value Co-Creation", Management of Engineering & Technology, PICMET, pp.2354–2360, 2008.

Sawatani, Y, and Niwa, K., "Service systems framework focusing on value creation: case study", International Journal of Web Engineering and Technology, Vol. 5, No. 3, pp.313-326, 2009.

Sundbo, J., "Management of innovation in services", The Service Industries Journal, Vol. 17, No. 3, pp. 432–455, 1997.

van der Aa, W. and Elfring, T., "Realizing innovation in services", Scandinavian Journal of Management, Vol. 18, pp.155-171, 2002.

Vargo, Stephen L., and Lusch, Robert F., "The Four Service Marketing Myths: Remnants of a Goods-Based, Manufacturing Model", Journal of Service Research, Vol. 6, No. 4, pp. 324-335, 2004a.

Vargo, Stephen L., and Lusch, Robert F., "Evolving to a New Dominant Logic for Marketing", Journal of Marketing, Vol. 68, No. 1, pp. 1-17, 2004b.

Vargo, Stephen L., Lusch, Robert F., and Akaka, M.A., "Advancing Service Science with Service-Dominant Logic", In P. P. Maglio, Cheryl A. Kieliszewski, and James C. Spohrer (Eds.), Handbook of Service Science, pp. 134-156, 2010.

Vargo, Stephen L., Maglio, Paul P., and Akaka, M. A., "On value and value co-creation: A service systems and service logic perspective", European Management Journal, Vol. 26, pp. 145-152, 2008.

von Hippel, E., "Lead Users: A Source of Novel Product Concepts", Management Science, Vol. 32, No. 7, pp. 791-805, 1986.

von Hippel, E., "Sticky information and the locus of problem solving: Implications for innovation", Management Science, Vol. 40, No. 4, pp. 429–439, 1994.

von Hippel, E., The Sources of Innovation, Oxford University Press, 1998.

von Hippel, E., "Perspective: User Toolkits for Innovation", Product Innovation Management, Vol. 18, pp. 247-57, 2001.

CHAPTER 6

Social Networks for Outsourcing and Developing a Firm's Creativity

Maria Colurcio*, Marco Tregua**, Monia Melia*, Angela Caridà*

*University of Magna Graecia
Catanzaro, Italy
** University Federico II
Napoli, Italy
mariacolurcio@unicz.it,
marco.tregua@unina.it,
monia.melia@unicz.it,
angela.carida@unicz.it

ABSTRACT

Social networks are leading to a fruitful involvement of collective creativity in firm's innovation processes. We propose a review about creativity, resource integration and social networks in order to define a framework, based on founding elements and conditions, to channel such collective creativity.

This study offers two main findings, the first relates to the organization of innovative processes within companies: internet-based collective creativity establishes an architecture of new sources, tools and practices. The second implication is linked to the viral effect that arises in the social media environment and concerns the company's capacity to absorb and manage distortional effects within the users' shared information or brand experience.

Keywords: creativity, social networks, resource integration, open innovation

1. Introduction

Recently, due to the awareness of the large creative potential offered by the interaction of a multitude of individuals through the internet, special attention has been paid to collective creativity (Boulaire, Hervet and Graf, 2010). Social interaction through virtual platforms (Social Networks) enables collective creative processes that companies can lead and develop (Hargadon and Bechky,

2006). Low entry barriers and the constant addition of new members allow for the continuous contribution of new ideas and knowledge. In this way, a continuous resource exchange and integration occurs and all members can share and use these types of mixed resources (Colurcio, Caridà and Melia, 2011).

From this perspective, the primary purpose of our study is to better understand the ways in which companies absorb creative ideas and insights that emerge from social interactions on the internet. In particular, we aim to frame a conceptual model to facilitate and convey companies' collective creative potential.

2. Creativity in the social-network environment

Numerous diverse and occasionally incompatible definitions of creativity exist in the literature (Kozinets, Hemetsberger and Schau, 2008).

In the firm context, creativity is closely related to innovation (Hunter, Bedell and Mumford, 2007; Amabile et al., 1996), as it concerns the art of discovering new solutions and new forms of action in engaging the market: *"Creativity must be employed to solve problems more effectively than competitors are able to do"* (Baccarani, 2005, p. 92). Processes that channel organizational creativity are varied, and they depend on both individual and environmental factors (Perry-Smith and Shalley, 2003; Colurcio, 2005) as well as the way in which each of these components interacts with the others (Bharadwaj and Menon, 2000; Baccarani, 2005).

Since their introduction, social networks such as MySpace, Facebook and Twitter have attracted and retained numerous users who have integrated social-network relationships into their daily practices (Boyd and Ellison, 2007).

Using these web-based services, users develop a personal webpage accessible to other users that is dedicated to the exchange of personal information, experiences and communication (Constantinides and Fountain, 2008).

Social networks provide individuals with relational opportunities from their own desk: they offer a means of presenting oneself, building one's social networks and maintaining connections with others (Ellison, Steinfield and Lampe, 2007). Social networks revolve around the individual because they offer firms the opportunity to activate a rich interaction and experience of knowledge-sharing with consumers.

In a social media environment, creativity becomes collective (Hargadon and Bechky 2006) because social interactions enable experience-sharing and matching that individual reflection alone could not have produced (Kozinets, Hemetsberger and Schau, 2008).

Crowdsourcing in one of the most common business practices that draws upon users' collective creativity (Howe, 2006); it is used to determine valuable contributions because the expertise is distributed among various group members (Peppler and Solomou, 2011).

The role of social network platforms to enhance collective creativity has been noted by many authors (Boulaire, Hervet and Graf, 2010; Peppler and Solomou, 2011; Greenhow, Robelia and Hughes, 2009) who have highlighted the convergence between individual, collective and organizational creativity.

In some cases, community members become partner with the firm/brand to develop new ideas and products that satisfy their needs and desires (Pitta and Fowler, 2005).

> **Nel Mulino che Vorrei—Barilla**
> This is an open online ideas-sharing platform, where a firm tests its ability to listen rather than to speak. Community members, supported by a tutor, can propose, examine and vote on ideas within the community. Moreover, they can follow in real time the firm's progress toward realizing the ideas. Two years since the social platform's launch, the brand has gained important insights due to members' engagement: 5,085 ideas proposed, 5 realized projects, 101,122 votes on ideas and 9,771 posts on ideas.

The infrastructure of social networks may channel creativity and imaginative potential for the purpose of play. Barbie.com, offers an interesting example of a virtual world where Mattel-Barbie immerses users into a magical space. Here, users express their creativity by becoming a stylist, creating new doll models in a way that is coherent with the Barbie lifestyle.

Facebook studio offers another fitting example of how social networks channel creativity. It is a new application for firms, creative developers and communications/marketing professionals.

> **Facebook studio**
> Launched in May 2011, this platform offers examples, start-up ideas and tools for advertisers and social-media marketers. Its goal is for users to learn, share creative projects, find inspiration in great marketing campaigns from around the world and identify resources to strengthen their creative approach to advertising.
> Different sections of the platform contribute to these purposes. For example, the learning lab is an educational space where it is possible to locate product descriptions and to share experiences and ideas with a community of experts.

Social networks also contribute to reducing the barriers to artistic participation that often accompany important creative businesses, such as the music industry. Examples such as You Tube, the best-known video-sharing platform, allows any internet user to watch video clips uploaded by other users, whether individuals or businesses (Boulaire, Hervet and Graf, 2010). YouTube provides users with their own personal internet space, along with the possibility of providing others with a window onto their personal space and thus onto their lives (Pace, 2008).

In the same way in Italy, Sanremo, the most famous and traditional Italian music event, experimented with developing a social network for the selection of a young Italian singer to participate in "Sanremo Festival".

> **Sanremo Social**
> Young and unknown singers who uploaded their own original song and a video had the opportunity to showcase their musical abilities before the online community in hopes of being selected to participate in the Sanremo Festival. Among the 1,174 songs uploaded, fans selected 60 singers to participate in the Sanremo Social Day, when the six Festival finalists were chosen. The winner of the Sanremo Festival was selected by social network members and television viewers via televoting. In only the first three weeks, Sanremo Social attracted 800.000 visits and 150.000 "likes."

Social networks ensure a more efficient process of screening new proposals, thus saving firms time and expense. Moreover, it contributes to increasing satisfaction and commitment among younger generations in the case of a traditional event such as the Sanremo Festival.

3. Resource integration: a key to enhancing creativity

Resource integration is multifaceted topic that encompasses such factors as value creation, new product development, customer engagement and customer participation.

This topic primarily focuses on the customer (Grönroos, 2004, 2011, Vargo and Lusch, 2004, 2008, Brodie, Hollebeek and Smith, 2011), as the resource-integration process is understood as a necessary step to facilitate usage and value creation. This result can be achieved if customers have the necessary type and amount of resources, viz., if firms and other parties can be aligned (Grönroos, 2004) to establish fruitful relationships.

In this way, firms and stakeholders can be classified as resource-providers (Lusch, Vargo and Tanniru, 2010) as well as those who benefit from the resource-integration process.

Many authors had attempted to classify the resources to be used by all stakeholders. All agreed to the presence of different types of resources related to behavior, personal context and the availability of specific tools.

All of these classifications are useful in understanding how all parties can take part in the resource-integration process in different ways and with different results. At the same time, it is important to emphasize the relationship between the available resources and the way in which an actor can contribute to a multifaceted result. To enhance the mutual benefit from resource integration, particular attention to the network of relationships is necessary, with particular reference to the creation of a setting in which parties can interact even if they come from different backgrounds. In this context, participants can act freely (as Virgin has done in calling for "big" new ideas connected to their business areas) or on the basis of a range of choices that the firm provides (as when Audi and Vitaminwater asked for suggestions for new models or products).

> **Fold-it: users-scientists**
> Thanks to an online game involving thousands of people all over the world, the 3D structure of a protein was determined in just three weeks. The game had been created by a group of computer scientists in Washington.
> Non-professionals, biologists, biochemists and other scientists took part in this process, in which a playful environment allowed for collective creativity. The resulting discovery will be applied in the contexts of the pharmaceutical industry and in future research.

The usefulness of resource integration can be verified in the results achieved by several firms (Dholakia et al., 2009, Mahr, Lievens and Blazevic, 2010). It can also be seen in theoretical findings published recently (FP9 by Vargo and Lusch, 2008) and in the past, such as Penrose (1959), who referred to a *"collection of resources"* instead of factors of production.

More and more firms are now developing channels to share resources (SDO by Karpen, Bove and Lukas, 2009). This orientation is itself a resource (Hartmann and Ots, 2010) and the means by which to support all resource integrators in achieving greater value-in-context (Chandler and Vargo, 2011).

Historically, firms focused on their core competencies to gain a competitive advantage and to hone their skills, while they outsourced all other resources, such as knowledge, to network actors who were directly and indirectly connected. As a result of this approach, firms are extending the control of critical resources (Möller and Halinen, 2000) and creating a basis upon which to share and apply a wider set of resources to attain density (Normann and Ramirez, 1993). These concepts can be applied both to service production and to innovation. The contexts that firms create intensify (Muñiz and Schau, 2011) actors' involvement. They also respond to the need for sharing that arises from the uniqueness of knowledge (e.g., customers' preferences in service recovery or for market acceptance, as stated by Hoyer et al., 2010) and the resources that are controlled by outside actors whose innovation potential is higher than the firm's.

> **Ford: Real-World Challenges**
> Ford challenges users to test its new parking technology and encourages them to submit testimonials through social networks to prove the drivers' confidence about the new parking technology's performance.

4. A framework to develop creativity through social networks

The literature review and the empirical study we have conducted allow us to propose a framework to support companies in channeling creativity through social networks (Figure 1).

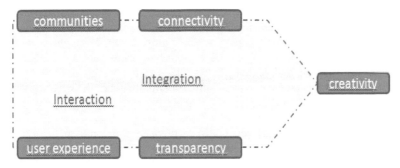

Figure 1 – A framework to channel creativity in social network environments

The framework is based on four founding elements and depends on the existence of two specific conditions—the interaction between the subjects (users/users and users/brand) and the integration of resources between the members—to channel collective creativity to produce useful results (e.g., new relationships, new ideas, rejuvenation of the brand, etc.). The conditions are based upon the necessity that participation occurs and that it is based on dialogue, as creativity may be stimulated and fostered through the viral process of continuous posting and sharing of information, emotions, sensations, and ideas.

Finally, we can say that companies have successfully channeled creativity when they can identify specific new ideas that will be converted into new products (e.g., things, projects, products, relationships, applications, etc.).

Founding elements

The founding elements are the basis of the framework and require the resources and engagement of all parties in the social relationship. On the company side, the establishment of a social media environment requires clear objectives, dedicated resources and particular competencies to design an attractive and "polysensorial" environment.

Communities of practice. A community of practice is not simply a group of friends or a network of personal connections but rather an aggregation of individuals or business partners with an identity defined by a shared domain of interest (Albors, Ramos and Hervas, 2008; Porter and Donthu, 2008).

Connectivity. Connectivity is the ability to link or connect to the internet, which provides access to global online information resources through a computer, laptop, or mobile device. It fosters the shift from a fully connected society to a hyper-connected one, wherein the number of network connections surpasses the number of humans connected to the network (Karakas, 2009).

Transparency. Transparency is the informational symmetry between the parties in the relationship (Prahalad and Ramaswamy, 2004). It is the fundamental condition to foster collaboration between firms and social network users and to develop the value co-creation processes.

User experience. A positive user experience refers to something real and authentic that excites or intrigues users by engaging their senses and emotions (Schmitt, 2009). It emerges when users sense a feeling of pleasure, satisfaction and happiness (Ravald, 2010).

Conditions

Interaction. Interaction is an important driver for the development of the network's experience and value (Ballantyne and Varey, 2008). Through actor-to-actor (A2A) interaction, a dialogue is established, knowledge and other resources are transferred and learning takes place (Gummesson and Mele, 2010). Interaction fosters the involvement and the commitment of the network's actors, thus enabling the development of collaborative creativity (Peppler and Solomou, 2011).

Resource integration. Resource integration is the incorporation of an actor's resources into the processes of another actor in accordance with their expectations, needs and capabilities. It implies a social and cultural process that enables an actor to become a member of a network (Gummesson and Mele, 2010).

5. Conclusions and main implications

Through online social networks, collective creativity may be channeled to very different ends: utilitarian, commercial, ludic as well as social (Boulaire, Hervet and Graf, 2010). We argue that social networks may convey collective creativity, thus enabling a process of crowdsourcing for the development of innovation.

Indeed, according to Davenport and Prusak (1998), knowledge-creating activities take place between people by means of connections and conversations.

Interconnectedness is a crucial property for knowledge integration (Ravasi and Verona, 2001), which joins customers and other stakeholders' resources in innovative value-creating activities. Numerous studies have found that most innovations emerge through a creative process that is not the result of a single inventor but rather of collaborative processes whereby many individuals contribute their individual knowledge, experiences, and strengths (Sawhney, Verona and Prandelli, 2005). This approach, defined as open innovation (Chesbrough 2003; von Hippel 2005), is a paradigm that assumes that firms can and should draw upon external as well as internal ideas (Bücheler and Sieg, 2011). In this model, the integration of customers into the creative process is one of the greatest resources for innovations (Prasarnphanich and Wagner, 2009).

According to the resource-integration approach (Vargo and Lusch, 2008 Gummesson and Mele, 2010; Colurcio, Caridà and Melia, 2011), social media environments foster collaboration between the communities' members and the firms through the sharing of ideas, experiences, knowledge, emotions, etc.

Connectivity and social-network infrastructure facilitate access to a great quantity of information, fostering the development of dialogue and collaboration in a peer-to-peer relationship between the consumer and the firm. According to this view, transparency, to which firms have been opposed in the past (Prahalad and Ramaswamy, 2004), now represents one of the fundamental conditions to enhance collective intelligence (Albors, Ramos and Hervas, 2008). Transparency also facilitates a collaborative approach to creativity that allows a crowd to identify the best idea among a large pool of potential innovations (Di Gangi and Wasko, 2009).

In terms of managerial implications, this study offers two main findings for reflection. The first relates to the organization of innovative processes within companies: internet-based collective creativity establishes an architecture of new sources, tools and practices and provides the opportunity to save time and resources. However, it requires devoting competencies and awareness to channeling collective insights. The channeling process of collective creativity must be well managed and organized.

The second implication is linked to the viral effect that arises in the social media environment and concerns the company's capacity to absorb and manage distortional effects within the users' shared information or brand experience.

References

Albors J., Ramos J. C. and Hervas J. L. (2008). New learning network paradigms: Communities of objectives, crowdsourcing, wikis and open source. *International Journal of Information Management*, 28, 194-202.

Amabile, T., Conti, R., Coon, H., Lazenby, J. and Herron, M. (1996). Assessing the work environment for creativity. *Academy of Management*, 39, 1154-1184.

Baccarani C. (2005). What do you think creativity is and where can we find it? *The Asian Journal of Quality*, 6, 2, pp. 90-104.

Ballantyne D. and Varey R.J. (2008). The service dominant logic and the future of marketing. *Journal of the Academy Marketing Science*, 36, 1, pp. 11-14.

Bharadwaj, S. and Menon, A. (2000). Making innovation happen in organizations: individual creativity mechanisms, organizational creativity mechanisms or both? *Journal Of Product Innovation And Management*. 17 (6), 424-434.

Boulaire C., Hervet G. and Graf R. (2010). Creativity chains and playing in the crossfire on the video-sharing site YouTube. *Journal of Research in Interactive Marketing*, 4, 2, pp. 111-141.

Boyd D. and Ellison N. (2007). Social network sites: definition, history, and scholarship. *Journal of Computer-Mediated Communication*, 13, 1, pp. 210-230.

Brodie, R. J., Hollebeek, L. D. and Smith, S. D. (2011). Engagement: An important bridging concept for the emerging S-D logic lexicon. *The 2011 Naples Forum on Service*, 14-17 June, Capri.

Bücheler T. and Sieg J. H. (2011). Understanding 2.0: Crowdsourcing and Open Innovation in the Scientific Method. *Procedia Computer Science*, 7, 327-329.

Chandler, J. D. and Vargo S. L. (2011). Contextualization and value-in-context: How context frames exchange. *Marketing Theory*, 11, 1, pp. 35-49.

Chesbrough, H. (2003). *Open Innovation: The New Imperative for Creating and Profiting from Technology*. Harvard Business School Press, Boston; MA.

Colurcio, M. (2005). La generazione delle idee. In *"La gestione dei percorsi di innovazione"*. Giappichelli Editore, pp. 73-109.

Colurcio, M., Caridà, A. and Melia M. (2011) Virtual Brand communities to integrate resource and experience. *The 2011 Naples Forum on Service*, 14-17 June, Capri.

Constantinides E. and Fountain S. J. (2008). Web 2.0: Conceptual foundations and marketing issues. *Journal of Direct, Data and Digital Marketing Practice*, 9, 3, pp. 231–244.

Davenport T. and Prusak L. (1998). *Working Knowledge*. Harvard Business School Press, Boston, MA.

Dholakia, U., V. Blazevic, C. Wiertz and R. Algesheimer (2009). Communal service delivery: how customers benefit from participation in firm-hosted virtual P3 communities. *Journal of Service Research*, 12(2), 208-226.

Di Gangi P. M. and Wasko M. (2009). Open innovation through online communities. Knowledge Management and organizational Learning. *Annals of Information Systems*, 4, 3, pp. 199-213

Ellison N., Steinfield C. and Lampe C. (2007). The benefits of facebook friends: social capital and college students' use of online social network sites. *Journal of Computer-Mediated Communication*, 12, 4, pp. 1143–1168.

Greenhow. C., Robelia, B. and Hughes, J. (2009). "Web 2.0 and classroom research: what path should we take now?". *Educational Researcher*, Vol. 38 No. 4, pp. 246-59.

Grönroos, C. (2004). The relationship marketing process: communication, interaction, dialogue, value. *Journal of Business & Industrial Marketing*, Vol.19 No.2, pp 99-113.

Grönroos, C. (2011). Value co-creation in service logic. A critical analysis. *Marketing Theory*, Special Issue, Vol.11.

Gummesson E. and Mele C. (2010). Marketing as Value Co-creation through network interaction and resource integration. *Journal Business Marketing Management*, 4, pp. 181-198.

Hargadon, A. B. and Bechky, B.A. (2006). When collections of creatives become creative collectives: A field study of problem solving at work. *Organization Science*, 17, pp. 484-500.

Hartmann, B. and Ots, M. (2010). What The Heck Is A Mash-Up?: Consumer Generated Media, Value Creation And Resource Integration. *39th EMAC Conference Proceedings: The Six Senses: The Essentials of Marketing*. June 1-4 Copenhagen: CBS Library.

Howe, J. (2006). The rise of crowdsourcing. *Wired Magazine*, (14), pp.1-4.

Hoyer, W., Chandy, R., Dorotic, M., Krafft, M. and Singh, S. S. (2010). Consumer co-creation in new product development. *Journal of service research*, 13, pp.283-296

Hunter, S.T., Bedell, K.E. and Mumford, M.D. (2007). Climate for creativity: a quantitative review. *Creativity Research Journal*, Vol. 19 No. 1, pp. 69-90.

Karakas F. (2009). Welcome to World 2.0: the new digital ecosystem. *Journal of Business Strategy*, 30, 4, pp. 23-30.

Karpen, I. O., Bove. L. L. and Lukas. B. A. (2009). Empirically Investigating Service-Dominant Logic: Developing and Validating a Service-Dominant Orientation Measure. *Australian & New Zealand Marketing Academy Conference*

Kozinets, R., Hemetsberger A. and Schau, H. J. (2008). The Wisdom of Consumer Crowds: Collective Innovation in the Age of Networked Marketing. *Journal of Macromarketing*, Vol.28, 2, pp. 339-354.

Lusch, R. F., Vargo S. L. and Tanniru M. (2010). Service, value networks and learning. *Journal of the Academy of Marketing Science*, 38 (1), pp.19-31.

Mahr, D., Lievens, A. and Blazevic, V. (2010). The value of customer co-creation during the innovation process. *39th EMAC Conference Proceedings: The Six Senses: The Essentials of Marketing*. June 1-4 Copenhagen: CBS Library.

Möller, K. and Halinen, A. (2000). Relationship marketing theory: its roots and direction. *Journal of Marketing Management*. Vol. 16., No. 1-3 pp. 29-54.

Muñiz A. M. Jr. and Jensen Schau, H. (2011). How to inspire value-laden collaborative consumer-generated content. *Business Horizons*, 54, pp.209-217.

Normann, R. and Ramirez, R. (1993). From Value Chain to Value Constellation: Designing Interactive Strategy. *Harvard Business Review*, Vol. 71, 4, July/August.

Pace S. (2008). YouTube: an opportunity for consumer narrative analysis? *Qualitative Market Research: An International Journal*, Vol. 11 No. 2, pp. 213-226.

Penrose, E. T. (1959). *The Theory of the Growth of the Firm*. London: Basil Blackwell and Mott.

Peppler K.A. and Solomou M. (2011). Building creativity: collaborative learning and creativity in social media environments. *On the Horizon*, 19, 1, pp. 13-23.

Perry-Smith, J.E. and Shalley, C.E. (2003). The social side of creativity: A static and dynamic social network perspective. Academy of Management Review, 28 pp. 89-106.

Pitta D.A. and Fowler D. (2005). Online consumer communities and their value to new product developers. *Journal of Product & Brand Management*, 14, 5, pp. 283-291.

Porter C.E. and Donthu N. (2008). Cultivating Trust and Harvesting Value in Virtual Communities. *Management Science*, 54, 1, pp. 113-128.

Prahalad C.K. and Ramaswamy V. (2004). Co-creation experiences: the next practice in value creation. *Journal of Interactive Marketing*, 18, 3, pp. 5-14.

Prasarnphanich P. and Wagner C. (2009). The role of wiki technology and altruism in collaborative knowledge creation. *Journal of Computer Information Systems*, 49, 4, pp. 33-44.

Ravald A. (2010). The customer's process of value creation. *Mercati e Competitività*, 1, pp. 41-54.

Ravasi D. and Verona G. (2001). Organizing the process of knowledge integration: the benefits of structural ambiguity. *Scandinavian Journal of Management*, 11, pp. 41-66.

Sawhney M., Verona G. and Prandelli E. (2005). Collaborating to Create: The Internet as a Platform for Customer Engagement in Product Innovation. *Journal of Interactive Marketing*, 19, 4, pp. 4-17.

Schmitt B.H. (2009). The concept of brand experience. *Journal of Brand Management*. 16, pp.417–419.

Vargo, S.L., Lusch, R.F. (2004). Evolving to a New Dominant Logic for Marketing. *Journal of Marketing*, Vol.68, pp.1-17

Vargo S.L., Lusch R.F. (2008). Service Dominant Logic: continuing the evolution. *Journal of the Academy of Marketing Science*, 36, 1, pp. 1-10.

von Hippel, E. (2005). *Democratizing innovation*. Cambridge, MA: MIT Press.

CHAPTER 7

Creativity and Learning in the Practices of Service Innovation

Tiziana Russo Spena, Cristina Mele

University of Napoli "Federico II" Italy
russospe@unina.it; crimele@unina.it;

ABSTRACT

This paper investigates the role of learning and creativity in enhancing the practice of co-creating service innovation. The analysis focuses on the contributions of both the network of social relationships and technological support. We identify four categories of practices: 1) engaging, 2) exploring, 3) exploiting and 4) orchestrating. By adopting a practice-based approach, service innovation results from ongoing creative and learning practices within web-based environments.

Keywords: innovation, practice, creativity, learning

1 SERVICE SYSTEMS AND SERVICE INNOVATION

A service system is the basic unit of service science. According to Mele and Polese (2011), a service system has four dimensions: (i) customers, (ii) people, (iii) information, and (iv) technology. The interactions between these key dimensions shape two kinds of nets: (i) a social network, which is the social relations that affect people and customers; (ii) a technological network, which is the Information and Communication Technology (ICT)'s patterns and tools fostering people's and customers' participation in the system.

People and customers create value and trigger the value process. Individuals interact, learn, create knowledge, and apply competences by taking actions consistent with certain goals and values. The dense and articulated relationships that characterize every individual—that is his/her social relationships —contribute to a service system's performance. An ICT is a set of nerve systems vital to learning and responding in a value-creating network (Lusch et al. 2010). It fosters the development of interaction and relationships, including inter-systems and intra-systems. Service providers and clients work together with support technologies, such as web services, to ensure a matching of practices within the networks of actors. Service systems manage to integrate the social and ICT nets as they form the basis for a greater value-creating network aimed at creating mutual value through the continuous development of value propositions.

The notion of value co-creation is one of the fundamental ideas of service-dominant (S-D) logic and service science, both of which explicitly recognize the importance of collaboration in sharing and integrating resources. The lever to be engaged in constant value creation is continuous service innovation.

The literature on service innovation is very rich in different conceptualizations. By breaking free from concepts of service innovation focused on services as outcomes, we adopt a conceptualization based on the different meaning of service given by S-D logic (Vargo and Lusch 2008). Thus, innovation is concerned with the development of new competencies or a new combination of existing competencies for the provision of new or increased benefits to one or more parties (Mele et al. 2009). Innovation is therefore developed in terms of finding new ways of participating in resource-integration or of co-solving customer problems (Michel et al. 2008 p.50). Consistent with this perspective, innovation is a process that is continuous, systemic and based on complex interactions between actors, activities and heterogeneous resources (personal, social, technological, etc) (Mele et al. 2010). This process leads to the application of 'open innovation' driven by 'service-centric thinking' (Chesbourough, 2011). Innovation is an open and democratized process (von Hippel, 2005), as a single firm does not possess 'enough knowledge and sufficient human resources to create the innovations that are needed to compete globally' (Lusch et al. 2010, p. 11). The innovation process is thus developed through continuous interactions between a range of stakeholders who integrate their resources to create a common value. This perspective considers stakeholders to be co-innovators (Mele et al. 2009) and new service development to be a process of co-creation. Based on the work by Frow et al., (2010) a recent study by Russo Spena, Mele (2011) identifies five "Co's" in the co-creation of innovation surrounded by various actors (such as customers, suppliers, users, experts, intermediaries and other partners): co-ideation, co-evaluation, co-design, co-test and co-launch. Each of these phases is considered to result from a dynamic and ongoing interaction process involving a group of actors who are interrelated via a dense network.

This paper investigates the role of learning and creativity in enhancing the practices of co-creating service innovation. The analysis focuses on both the contribution of the network of social relationships and technological support. We identify four categories of practices: 1) engaging, 2) exploring, 3) exploiting and 4) orchestrating.

The paper is organized as follows: first, we provide a brief review on creativity and learning and the practice-based approach, and then, we outline the research process and present the four practices.

2 CREATIVITY AND LEARNING

Creativity and learning are often intertwined concepts in many scholars' studies of innovation (Hirschman 1980; von Hippel 2005; Amabile 1997). Learning can be defined as the process of absorbing and acquiring new knowledge (Cohen and Levinthal, 1990; Nonaka and Takeuchi 1995.), while creativity is defined as the generation of new ideas from that knowledge (Amabile 1997). Although there are various ways to innovate, developing and applying knowledge is regarded as the key to maintain long-lasting innovative ability, as well as (Nonaka and Takeuchi 1995) to continually build firms' capacity to generate and discover applications (Howard et al. 2008). Learning and creativity are considered the double face of the innovation process by which knowledge is developed and transformed into an attractive value proposition for users and customers (Amabile 1997, Spohrer and Maglio 2008). For service innovation, a crucial role is performed by knowledge defined a 'meta-resource' (Mele and Polese 2011)—that is, a primary factor that is capable of activating the development of other resources. From this perspective, the service innovation is understood to be primarily a learning process mobilizing knowledge in and through an organization.

A wide debate on how creativity links with learning development has flourished into the field of psychology and cognitive science literature (Howard et al. 2008). The view was that creativity is a part of the general process by which people use and acquire new knowledge. It was the 'cognitive approach' to creativity (Amabile, 1996) that first put into focus the domain-relevant knowledge aspect of creativity as it is referred to the sequence of cognitive activities that generates novel and appropriate productions in a certain problem context (Lubart 2000). According to the cognitive-approach (Gardner 1982; Finke et al 1992), creativity is mainly perceived to be synonymous with 'problem-finding, problem-solving' and 'thinking skills'. This approach emerges as a more compelling perspective, revealing the potential for fostering creativity (each human has the potential to be creative) through education and training, especially when augmented by the appropriated use of new technological tools (Lubart 2000, Zeng et al 2009, Prandelli et al. 2009).

Other authors linked particular types of thinking, labeled 'lateral', 'divergent' or 'intuitive,' with creativity and considered them to be providing more fertile ground for creativity. According to De Bono (1990), creativity involves critical thinking (*i.e.,* observing laterally the information that has been available to everyone else). In other words, creativity is the ability to perceive, in any event, diverse aspects that would fit into different patterns.

Within the design literature (Ulrich and Eppinger 2000; War and O'Neill 2005), the outcomes of critical thinking were demonstrated when groups of people engaged in the creative process together, offering multiple perspectives and providing opportunities for people to take different approaches to problem solving.

Other scholars within organizational learning and knowledge management research note that many creative activities involve social and collaborative processes spanning functional boundaries within and between organizations. If creative insight would take place within the human mind, these perspectives stress that it is highly soaked with a long process of social interaction (Gherardi and Nicolini 2006). Additionally, recognizing the importance of environment or context, communities and networks of practice (Lave and Wenger 1991) are fertile venues for research. These studies explore how social factors promote creativity in a collaborative contest. Such communities have been considered mechanisms to catalyze situated and distributed knowledge (Paavola et al. 2004) and to extend learning processes far beyond the companies' boundaries (Prandelli et al. 2008).

Social and collective creativity flourish as a master discipline, promoting examples of the creation of valuable and useful ideas, procedures or processes by individuals working together within a complex social system. This interpretation presupposes a conception of creativity and learning as a mainly collective and situated activity as well as a social conception of knowledge and knowing (Paavlona et al. 2004).

3 A PRACTICE-BASED APPROACH TO SERVICE INNOVATION

Service systems are a collection of practices performed by people through the use of technology (Orlikowski 2000; Schatzki 2005). The idea of these practices is not unknown in the mainstream innovation literature in terms of well-codified internal routines. However, we advance this traditional vision of practice in order to adopt a wider meaning, in line with what Schatzi (2005) labels as the practice turn. In the last decade, the term practice has been used as a lens for the understanding of social and organizational phenomena with a renewed view. Although there is not a unified theory of practice but an array of theoretical perspectives, the practice perspective is an epistemological choice to examine phenomena in organizations and society. Central to the practice perspective is the acknowledgement of the social, historical and structural contexts in which actions take place. Contextual elements are thus considered to shape how individuals learn and how they acquire knowledge and competence. The practice lens joins the individual and collective dimensions, human and technological elements, outlining doing and knowing.

Corradi et al. (2010) well synthesize the renewed interest in practice theories:

> One reason for the renewed interest in practice theories in organization studies is linked to the search for a non rational-cognitive view of knowledge. Central to the practice perspective is acknowledgement of the social, historical and structural context in which knowledge is manufactured. Practices allow researchers to investigate empirically how contextual elements shape knowledge and how competence is built around a contingent logic of action (p. 267).

The adoption of a practice-based approach shifts our investigation from examining creativity and learning by adopting a cognitive and individual vision to a social and situated perspective. "Learning is not a phenomenon that takes place in a person's head rather it is a participative social process" (p.267). Therefore, knowledge is considered a form of competent reasoning, and thus, it is embedded in ongoing practices. It is a 'collective good' enacted in the collective actions through practices in communities (Schau et al. 2009).

Capturing knowledge therefore requires an appropriate arrangement and disarrangement of the practices and constituents that are the net of actions, people, context artifacts and tools. In this process, a crucial role can be performed by technology. Starting from the assumption that technologies have two dimensions-that of the artifact and that of its use (what people do with the technological artifact in their practices)-the practice lens allows the ability to examine "how people interact with a technology in their ongoing practices, enact structures which shape their emergent and situated use of that technologies" (Orlikowski, 2000, p. 404).

In sum, knowledge, learning and technology are key factors in service innovation. However, the practice-based approach in innovation has not yet been fully employed, with the exception of few studies (Dougherty 2004, Russo Spena and Mele 2011). The main insight coming out from these works is that the practice lens allows for analyzing innovation as an array of factors -namely, subjects, actions, tools, and know- how-rather than the mere innovation output.

4 RESEARCH PROJECT

This paper is part of an ongoing research project aimed at analyzing service innovation according a practice-based view. In line with other studies adopting the practice approach (Schau et al. 2009), our project's method moves the unit of analysis away from the individual consumer and the single firm -or the single brand- to the practices across individuals and networks. Some findings have already been published (Russo Spena and Mele, 2011).

In this paper, we analyzed a specific encounter contest: "the websites." The literature on open innovation and communities stresses the role of the Internet in shaping innovation as a socially interactive and open process that emerges through the collaboration of empowered users. During an eighteen-month period, we analyzed the innovation practices occurring within a Web contest of twenty-one companies. By following the suggestions by Kozinets (2002), we carried out netnographic research within websites and related forums. Our data include an analysis of content websites and, messages posted by users and participants and naturalistic observation of users' activities.

In analyzing how the social and ICT networks shape service innovation through creativity and learning, we identified four categories of practice: 1) engaging, 2) exploring, 3) exploiting, and 4) orchestrating. Each of the practices feed learning processes and is performed by employing knowledge and creativity. In the following section, we present the practices and use cases as examples (table 1 in appendix).

5 THE PRACTICE OF ENGAGING

Service systems widen their ICT net through the development of websites as context for innovation. The web is a tool to connect people and develop a social net aimed at innovation.. The open innovation platform has the goal of finding talented people who are skillful at implementing improvements, solving problems and developing projects. The approach to co-create innovation with external people complements traditional research and development activity. This approach is a way to find partners and develop social networks for building new business opportunities. The practice of engaging is based on an array of actions performed by service systems, namely, 1) soliciting, 2) socializing and 3) expanding.

Soliciting consists of letting people know about the innovation project in order to involve them. It involves attracting a number of supporters able to interpret and recognize the project as interesting. Service systems design the website to catch the right people who want and have the interest and competences to participate. Individuals must register in order to access and have the ability to contribute with their creativity in terms of ideas and skills.

The crucial step in sharing knowledge and creating new knowledge by enabling learning processes is *socializing*. The social net has its core in the innovation project community, which is the source and the medium for socialization. People participating in the innovation project can take part in the community. Creating a common ground is a way to transform separate understandings into coherent views of solutions. People collectively make sense of new ideas or insights (Nonaka and Takeuchi, 1995), and ongoing articulation is central to this collective sense making.

In *expanding* the "range of possibility", service systems listen beyond the community sphere hosted on the company's web sites. They move toward the social virtual spaces (Facebook, Twitter, Ovi, etc) where the brand or interesting topics are discussed (by customers and non-customers). Service systems need to expand their peripheral vision beyond their own customers and fans by making an effort to gather ideas from as broad an audience as possible.

In sum, the practice of engaging focuses on combining a heterogeneous group of subjects in pursuit of the same goal. The ICT net is a means to build and strengthen the social net by connecting individuals, companies, competitors, venture capitalists, research centers, universities, design institutes and intermediaries. Service systems connect not only people but also ideas and resources (artifacts, tools, language, etc). Engaging people aims to foster creativity that is geared toward producing new ideas and solutions in all the phases of the co-creation process. The higher level of the practice of engaging, the higher the potential exists for new ideas and proposals. Additionally, the higher level of the practice of engaging, the higher the possibility of co-creation in the innovation development.

6 THE PRACTICE OF EXPLORING

Within service systems, innovation mainly concerns creating variety in experiences, which is associated with broadening existing knowledge bases toward new and unexpected creativity paths (Zeng et al. 2009). The practice of exploring feeds innovation, including activities such as searching, creating variation, risk taking, experimenting and playing (March 1991). The ICT net puts in practice these activities by allowing the continuous dynamic of knowledge creation and reconstitution through human and non-human interactions (Orlikoswki 2000). The practice of exploring concerns three activities: 1) generating, 2) articulating and 3) discussing. These three activities describe the attempt to expand the knowledge and creativity dispersed in service systems and to tap them into the constitutive activities of innovation processes.

The *generating* aspect consists of giving off users' creativity that involves the identification and generation of opportunities, fresh ideas and novel concepts. To produce a large number of ideas, service systems utilize different tools and methods through the web nexus. Service systems engage users to perform collaborative work not only on simple ideas or concepts' generation but also in advancing business and innovation to improve competitiveness. Most of the generation tasks are defined within problem finding contexts and open-ended structures within the web nexus as a stage for creative solutions.

Through *articulating* actions, the users' activities are organized around collaborative knowledge advancement through shared aims. The users are challenged to elicit their creativity and integrated it into the co-design process promoted by service systems. In many cases, users work on an authentic new project, involving skills, knowledge and competencies often at the level of those used by professionals and experts in the design field. Articulating actions invest users with more compelling activities in order to externalize tacit knowledge and to explicate and formalize the connections between their ideas. In many cases, users are brought in contact with specific problem finding-solutions by creating and working with specific knowledge artifacts to store, visualize and improve knowledge and creativity expression. These systems accompany learning by doing and learning by interaction processes for the user and are the primary instrument through which users should be challenged and motivated to learn aloud from the activities that they perform and to create a positive design experience. To advance this experience, the open source mechanism for development is also used. This mechanism challenges users knowledge and creativity not only in developing a requested concept or product but also in advancing technology frontiers by using open source software on which users can work within true design freedom.

Often, the interactive multimedia tools support the users not only by contributing to their original idea, design and picture but also in *discussing* and evaluating the ideas created by others (in many cases also proposed by companies) or affecting the others' contributions. Through discussing activities, the most promising ideas and concepts can be selected among those ideas that inspire the group. This discussion provides a foundation for subsequent creative ideas. The produced ideas, designs, and thoughts play a prominent role in interaction and should provide a useful foundation for other users' creativity.

In sum, the practice of exploring enables users to place creativity in the context of change and experimentation to support the dynamics of knowledge creation and regeneration in a community's way of doing and meaning (Paavola et al. 2004). The ICT net prompts the open-ended nexus of creativity around collaborative knowledge creation. Through the interaction in generating, articulating and discussing, service systems produce the transformations of users' shared ideas, objects and thoughts. The practices of exploring promote the creation and regeneration of a shared understanding, which enables new creativity to emerge.

7 THE PRACTICE OF EXPLOITING

Exploitation represents the other side of service innovation, complementing the learning and creativity processes of group and institution (March 1991). Through the ICT net, the service systems perform an array of creative activities and draw on existing competences. These activities describe the attempt to combine and institutionalize knowledge and experience in new value propositions, norms and routines to advance to the forefront of the innovation process. Specifically, the practice of exploiting involves three activities: 1) experimenting, 2) appropriating and 3) disseminating.

Experimenting includes the intentional introduction and application of users' knowledge to validate and suggest refinements and improvements to make companies offer better value propositions. In this stage of the innovation process, users support learning and creativity with working on "real" versions that can scale up and grow into more advanced outputs. Many companies posit the dynamic of experimentation to be the heart of their innovation process. The feedback from skilled users allowed companies to assess the extent to which new proposals satisfy clients' needs. This assessment, in turn, will help identify room for further improvement. The website allows users of various backgrounds to provide input. By increasing the number of different views , new solutions to problems (especially concrete problem statements to improve service design) can be found.

Most definitions of creativity suggest that the creative act is also the recombination of materials in a novel way (Zeng et al .2009) In this way, *appropriating* performs activities to support creative work to let users make associations in order to create their novel commercial concepts. In the web domain, the mass-customization approach is widely used to enhance creativity through personalization, and personalization has been identified as one of the key dimensions of value creating activities. Users enjoy creating value by themselves, for themselves, or for others users by tapping into their own knowledge and skills complemented by the company. These creations for self and for others practices involve an extension of the consumer's proactive role widening his capacity to put their creativity to work (Nuttavuthisit 2010). In some cases, the users' creativity and learning skills can be challenged within a real business innovation experience.

The success of new ideas depends on building acceptance from a wider audience. This acceptance occurs if the widest share of prospective clients is persuaded to buy the new output because of effective networking of *disseminating* activities. Online activities such as user-generated communication contests, web-enabled word of mouth, viral messaging, and virtual communities are all tools that elicit users' creativity to share brand good newness. These activities inspire others to use of the new solutions or turn them in to real evangelists preaching and spreading the appreciated features of newness (McAlexander et al 2002). Viral communication promoted by protagonist users grouped around a specific interest and values are expected to influence the purchasing expectations and trends (McAlexander et al 2002).

In sum, the practice of exploiting describes the activities through which users' creativity and knowledge work in a context of use, refinementt and dissemination to support the dynamics of knowledge transmission, thus allowing for routine emergence in a social way of doing and sensing. Through interaction in experimenting, appropriating, and disseminating, users' creativity joins around collective sensing and language. Service systems promote users' alignment about perceptions of what matters and the manner in which they attend to things.

8 THE PRACTICE OF ORCHESTRATING

Service systems frame innovation practices as a problem of value creation. They strive to govern learning and creativity as a way to create and exploit value creation-driven knowledge. Such knowledge is practice-based and is produced continuously in situated action, as people draw on their presence in social settings, on their cultural background and experience, and on conscious and sensory information (Tyre and von Hippel, 1997; Orlikowski, 2000). The ability to put know-what and know-how into practice depends upon the contest, the tools, and the skills service systems enable. These skills include tapping into collective knowledge held by a community. Practice-based knowledge is indeed collective, since no person can know all the heuristics or principles involved, or possess all necessary experience. Competent practitioners need to know how to access, interact, and participate in the community (Wenger, 1998). Thus, service systems develop practices of orchestrating to design the contest and set the conditions to allow a better flow of the other practices—engaging, exploring and exploiting. In doing so, the orchestration of the ICT and social nets foster the service innovation and its experience. The practice orchestrating involves three activities: 1) framing, 2) mobilizing, and 3) strengthening.

Framing is about the design of the innovation context in order to foster knowledge mobility and sharing. The aim is to create a virtual environment for the exploration and exploitation of creativity oriented to specific scopes (new proposals). It also consists in the proper arrangement of the resources, such as tools, images, and artifacts, thus fostering creativity and learning. This process aims at *mobilizing* creativity, that is, allowing people's knowledge to emerge and learning process to flow. Service systems try to capture and represent individual and collective knowledge by using specific means, such as text documents, drafts, videos with informal language, and text- and relation-based code.

The building up of social and ICT nets as key factors in enabling service innovation is defined as *strengthening*. Web-enabled innovation needs a form of orchestration of the community and other involved actors to exploit its potential and to transform ideas and contributions into new solutions to be marketed.

In sum, the practice of orchestrating is about reflecting on how the practice and the work can be organized to mobilize and use knowledge in order to spur creativity and learning. Service systems perform the role of orchestrators of innovation initiation and appropriation.

9 CONCLUSIONS

This paper investigates the role of learning and creativity in enhancing practices of co-creating service innovation. By adopting a practice-based approach, service innovation results from ongoing creative and learning practices within the web-based environment. Creativity is co-constructed in social learning interactions and within social settings. The ICT net interweaves with the social net and supports social relationships in their value aims and practices. Both nets support and foster learning processes and knowledge use, creation and institutionalization.

Four practices of innovation are identified at the foundations of innovation in service systems: 1) engaging, 2) exploring, 3) exploiting and 4) orchestrating. By deploying practices of engaging, exploring, exploiting and orchestrating, all participants learn from and contribute to the knowledge of the service systems. Additionally, all participants encourage other members to improve and elaborate on ideas, thoughts and output). The web is the tool to connect people and develop a social net aimed at innovating through collective knowledge and creativity. New knowledge is generated through relationships (Prandelli et al.2008) and new collective knowledge allows the creativity process to emerge and progress toward a shared aim. The interactions between networks of actors allow for the exploitation and exploration of complementarities and interdependences, reinforced by the web technology endowment. Practices are centripetal forces that make the base for utilizing the knowledge of the virtual environment in order to foster creativity and innovation. Most innovations are a mixture of emergent adapted and adopted practices, which become sharpened over time by interaction within social interconnected activities. Thus, a key challenge of service innovation for service systems resides in their ability to interweave the social and ICT nets in order to foster practices for open innovation as they enable users' knowledge and creativity. Collaborative users who create and learn from (and with) others in the work place of innovation (the web setting) call for managing a new kind of web interactions aimed at co-creating service innovations. There is clearly a need for more research into how complex creative service systems can foster virtual working and involve dispersed and often mobile transient groups.

REFERENCES

Amabile, T. 1997. Motivating creativity in organizations: On doing what you love and loving what you do. *California Management Review* 40:39-58.

Chesbrough, H. 2011. *Open Services Innovation: Rethinking Your Business to Grow and Compete in a New Era*, New York, John Wiley & Sons, Inc.

Cohen, W.M. and D.A. Levinthal. 1990. Absorptive capacity: A new perspective on learning and innovation. *Administrative Science Quarterly* 35(1):128-152.

Corradi, G., S. Gherardi, and L. Verzelloni. 2010. Through the practice lens: Where is the bandwagon of practice-based studies heading?. *Management Leraning*, 41(3): 265-283.

De Bono, E. 1990. *Six Thinking Hats*. England: Penguin.

Dougherty, D. 2004. Organizing Practices in Services: Capturing Practice-Based Knowledge for Innovation. *Strategic Organization* 2(1): 35-64.

Finke, R.A., T.B. Ward, and S.M. Smith. 1992. *Creative cognition: Theory, research and applications*. Cambridge, MA: The MIT Press.

Frow, P., A. Payne, and K. Storbacka. 2010. Co-creation: A Framework for Collaborative Engagement, Proceedings of *18th International Colloquium in Relationship Marketing*. Henley Business School, UK, 27 - 30 September, 2010.

Gardner, H. 1982. *Art, Mind and Brain: A Cognitive Approach to Creativity*. New York: Basic Book.

Gherardi, S., and D. Nicolini. 2006. *Organizational Knowledge: The Texture of Workplace Learning*. Blackwell Publishing, Ma.

Hirschman, E.C. 1980. Innovativeness, novelty seeking, and consumer creativity. *Journal of Consumer Research*. 7(3): 283-295.

Howard, T.J., S.J. Culley, and E. Dekoninck. 2008. Describing the creative design process by the integration of engineering design and cognitive psychology literature. *Design Studies* 29: 160-180.

Lave, J.W. and E. Wenger. 1991. *Situated Learning: Legitimate Peripheral Participation.* Cambridge: Cambridge University Press.

Lubart, T.I. 2000. Models of the creative process: past, present and future. *Creativity Research Journal* 13(3/4): 295-308.

Lusch, R. F., S. L Vargo, and M. Tanniru. 2010. Service, value networks and learning. *Journal of the Academy of Marketing Science*, 38:19-31.

March, J.G. 1991. Exploration and Exploitation in Organizational Learning. *Organization Science* 2 (1): 71–87

McAlexander, H.J., J.W. Schouten, and H. F. Koenig. 2002. Building Brand Community. *Journal of Marketing* 66(January): 38-54.

Mele, C., M. Colurcio, and T. Russo Spena. 2009. Alternative Logics for Innovation: a call for service innovation research. *Proceedings of Forum on Service*, Napoli, 2009.

Mele, C., T. Russo Spena, and M. Colurcio. 2010. Co-creating value innovation through resource integration. *International Journal of Quality and Service Science* 2(1):60-78.

Mele, C., and F. Polese. 2011. Key dimension of Service Systems: Interaction in social & technological networks to foster value co-creation, in *The Science of Service System* Demirkan, H., Spohrer J., Krishna V., Springer.

Michel, S., S.W Brown, and A.S. Gallan. 2008. An expanded and strategic view of discontinuous innovations: Deploying a service-dominant logic. *Journal of the Academy of Marketing Science* 36(1):54-67.

Nonaka, I. and H. Takeuchi. 1995. *The knowledge-creating company.* New York, Oxford: Oxford University Press.

Nuttavuthisit, K. 2010. If you can't beat them, let them join: The development of strategies to foster consumers' co-creative practices. *Business Horizons* 53:315-324.

Orlikowski, W. J. 2000. Using Technology and Constructing Structures: A Practice Lens for Studying Technology in Organizations. *Organization Science* 11(4): 404-28.

Paavola, S., L. Lipponen, and K. Hakkarainen. 2004. Modeling innovative knowledge communities: A knowledge-creation approach to learning. *Review of Educational Research* 74:557-576.

Prandelli, E., M. Sahwney, and G. Verona. 2008. *Collaborating with Customers to Innovate: Conceiving and Marketing Products in the Networking Age.* Edward Elegar, London.

Russo Spena, T., and C. Mele. 2011. Co's in innovation: co-creation within a practice-based view. Proceedings of *Forum on Service*: 14- 17 June Capri.

Schatzki, T.R. 2005. Peripheral Vision: The Sites of Organizations. *Organization Studies* 26(3): 465-484.

Schau, H., A. Muñiz, and E.J. Arnould. (2009). How Brand Community Practices Create Value. *Journal of Marketing* 73(5):30-51.

Spohrer, J., and P. Maglio. 2008. The Emergence of Service Science: Toward Systematic Service Innovations to Accelerate Co-Creation of Value, *Production and Operations Management* 17(3):238-46.

Tyre, M. J., and E. von Hippel. 1997. The Situated Nature of Adaptive Learning in Organizations. *Organization Science* 8(1):71-83.

Ulrich, S., and D. Eppinger. 2000. *Product Design and Development*, Second Edition. McGraw-Hill, New York.

Vargo, S L., and R.F. Lusch. 2008. Why "service?. *Journal of the Academy of Marketing Science* 36: 25-38.

Vargo, S.L., 2009. Toward a transcending conceptualization of relationship: a service-dominant logic perspective. *Journal of Business and Industrial Marketing* 24 (5-6): 373-379.

von Hippel, E. 2005. *Democratizing Innovation*, MIT Press: Cambridge, MA.

Ward, T. B., S. M. Smith, and R.A. Finke. 1999. Creative cognition in *Handbook of Creativity*, Robert Sternberg ed., New York, NY: Cambridge University Press, p.189-212.

Wenger, E. 1998. Communities of Practice. Learning as a social system', *Systems Thinker*, http://www.co-i-l.com/coil/knowledge-garden/cop/lss.shtml. Accessed December 30, 2002.

Zeng, L., R.W. Proctor, G. Salvendy. (2009). Fostering Creativity in Service Development: Facilitating Service Innovation by the Creative Cognition Approach. *Service Science* 1(3):142-153.

Table 1 Practices of innovation

Practice	Actions	Contents	Examples
Engaging	Soliciting	"Sign in' to contribute to creativity"	On the main page of **Mystarbucksidea,** an individual can click on 'sign in' and create an account to become a current member. This will allow the individual to give ideas and make comments. Open innovation at **P&G** works both inbound and outbound and encompasses everything from trademarks to packaging, marketing models to engineering, and business services to design. P&G lists the needs company is looking for solutions to with the slogan "do you have what we need?"
	Socializing	Create common ground	**NelMulinochevorrei** hosts Facebook connection links that allow users to directly access without registration. This allows the users involved to become larger in number and sufficiently mix in their background and views. **Popcuts'** catchphrase is 'Be social. Go out and spread the word'. Fans do not keep the music to themselves. With MiniStores, people can showcase their Popcuts music on Facebook, MySpace, their blog, or anywhere.
Exploring	Expanding	Find out newness beyond peripheral vision	**Nokia's** Ideas Project is an online community for everybody from all around the world to brainstorm. Facebook, Ovi and Twitter links allow users to be in real time connection with companies, developers, experts, NGO's and all who can help in the realization of ideas . **Xerox** has fourteen blogs, a Twitter account (@XeroxCorp), a Facebook page titled "So, what DOES Xerox do?," A LinkedIn profile page, and a YouTube XeroxCorp Channel with over 120 video uploads.
	Generating	Identify and generate fresh ideas and novel concepts	A **MyStarbucksidea** contest obtains ideas on several topics including experience, atmosphere and location, ordering, payment, and pick-up in addition to compelling aims, such as building community, social responsibility and "Outside USA." **Electrolux** global design competition invites students and graduates of industrial design to compete for a six month paid internship and prize money by submitting ideas that develop home appliances.
		Generate solution within problem finding process	**BMW's co-creation lab** is a virtual meeting place allowing interfaces between external innovation sources and developers from the BMW Group. Everyone can submit online ideas, concepts, and patents on new technologies and services. **Fujisty Siemens'** innovation contest challenges the users to anticipate how date center function in the future and to find out what service will be required in the years ahead.

Phase	Activity	Example
Articulating	Elicit user creativity in a professional environment	The "App my Ride" contest by Volkswagen seeks Apps and ideas for possible Apps for a prototype of a future Volkswagen infotainment system. The goal of the contest is not only to gain innovative applications but also to involve interested users, coders and developers into the process of developing Apps. P&G Connect+Develop catches higher advanced expert in co-design practices. Through a highly selective registration process, participation on to design activities is open only to advanced users who can provide solutions (detailed in project and prototype) that have already been reduced to practice.
Articulating	Open-sourcing to user creativity to advance technology frontiers	LEGO released as open source firmware of the Lego Mindstorms NXT- the robotics toolset available on Lego platform for armchair inventors, robotics fanatics, and LEGO fans to build and program robots that do what they want. In the section "Featured Technologies" of Nokia's developer web site, open code mechanisms are used to develop new technology applied to mobile phone and related services and to connectivity application for computer.
......(Explori.) — Discussing	Discuss on users thoughts about specific or general topics	The Watch design community of Swarovski allows users to have the ability to comment on designs and provide suggestions for improvement. The Discussion section of Fiat Mio is a sensing platform to acquire knowledge on what people think about different topics. Each discussion starts with a short demand, and different solutions are presented. Visitors are invited to write a comment and to provide their opinions.
Experimenting	Build users creativity on the ideas of others	The Threadless design community allows hobbyists and professional graphic designers to interact in a variety of ways in discussing reviewing, and submitting concepts and using their interaction and responses to promote spontaneity and excitement.
Experimenting	Put skills to work to meet the targeted standard	Trough broadcast invitation on its web site Blizzard Entertainment recruits heavy hitters who access a Beta testing portal (a secure site) for playing the game where testers interact with one another. The company, by observing game's reactions and listening to feedback, refine features of game.
Experimenting	Refine and improve to make offering better.	Open Xerox Web portal hosts technology prototypes from the Xerox R&D labs, making them accessible to the external user community before the launching of a product offering.
Exploiting — Appropriation	Created users personal	The Nokiabeta labs community invites users to test pre-commercialized applications. By employing crowdsourcing, customer feedback and user testing, the firm can test-market its new applications and obtain ideas for further development. At NIKEiD, the customer becomes the designer of his own shoes. He can choose which area of the shoe to personalize, changing the color and fabric to what he desires, adding his personal look and feel to a selected

Disseminating	commercial concepts	item. If the customer does not want to have to design the shoes, they can choose from a range of pre-made designs or buy designs other customers have made.
	Set up users' personal business experience	**Spreadshirt** is a creative platform for custom clothing. The users can create and buy custom t-shirts and other personalized apparel and sell it through a free T-Shirt Shop that the users can easily open up. A special section of the web site provides instructions on how to setup shop and attract customers.
	Make protagonist the users	In 2009, the **Nestlé** campaign for the launch of 'Twitter Pulse' format allowed users to post tweets into an advertising unit launched on different sites with questions related to their product. Twitter could then respond to the question and their answer would appear in real-time on the advertising units. All the tweets were moderated by Nestlé before showing. By allowing their community to do the talking, they increased brand awareness and engagement.
	communication contests	For its Aggregated Services Router (ASR) launch, **Cisco** executes entirely online leveraging social media. The company used video to educate its audience and engage them in a gaming world. In addition, the firm assembled videos and images in a widget format encouraging video blogging and the spreading of information. Cisco seeded its Networking Professionals Technology Community Forum with launch related discussion topics and gave customers an "Ask the Expert" function.

Mobilizing			
....(Dissem.)		Evangelize users to spread the appreciated features of newness	The team at **Articulate** used internal and employee blogs and Twitter accounts to launch the new screen casting tool called Screen. Within a few days, the campaign received coverage from major industry blogs and mainstream outlets such as The New York Times. Key to the launch was the first review by the hugely popular ReadWriteWeb blog, which posted its review in conjunction with form official launch. @copyblogger (Brian Clark) tweeted the launch news to his 31,000+ followers. **Plantronics**, for the launch of its new product, CS70N, identified through a survey its Brand evangelists. Once selected, they were given access to online tools to help to facilitate their conversations about product.
Framing		Arrange the resources to let creativity emerges	**Lego Mindstorm NXT** succeeded with a co-design orientation framed by a strong focus on knowledge transfer and learning networking. Lego has collaborated with users in editing books to better explain the Lego Mindstorm NXT technology. In **Volkswagen**, contest users can test apps with the "AppPlayer". They have to download the software development kit, which includes the AppPlayer, as well as the instructions and the available data streams.
Orchestrating / ...(Exploi.)		Allow collective knowledge to emerge.	**Nokiabetalab** mobilizes people knowledge by posting beta and experimental application with the slogan "Try what you like. Say what you think". In the last **Fiat** project, more than 2 mio net-surfers of 160 counties over the world, gave more than 10.000

		ideas on the website www.fiatmio.cc.
Strengthening	Foster the creative engagement	On the website "Create your style", Swarovski gives users the support of online tutorials, beginners' guides, techniques, online design tools and other tools useful to create new designs with Swarovski elements. Users can also find an overview of all of the designs produced by the sparkling community.
	Use technology convergent	Threadless launched an iPhone application to let people participate in design challenges. "Create your style with Swarovski elements" introduced a free app for the iPhone and iPad to offer users a direct connection to inspiring ideas, sparkling designs and the latest crystal innovations. The Fiat Mio Project adopts the tool of Commons creative licenses. This method allows users to get, diffuse, and combine the ideas that will be developed by Fiat team, as well as to make available to whoever uses the technological knowledge generated by Fiat Mio.
	Feed community and social net	Threadless tries to widen people's interaction through the involvement of social networks (e.g., Facebook) and communities. The company has a wide community with several activities, namely, forum, tee-V, artists program, designer interviews, thread spotting, and street team.

CHAPTER 8

Service Design to Evoke Users' Enthusiasm —Proposal and Evaluation of a Museum Information Providing Service

Motoya Takahashi, Yuki Yasuma, Miwa Nakanishi

Keio University, Fac. of Science & Technology, Dept. of Administration
Engineering
Hiyoshi 3-14-1, Kohoku, Yokohama 223-8522, Japan
takahashimotoya@a5.keio.jp, yasuma@z3.keio.jp, miwa_nakanishi@ae.keio.ac.jp

ABSTRACT

In psychology, users' enthusiasm for products or services is considered to be a type of intrinsic motivation. From the viewpoints of emotion, cognition, and ability, Hunt suggested that enthusiasm is evoked when users' perceive an adequate gap between their own characteristics and those of an object. Our previous study reported objective methods for detecting intrinsic motivation and quantifying a psychological gap. In this study, we experimentally produced a service that exposed each user to an adequate psychological gap from his characteristics, and conducted a scientific evaluation. In particular, by focusing on the case of an information service at a museum, we correlated the observational and subjective data of the participants' reaction to the service. The results of the analysis of the participants' behavior, blink rate, variation in the regional cerebral blood volume suggest that users' enthusiasm can be increased to a greater extent by applying a scientific method to compute an adequate gap from each person's characteristics.

Keywords: Enthusiasm, Intrinsic motivation, Adequate gap, Museum, Information Provision Service

1 INTRODUCTION

In a mature society, it is important to study the aspects of users' minds as well as the functionality or performance of products and services in order to effectively attract the attention of users.

In psychology, users' enthusiasm for products or services is considered to be a kind of intrinsic motivation [1]-[10]. Intrinsic motivation involves curiosity and satisfaction. From the perspectives of emotion, cognition, and ability, it has been suggested that enthusiasm is evoked when users perceive an adequate gap between their own characteristics and those of an object [9] [10]. Our previous study described objective methods for the detection of intrinsic motivation and the quantification of psychological distance [11].

We hypothesize that products and services that have an adequate psychological gap elicit users' motivation to use them. In this study, we attempt to verify this hypothesis through a case study of a concrete service by applying a psychological gap to an information service.

2 EXPERIMENT

2.1 Outline of the Experiment

Different users have different characteristics. Accordingly, to create services that evoke enthusiasm, it is necessary to provide an effect that is appropriate for each user. In particular, by focusing on the example of a museum, we constructed an application to provide different sets of information by using an optical see-through head-mounted display (HMD), thus enabling each user to experience an adequate psychological gap. This application was developed according to the following three steps.

STEP 1. Identifying users' intrinsic characteristic

From the perspectives of emotion, cognition, and ability, the gap between a user's characteristics and those of an object cannot be observed directly. Moreover, developing a method to measure this gap would be difficult. Thus, we considered users' characteristics in relation to the multiple aspects of emotion, cognition, and ability as simply a kind of intrinsic characteristic.

Users were given thirty items that are of interest to young people (selected based on [12]; see Table 1), and their level of interest for two randomly chosen categories was scored by a paired comparison method. From the results, the category rated as most interesting was construed as the intrinsic characteristic of each user. For example, a user rating "Sports" as the most interesting category is regarded as having the intrinsic characteristic of "sporty."

Table 1 Items of interest given to the participants

Sports	Travel
Music	Art
Game	Comic,Animation
Outdoor Life	Eating/Drinking
Photo	SNS
Reading	Gamble
Movie	

STEP 2. Derivation of an adequate gap

In our previous study, it was reported that an adequate gap exists when the distance from the intrinsic characteristic is around 0.1–0.2 and each pair of categories compared is scaled by the fixed-Scheffe's method and standardized to 0–1. The gap also indicates the distance between the intrinsic characteristic and each category.

Therefore, this study selected a category falling in the gap for each user by using the derivation method described above.

STEP 3. Information

We produced annotation information related to the above-mentioned thirty items as a movie for each exhibit in the museum. The elements of the set of information items were provided to each user.

2.2 Experimental Environment

We built a mock museum booth with seven kinds of paintings. Fig. 1 shows the layout of the museum booth. The basic information on the seven kinds of paintings exhibited in the booth is summarized in Table 2. As an example, Fig. 2 shows an outline of annotation information for "Vulcan's Forge" (LucaGiordano, 1660).

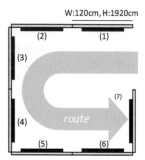

Figure1. Layout of the museum.

Table 2 Paintings exhibited in the museum booth

No:			
(1)	Self portrait	Henri de Toulouse-Lauterec	1880
(2)	Die Gesandten	Hans Holbein der Jüngere	1533
(3)	Vulcan's Forge	Luca Giordano	1660
(4)	The Tower of Babel	Pieter Bruegel de Oude	1563
(5)	The school of Athens	Raffaello Santi	1511
(6)	Waterfall	Maurits Cornelis Escher	1961
(7)	Ultima Cene	Leonardo da Vinci	1498

Figure2. Example of annotation information (for "Vulcan's Forge").

2.3 Experimental Procedure

After agreeing to participate in the experiment, participants first heard an account on the museum booth experience. The main instructions were as follows: (a) When you enter the museum booth, you must first appreciate each painting by walking along the regular route (see Fig.1). (b) After completing a circuit of the museum booth, you are free to walk inside and appreciate any painting you like. (c) You could leave at any time after appreciating the paintings. Next, a set of optical encephalography measurement equipment (OEG-16, Spectratech Inc.) was fixed to the participant's forehead and held there for 2 min while the subject was in a state of rest.

The participants wore an HMD (FR-G20, BROTHER) before entering the museum booth. After checking visibility using a test pattern, they switched the image on the HMD to a standby image and entered the museum booth. After entering, the participants viewed each painting while referring to the additional information presented on the HMD. The additional information corresponding to each painting was a movie of duration 1 min comprising multiple chapters. The participant could advance to the next chapter, return to the beginning of a chapter, or skip a chapter by using a handheld mobile controller. Only one participant could be in the museum booth at a given time. Fig. 3 shows a participant appreciating a painting. The participants exited the museum booth and then returned to the waiting room to note their reflections on a self-evaluation form.

2.4 Experimental Conditions

To examine whether providing information with an adequate gap from the participants' intrinsic characteristics enhanced participant enthusiasm in the museum booth experience, we set the following four experimental conditions on the basis of the results of STEP 1 described in 2.1.

Condition 1

Annotation information about the category a participant is most interested in is provided. In particular, the information is about the category the distance from the participant's inner characteristics is without.

Condition 2

Annotation information that has an adequate gap from the participant's intrinsic characteristics is provided, that is, information about the category whose distance from the participant's inner characteristics is 0.1–0.2.

Condition 3

Annotation information about the category that does not interest the participant much is provided, that is, information about the category whose distance from the participant's intrinsic characteristics is 0.6–0.7.

Condition 4

Annotation information about the category a participant is not interested in is provided, that is, information about the category whose distance from the participant's intrinsic characteristics is 0.9–1.0.

2.5 Data

Before the museum booth experience, each participant's face was captured by a digital video camera in the waiting room while relaxing for 2 min (Cyber-shot DSC-TX7, SONY) and the frequency of their blinks was determined. At the same time, the variation of regional cerebral blood volume (rCBV) was

Table 3 Self-evaluation form

Classification:	No:	Question items:
Self-determination	(1)	Were you absorbed in the experience?
	(2)	Do you want similar experience again? Or did you want more this experience?
Competence	(3)	This time, do you think you did a special experience?
	(4)	Do you think there was something to gain through this experience?
Unexpectedness	(5)	Did you feel this experience as a fresh?
	(6)	Does each painting which you appreciate stick out in your mind?
Non-displeasure	(7)	Did you satisfied with this experience?
	(8)	Did you appreciate to the last without tired or boredom?

measured by a 16-channel recorder fixed on the participant's forehead.

During the museum booth experience, a participant's movement within the museum booth was captured by a digital video camera (Cyber-shot DSC-WX1, SONY) positioned at the center of a 266-cm-high ceiling. The participant's face was also captured by an ultra-small video camera (AGENT CAM, Agent Camera) positioned above the frame of "The Tower of Babel." In addition, the participant's variation of rCBV was measured.

After the experiment, the participants filled out a self-evaluation form and rated their experience using a 0–6 points rating scale (see Table 3). Intrinsic motivation has four factors: Self-determination, Competence, Unexpectedness, and Nondispleasure. (see [13]). The self-evaluation form comprised eight questions, two for each factor.

2.6 Participants

The participants comprised 28 adults aged 20–26 years. After reviewing their consent forms, they were randomly divided into four groups of seven, which were allocated to each of the four conditions (see 2.4). The participants did not consume caffeine or alcohol for at least 24 h prior to the start of the experiment.

3 RESULTS

Our analysis of the behavioral, physiological, and psychological data yielded the following results.

3.1 Duration of the Experience

We analyzed the video data captured from the overhead camera to examine the duration of the participants' stay in the museum booth and the time spent appreciating each painting. Condition 2 had the longest average duration. This suggests that providing annotation information that had an adequate gap from the participant's intrinsic characteristics had a fixed effect on intrinsic motivation. Moreover, the finding that the average free time was the longest suggests that the information provided increased their spontaneous appreciation.

Thus, we consider that providing annotation information that has an adequate gap from the participant's intrinsic characteristics has the potential to increase intrinsic motivation.

3.2 Eye Blink Frequency

Previous studies have reported that eye blink frequency changes when emotions related to enthusiasm such as interest and those imparting satisfaction are enhanced [14]. Therefore, we analyzed the video data of the participants' relaxed faces captured before the museum booth experience and those obtained while appreciating "The Tower of Babel." Next, we calculated the difference between the eye blink frequencies for these two conditions, as shown in Fig. 4. The results showed that the average frequency was the highest for Condition 2.

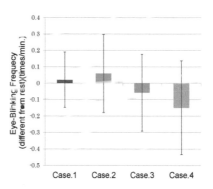

Figure 4. Difference in eye blink frequency

Thus, we consider that providing annotation information that had an adequate gap from the participant's intrinsic characteristics increased the participants' interest.

3.3 Variation in rCBV

In general, an increase in oxyhaemoglobin (OxyHb) represents cerebral activation. Therefore, we analyzed the data for variation in the concentration of OxyHb level of the participants' relaxed frontal cortex, both before the museum booth experience and while appreciating the paintings. Then, we calculated the difference between the average variation in OxyHb concentration for these two conditions. Figs. 5a, b, and c show the difference between the variation in OxyHb concentration in the medial prefrontal cortex (mPFC) for the conditions.

The results showed that the average variation in mPFC activation was highest for Condition 2. Previous studies have reported that patients with mPFC cannot act voluntarily [15]. Hence, the mPFC might be involved in self-motivation. Thus, we consider that providing annotation information that had an adequate gap from the participant's intrinsic characteristics stimulated them to voluntarily appreciate the pictures.

Figs. 6a, b, and c compare the variation in OxyHb concentration in the orbitofrontal cortex (OFC) for each condition. The results show that the average variation was the highest for Condition 2. Previous studies have reported that patients with OFC become affectively labile [16]; therefore, the OFC might be involved in emotions. Thus, we consider that providing annotation information with an adequate gap from the participant's intrinsic characteristics has the potential to elicit intrinsic motivation.

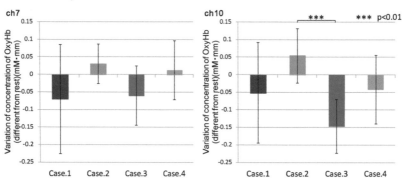

Figure 5a. Difference between the variation in the concentration of OxyHb (ch7)

Figure 5b. Difference between the variation in the concentration of OxyHb (ch10)

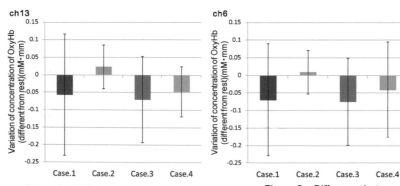

Figure 5c. Difference between the variation in the concentration of OxyHb (ch13)

Figure 6a. Difference between the variation in the concentration of OxyHb (ch6)

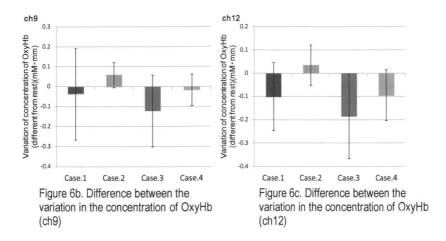

Figure 6b. Difference between the variation in the concentration of OxyHb (ch9)

Figure 6c. Difference between the variation in the concentration of OxyHb (ch12)

3.4 Self-evaluation

We analyzed the self-evaluation form that participants scored after the experiment. Fig. 7 shows the difference in the average scores on the "Self-determination", "Competence", "Unexpectedness", and "Nondispleasured" questions for each condition. The result showed that the score was relatively higher for Condition 2 than that for the other conditions. The "Competence" and "Unexpectedness" scores were particularly high. Thus, we

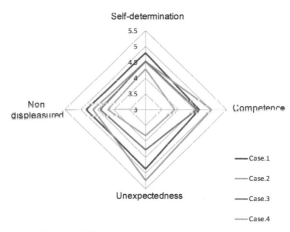

Figure 7. Difference in the score four aspects

consider that providing annotation information with an adequate gap from the participant's intrinsic characteristics has the potential to at least provoke intrinsic motivation, although not all aspects of Condition 2 had the highest scores.

4 CONCLUSION

In this study, we hypothesized that products and services with an adequate psychological gap increase users' motivation to use them. Therefore, we attempted to verify and model the hypothesis through a case study by applying a psychological gap to a concrete example of a museum information service. Analysis of the participants' behavior, blink frequency, variation in rCBV, and self-evaluation revealed the following results.

First, when we provided annotation information with an adequate gap from the participant's intrinsic characteristics (Condition 2), the average duration of the participants' stay in the museum booth was higher than that in the other conditions. This suggests that such information has the potential to provoke the participants' intrinsic motivation

Second, the participants' blink frequency was highest for Condition 2. In general, blink frequency has been a barometer of response to emotion. Thus, this also suggests that providing such information has the potential to provoke intrinsic motivation.

Third, the participants' variation in OxyHb concentration in the mPFC and OFC were the highest for Condition 2. In general, these brain areas are involved in self-motivation and expression of emotions. Thus, this supports the hypothesis that providing such information has the potential to provoke intrinsic motivation.

Finally, the results of the participants' self-evaluation showed that the evaluation of "Competence" and "Unexpectedness," which constitute intrinsic motivation, was the highest for Condition 2. Thus, this also suggests that providing such information has the potential to subjectively provoke intrinsic motivation.

The above results indicate that if annotation information with an adequate gap from the participant's intrinsic characteristics is provided, the participant's intrinsic motivation will grow more compared with the case that annotation information is provided without considering this factor. So we consider that the results supported the hypothesis.

In this study, users' emotions, cognition, and ability were together interpreted as users' intrinsic characteristics. Future studies will attempt developing a multidimensional approach.

REFERENCES

Berlyne, D.E. 1957. Conflict and choice time. *British Journal of Psychology*, 48: 106–118,.

Deci, E. L. and R. M. Ryan 1985. *Intrinsic Motivation and Self Determination in Human Behavior*. New York, NY: Plenum Press.

Deci, E.L. and R.M. Ryan. 2000. Self-determination theory and the facilitation of intrinsic motivation: Social development and well-being. *American Psychologist*, 55(1): 68–78.

Deci, E.L. 1975. *Intrinsic motivation*. New York, NY: Plenum Press.

Hall, R. J., B. L. Cusack. 1972. The measurement of eye behavior: Critical and selected reviews of voluntary eye movement and blinking. U.S. Army Technical Memorandum, 18–72.

Harlow, H.F. 1950. Learning and satiation of response in intrinsically motivated complex puzzle performance by monkeys. *Journal of Comparative and Physiological Psychology*, 43: 493–508.

Harter, S. 1987. Effectance motivation reconsidered: Toward a developmental model. *Human Development*, 21: 34–64.

Harter, S. 1981. A new self-report scale of intrinsic versus extrinsic orientation in the classroom: Motivational and informational components. *Developmental Psychology*, 17, 300–312.

Horlow, H.F. 1953. Mice, monkeys, men, and motives. *Psychological Review*, 60: 23–32.

Hunt, J. McV. 1963. *Motivation inherent in information processing and action.* In O.J. Harvey (Ed.) *Motivation and social interaction: Cognitive determinants.* New York, NY: Ronald. 35–94.

Hunt, J. McV. 1965. *Intrinsic motivation and its role in psychological development.* In D. Levine (Ed.) Nebraska symposium on motivation, 13: 189–282. Lincoln, NB: University of Nebraska Press.

Kertesz, A. 1994. *Frontal lesions and function: Localization and neuroimaging in neuropsychology,* Kertesz, A. Academic Press, San Diego, 567–598.

Levy, R., and B. Dubois 2006. Apathy and the functional anatomy of the prefrontal cortex-basal ganglia circuits, *Cereb Cortex* 16: 916–928.

Takahashi, M. and M. Nakanishi 2012. *Psychophysiological approach to evaluation of intrinsic motivation,* The 53th Conference of Japan Ergonomics Society, Fukuoka: Japan printing.

Ministry of Internal Affairs and Communication: Survey of Time Use and Leisure Activities.: Table 60-1. Participants in Hobbies and Amusements by Sex, Usual Economic Activity and Age. 2006.

Yasuma, Y., M. Nakanishi, and Y. Okada 2010. Can "tactile kiosk" attract potential users in public. Proceedings of the 3rd International Conference on AHFE (Applied Human Factors and Ergonomics) in Miami, USA, on CD-ROM.

CHAPTER 9

Ergonomics at Home: Contribution to the Design of a Smart Home Lighting Service

Germain Poizat[1], Myriam Fréjus[2], & Yvon Haradji[2]

[1]Université de Bourgogne
Dijon, France
germain.poizat@u-bourgogne.fr
[2]EDF Recherche et Développement
Clamart, France
{myriam.frejus}{yvon.haradji}@edf.fr

ABSTRACT

Pervasive technologies suggest new horizons for developing innovative products and services for domestic settings. This paper on lighting service shows how activity analysis can contribute to the design of smart home services. Preparing dinner, programming a video recorder, removing clothes from the washing machine, lighting a floor lamp and turning off the ceiling light: all are ordinary daily acts. Yet it is precisely by observing these ordinary activities that it becomes possible to determine the most useful services to design, the functionalities to define, and even the interaction modes to encourage.

Keywords: activity, design, context-sensitive services, smart home, lighting.

1   INNOVATIVE SERVICES FOR THE HOME

Since Weiser's introduction of the concept of ubiquitous computing (1991), the design of pervasive technologies for the domestic environment has become an active area of research and has generated a growing body of industrial initiatives (e.g., Aldrich, 2003; Edwards & Grinter, 2001; Intille, 2002; Howard et al., 2007). These pervasive technologies (context-aware, invisible in use, interacting with people in a natural way) suggest new horizons for developing smart home products and services.

Much of the research has dealt with the development of care services for the elderly. In this area, pervasive technologies have addressed a real problem of society: home support for seniors who want to continue living independently. These technologies have been used to design systems to (a) identify and prevent emergency situations, (b) monitor medical conditions and inform on changes in health status, and (c) provide assistance for daily activities. Other systems have also been devised to combat the isolation of many of the elderly. For example, the aim of systems like home "telepresence" is to encourage and maintain family ties (e.g., Mynatt & Rogers, 2002). Although these novel applications of pervasive computing technology are interesting, the fact remains that the majority of these solutions have been developed almost exclusively from a technocentric approach. For this reason, they often do not reflect how people actually conduct their lives at home. Very few of the support systems for elderly people today have been designed with issues of real daily activity and privacy in mind.

Pervasive technologies can also provide solutions for another priority today: the need for efficient energy use and reduced energy consumption. In this area, pervasive technologies have been associated with smart metering (e.g., Meyers et al., 2010) to develop new ways of think about services for energy management. The objective is to develop context-aware systems of energy management that reduce energy consumption while also addressing issues of comfort and safety (Frejus & Guibourdenche, 2012). Two types of energy consumption are particularly important in the development of energy management services: home heating and lighting. As part of this presentation, we will thus specifically focus on the design of a smart home lighting service.

2   A NEED FOR STUDIES IN DOMESTIC SETTINGS

The ordinary activities of families remain understudied, yet the design of context-aware services adapted for domestic use is dependent on a good understanding of the rhythms, content, and organization of daily living activities in the home (e.g., Edwards & Grinter, 2001). Like Abowd & Mynatt (2000), we assume that studies in domestic settings are vital to inform design.

In recent years, many ethnographic studies have explored domestic activities in order to guide innovation in the design of pervasive systems for the home (e.g., Cordelois, 2010; Grinter et al., 2005; Tolmie et al., 2007). Many of them have focused on the social organization of home activities, and some have specifically centered on domestic routines (e.g., Tolmie et al., 2002). These researchers assume that understanding domestic routines is important for two reasons. First, a keen understanding of routines will improve the design of systems to support them. Second, it is hypothesized that understanding the characteristics of routines will aid in the development of "unremarkable" systems.

Routines are generally invisible to those involved in them, and the idea guiding the work of several authors has been to design computational resources that can be unremarkably embedded into routines so that they are invisible in use. Other authors (e.g., Crabtree & Rodden, 2004) have particularly focused on the spatial and temporal distribution of communication in domestic settings. They looked at how householders organize both their homes and themselves to manage the flow of incoming and outgoing communications. Two main findings are noteworthy: how householders distribute communication media around the home and how they construct organizational sites to promote awareness and facilitate coordination. The authors identified the locations that the inhabitants habitually exploited in their communication activities. These places were considered as "prime sites" for situating pervasive computing in the domestic environment. In other words, these studies helped locate the best places to install pervasive technologies to meet people's needs.

Our study extends the above-cited work in that it does not further explore domestic routines but instead examines both individual and collective activity in the home from within the theoretical and methodological framework of the course of action (Theureau, 2003). This framework was originally developed in the field of French-language ergonomics, and it provides analytic methods well-suited to address questions of design. First, the real activity of inhabitants in authentic living situations can be analyzed in detail using this framework. Second, activity can be described at the level that is meaningful for the actors (i.e., activity that can be shown, told, and commented on). Thus, compared with ethnographic approaches, it offers the possibility of collecting detailed information on situations as they are experienced by the inhabitants. Last, this framework from ergonomics research is a powerful tool for determining which contextual factors need to be incorporated into pervasive systems because it takes into account the environment and the situation as the occupants experience it (e.g., Salembier et al., 2009). This dimension is fundamental to ensure meaningful interactions with context-sensitive services.

The aim of this paper is to show how activity analysis contributes to the design of smart home services by focusing on lighting service. It is precisely by observing everyday home activities that it becomes possible to determine the most useful services, the essential functionalities, and even the interaction modes to encourage. This is especially true when the goal is to design home services based on pervasive technologies: a detailed knowledge of domestic activity is essential to ensure that these systems become "invisible" for users (e.g., Tolmie et al., 2002). Only those technological systems that respect the action and reasoning logics of the inhabitants, as well as the cooperation/coordination modes between them, will be judged useful, useable, and appropriate by them. Our aim is thus to contribute to the design of an efficient lighting system both from the users' point of view (appropriate, offering comfort, ease of use, assisted activities, etc.) and from the demand side management (reducing energy consumption, management of the load curve).

3   OBSERVATION OF HOME ACTIVITY

Our approach was to design methods for the extended observation of domestic activity, with no special focus on lighting.

3.1   Participants and procedure

Six families agreed to participate in this study. They were recruited from the local telephone book and all were financially compensated. Half of them had very young children and the other half had teenagers to increase the complexity and variety of the activities observed.

To gain an initial understanding of each family's daily living activities and to become familiarized with the lay-out of the homes, we began by conducting semi-structured interviews with the participants (mainly the parents) on the premises. Interviewees were asked to describe their typical weekday and weekend activities at home and their time schedules. This information helped us determine the optimal time for video recording. A home tour with the participants was also conducted in order to choose the best places for installing the recording equipment. This phase was also a way to build a trusting relationship with the participants before beginning data collection.

3.2   Data collection

Two types of data were gathered: (a) continuous audiovisual recordings of the participants' actions and (b) verbalizations during post-recording interviews.

Audiovisual recordings

We used a system for recording based on a proposal by Relieu et al. (2007). The recorded data were collected from eight mini-cameras and eight small microphones. The cameras were able to capture images in low light and were set to record wide-angle views of the areas of interest. Each camera was connected to a digital pocket recorder which allowed multiple shots to be pre-programmed and automatically started. The recording system also offered the families the opportunity to monitor the recording using a switch to temporarily suspend the recording. They were also able to ask us to erase undesirable fragments. The recording system remained in the homes for one week. The installation of the recording equipment was fitted to the structure of each home. The participants gave prior agreement to the camera positions. In most cases, activity in the following areas was recorded: kitchen, living room, dining room, office, children's rooms, hallway, landing, entrance hall, laundry room and garage. Each household was continuously recorded for about six hours a day for two weekdays and one day on the weekend.

88

Verbalization data

The verbalization data were gathered from self-confrontation interviews with the householders (i.e., adults, teenagers). In general, the self-confrontation interview is a procedure during which an actor is confronted with the audiovisual recording of his or her activity and is asked to explain, show and comment on the elements of this activity that are personally meaningful, in the presence of an interlocutor (Theureau, 2003). We conducted these collective interviews because of the organizational constraints linked to conducting the interviews in the homes. The researcher and inhabitants of a given home viewed the audiovisual· recordings together one week after collection. During the interviews, the researcher and the inhabitants could alternate between two viewing modes: the focused mode monitored activity in one room and the extended mode presented a view of all parts of the home. The inhabitants were invited to describe and explain their activity step by step while viewing the recording. The researcher's prompts served primarily to help them to re-experience their past activity and reconstruct their individual and collective experience.

3.3   Data processing

The data were processed from within the theoretical and methodological framework of the course of action to account for the level of activity that was meaningful for the inhabitants. The first step consisted of generating a summary table of the data collected from each recording. The data were presented by mapping two levels of data together. The first level pertained to the data recorded in situ: it contained the descriptions of the actions and communications of the inhabitants. The second level pertained to the data recorded during the interview: it contained the verbatim transcriptions of the prompted verbalizations. The second step was to reconstruct the condensed narrative of each inhabitant for each recording. The condensed narratives were constructed by breaking down the verbalizations into meaningful units organized in several hierarchical ranks (Theureau, 2003). The assumption was that the sequence of these units organized into ranks would reflect the dynamics of the activity from the actor's point of view. The condensed narratives of all the inhabitants were then mapped chronologically to reflect the collective activity in a given home.

4   DAILY DOMESTIC ACTIVITIES AND LIGHTING: EMPIRICAL RESULTS

The following section presents some of the results highlighting important issues in the design of a smart lighting service.

4.1 Lighting is not always useful

The analysis of activity revealed that lighting is not always necessary from the inhabitant's point of view. Entering a room did not necessarily imply activating the light switch, even if this had been observed a few minutes earlier. Everything depended on the activity that was to be conducted in the room. As an illustration, we observed a situation in which a father came into his kitchen without turning on the light. He explained during the interview that he had expected to remain in the kitchen for a very brief time, just to take a snack. He therefore judged that the ambient light was sufficient to proceed to the table. In this case, a light automatically switching on would have been seen as superfluous for the inhabitant.

4.2 Lighting is not always usable

The results also showed various situations in which it was very difficult for the participants to use conventional light switches. For example, we repeatedly observed that the participants had problems turning on lights when they had their arms full (e.g., carrying a child to bed or carrying a laundry basket). In such cases, it would be appropriate to offer the option of temporarily automatic lighting.

4.3 Not turning off the light is not always "forgetting"

Several participants mentioned during the self-confrontation interviews that they sometimes neglected to turn lights off. However, the empirical data enabled us to gain greater insight into the meaning and emergence of such "forgetfulness" in the home. These data also revealed that the lights were not always forgotten because, in fact, the participants sometimes left the lights on deliberately. The parents of young children were usually concerned with safety and peacefulness when putting children to bed. They therefore often left the lights on in hallways and unoccupied rooms to allay their children's fears about the dangers that might be lurking in "dark places." What might have appeared as a waste of energy was in fact an overriding concern on the part of parents for their children's wellbeing.

4.4. Lighting the way in advance

The results showed that the inhabitants switched on the light in anticipation of an upcoming individual or collective activity. For example, we observed a mother turn the dining room light on as she headed toward the kitchen to prepare a meal. In fact, she then explained in the interview that not only was she expecting to return to the dining room to set the table, but she also was expecting other family members to soon enter the room. She took advantage of walking by the light switch to turn on the light as she left. In such a case, an automatic system for switching off the lights on the basis of presence/absence detection in order to reduce energy waste and ensure economical use would likely be perceived as a disturbance by the inhabitants.

4.5 Lighting is often adjusted

The results repeatedly emphasized that the participants modulated the lighting depending on their current activity. For example, we observed that they were fairly systematic in the lighting they used in the kitchen during meal preparation. In this case, it might be appropriate to use "scenarios" to control an entire set of lights and configure the set, for example, into a single act of lighting "for cooking." Conversely, we also observed situations that pointed to the limits of operating on the basis of scenarios, notably because the scenarios were often based on the actions of a single user. For example, we observed an episode in which all the family members had settled on the couch and were viewing a TV program. The father had turned off the living room lights to maximize the TV image. But some time later in the evening, the mother walked toward the light switch and switched a light on in order to read while still following the TV program. In this type of case, scenarios may be ineffective because sometimes the lighting in a given space ultimately depends on an ongoing negotiation among the inhabitants.

5 CONTRIBUTION TO THE DESIGN OF A SMART LIGHTING SERVICE

Our study identified several modes of interaction to address the diverse contexts for the occupants' activities. The first mode is to activate the automation. This mode of interaction is relevant in relatively simple contexts where the goal is to respond in a timely and localized way to safety concerns or to forgotten lights. In this case, a simple solution is to install an automatic system in very specific places (e.g., staircase, basement) for switching off the lights on the basis of presence/absence detection. However, it would be irrelevant to install this system throughout the home because it is too limited with regard to the wide diversity of contexts for the occupants' activities.

The second mode of interaction is an intermediate solution, allowing the occupants to temporarily enable automation. A technical solution might be to install presence/absence sensors throughout the home that can be activated on demand (e.g., an "auto" setting on a conventional light switch). Once activated, this setting would switch all the lights in the home to automatically operate on the basis of presence/absence detection. This mode seems to be a good compromise because it provides a response to lighting needs that are brief, simple, and often while the inhabitant is moving from room to room: the auto setting would temporarily light spaces following an inhabitant's movements. However, this solution would require prior negotiation among the occupants to ensure that all would accept potentially being disturbed by the lighting for a relatively short duration.

The third mode of interaction is to combine parametrization and control by the occupants. This mode of interaction is adapted to specific contexts with clearly defined lighting needs (e.g., cooking, watching TV, etc.). In these contexts, one solution might be to provide occupants with the opportunity to configure and enable a variety of lighting scenes, which could easily be done by accessing the central system with a computer or tablet. However, expecting occupants to set the parameters for a high number of lighting scenes would most likely render this solution ineffective. To activate these lighting scenes, two solutions stand out: the addition of a third setting, labeled "scenario", on the conventional light switches in specific locations offering a maximum of three lighting scenes to choose from, and the addition of a function on the "target" device, depending on the scenario (e.g., a television for "watch a movie").

The fourth mode of interaction is to facilitate the occupants' control of lighting in any place and at any time, notably by increasing the opportunities for remote control. In our opinion, the concrete objective should be to provide a control device as near as possible to the current activity at all times (without this having to be determined in advance) and to adapt the mode of interaction to the probable activity of the occupants. One solution would be to take advantage of other future smart appliances with new communication capabilities that will emerge in the domestic environment. These could form both a network of relevant information on the context and activities underway and a support to interactions with lighting easily at hand. This mode of interaction in our opinion should be at the heart of the system architecture because it can meet all the contexts of activity not covered by the previous solutions and thus provides added value to a smart lighting system.

6   CONCLUSION

In summary, activity analysis seems essential to the design of smart services for domestic settings. In the present study, activity analysis identified the action logics and even the modes of coordination among household members. It was thus possible to build an analytic empirical model to guide design. In the framework of this study, the analysis of activity also allowed us to identify areas of difficulty and the situations in which interactive and context-aware systems can be of aid to the inhabitants. We identified certain needs regarding the modes of interacting with future lighting systems and were able to make some design proposals. Last, the activity analysis enabled us to compare the context that is potentially perceptible by the system (depending on the preferred technical solution) and the situation as experienced by the inhabitants. This is particularly interesting for the design of pervasive technologies in the domestic environment. The comparison helped to define the "relevant context" (e.g., Salembier et al., 2009) for a system of lighting management and to specify the architecture of future systems. Several technical solutions are currently being tested to help develop solutions for lighting control that is localized and adapted to the unfolding activity in domestic settings. The analysis of household activity has significantly contributed to specifying the system architecture (Dominici et al., 2011), as particular attention has been paid to the difficulties in recognizing domestic activity. The results of our study are currently being used to formulate minimalist, credible and relevant scenarios that will serve to assess the relevance of the behaviors of various prototype systems.

92

Further studies are obviously needed before a smart home lighting system can be finalized. The next step is building mock-ups incorporating the various solutions and testing them in real or realistic situations. This phase seems essential as preparation for installing these new services, because the testing process will pinpoint the important questions that need to be raised about the appropriation/individuation of the devices (e.g., Haradji et al., 2010). In addition, ongoing studies using these mock-ups should be combined with dynamic pricing models of electricity. This will be particularly important in achieving efficient home lighting from both household members' points of view and a demand control perspective.

REFERENCES

Abowd, G. and Mynatt, E. 2000. Charting Past, Present, and Future Research in Ubiquitous Computing. *ACM Transactions on Computer-Human Interaction* 7: 29-58.
Aldrich, F. 2003. Smart Home: Past, Present and Future. In. *Inside the Smart Home*, ed. R. Harper. London: Springer-Verlag, pp. 17-39.
Cordelois, A. 2010. Using digital technology for collective ethnographic observation: an experiment on "coming home". *Social Science Information* 49: 445-463.
Crabtree, A. and Rodden, T. 2004. Domestic routines and design for the home. *Computer Supported Cooperative Work* 13: 191-220.
Dominici, M., Fréjus, M., Guibourdenche, J., et al. 2011. Towards a system architecture for recognizing domestic activity by leveraging a naturalistic human activity model. Proceedings of the International Conference on Automated Planning and Scheduling. Freiburg, Germany.
Edwards, W. K. and Grinter, R. E. 2001. At home with ubiquitous computing: seven challenges. Proceeding of the Ubiquitous Computing International Conference. Atlanta, Georgia.
Fréjus, M. and Guibourdenche, J. 2012. Analysing domestic activity to reduce household energy consumption. *Work : A Journal of Prevention, Assessment and Rehabilitation* 41: 539-548.
Grinter, R. E., Edwards, W. K., Newman, M., et al. 2005. The work to make the home network work. Proceedings of the European Conference on Computer Supported Cooperative Work, Paris, France.
Haradji, Y., Poizat, G., and Motté, F. 2011. Activity-Centered Design: An Appropriation Issue. *Communications in Computer and Information Science* 173: 18-22.
Howard, S., Kjeldskov, J. and Skov, M. B. 2007. Pervasive computing in the domestic space. *Personal and Ubiquitous Computing* 11: 329-333.
Intille, S. S. 2002. Designing a home of the future. IEEE Pervasive Computing 2: 80-86.
Meyers, R. J., Williams, E. D., and Matthews, H. S. 2010. Scoping the potential of monitoring and control technologies to reduce energy use in homes. *Energy and Buildings* 42: 563-569
Mynatt, E. D. and Rogers, W. A. 2002. Developing technology to support the functional independence of older adults. *Ageing International* 27: 24-41.
Poizat, G., Fréjus, M., and Haradji, Y. 2009. Analysis of collective activity in domestic settings for the design of Ubiquitous Technologies. Proceeding of the European Conference on Cognitive Ergonomics. Otaniemi, Finland.

Relieu, M., Zouinar, M., and La Valle, N. 2007. At Home with Video Cameras. *Home Cultures* 4: 45-68.

Salembier, P., Dugdale J., Frejus, M., et al. 2009. A descriptive model of contextual activities for the design of domestic situations. Proceeding of the European Conference on Cognitive Ergonomics, Otaniemi, Finland.

Theureau, J. 2003. Course of Action analysis & Course of Action centered design. In. *Handbook of cognitive task design*, ed. E. Hollnagel. Mahwah, NJ: Lawrence Erlbaum, pp. 55-81.

Tolmie, P., Crabtree, A., Rodden, T., et al. 2007. Making the Home Network at Home: Digital Housekeeping. Proceedings of the European Conference on Computer Supported Cooperative Work. Limerick, Ireland.

Tolmie, P., Pycock, J., Diggins, T., et al. 2002. Unremarkable computing. Proceedings of the ACM CHI Conference on Human Factors in Computing Systems Conference. Minncapolis, Minnesota.

Weiser, M. 1991. The computer for the 21st century. *Scientific American* 265: 94-104.

CHAPTER 10

Designing a Mobile-based Banking Service: The MOBSERV Project

Simões, Anabela*, Guerreiro, Sérgio**, Ferreira, Ana*, Rôla, Susana*, and Freire, Graça**

Instituto Superior de Gestão / CIGEST
Universidade Lusófona de Humanidades e Tecnologias / CICANT

ABSTRACT

This paper is centered on the work under development in the frame of the MOBSERV project, which is a national funded project aiming at developing a mobile banking service. Due to an increasing usage of mobile technologies, banking services are experiencing major changes in order to satisfy customers' demands and expectations. This challenge is imposing a large effort for reinventing new approaches compliant with user's expectations and also to design solutions that are adaptable to the fast services demand change cycle. Therefore, the MOBSERV project aims at responding to these new market needs by means of developing an innovative service to provide secure access to new types of mobile banking services. The development of this innovative service highlights the need for identifying the diffusion elements based on attitudes and behaviors of potential users regarding the adoption of the targeted mobile banking services in Portugal. Therefore, focus groups (FG) interviews have been carried out in order to collect perceptions, opinions, beliefs and attitudes of potential users regarding the system under development and thus, identify factors behind the resistance to the use of the service, around issues such as the technological, product and service experience, personal, cultural and sociological characteristics, as well as dissemination models. Preliminary results of FG interviews that have been carried out are presented in this paper together with the detailed methodology and further steps in the project.

Keywords: Mobile banking; Focus Groups; Adoption; Resistance; Behavior.

1   INTRODUCTION

Banking services are actually suffering major changes due to the continuously emergence of the mobile devices usage as well as the profound technological developments that have occurred in operating systems, computer networks and human-machine interfaces. A bank customer is no longer satisfied with home banking services accessed at Internet. Customers actually demand new types of mobile banking services supported on tablets, smart devices or mobile phones. This challenge is demanding a large effort for reinventing new approaches compliant with user's expectations and also to design solutions that are adaptable to the fast services demand change cycle. In fact, the high capabilities offered by the new mobile devices are pushing banks to reinvent interfaces and to redefine services. Some qualities, such as real-time services where the user is decoupled from a physical location, are actually already identified and pursued by banks. However, the approach to this problem should be founded in the holistic study of the constructional parts of a mobile banking service rather than in its functional or behavioral usage by the actors. Constructional solutions to mobile banking services using a design approach (Hevner et al., 2004; Winter, 2008) enables a comprehensive service description allowing for a consequent faster adaptability when needed. Therefore, the technological development, together with the results of the survey carried out by Cruz, Laukkanen & Muñoz (2009) and Cruz, Neto & Laukkanen (2010) regarding the adoption of mobile banking services, encouraged the design of the MOBSERV project and its submission to the Portuguese authority for science funding.

The MOBSERV project aims at responding to these new market needs by means of developing an innovative service to provide secure access to new types of mobile banking services. The service under development in the frame of MOBSERV is innovative, which highlights the need for identifying the diffusion elements based on attitudes and behaviors of potential users regarding the adoption of the targeted mobile banking services in Portugal. Therefore, focus groups (FG) interviews have been carried out in order to collect perceptions, opinions, beliefs and attitudes of potential users regarding the system under development and thus, identify factors behind the resistance to the use of the service, around issues such as the technological, product and service experience, personal, cultural and sociological characteristics, as well as dissemination models. This paper is centered on the preliminary results of the FG interviews that have been carried out with participants of two categories of age.

2   FOCUS GROUPS

In order to collect data about attitudes and behaviors of potential users regarding the adoption of mobile banking services in Portugal, Focus Group interviews have been carried out in the frame of the MOBSERV project. This technique will also allow identifying factors behind the potential resistance to the use of this type of service and related issues, such as technological, product and service experience, personal, cultural and sociological characteristics, as well as dissemination models.

Focus groups are a special type of discussion group aiming at better understanding how people feel or think about an issue, product or service (Krueger & Casey, 2009). Discussions are conducted in an unstructured and natural way where respondents are free to give views from any aspect. They are asked about their perceptions, opinions, beliefs and attitudes towards a product, service or concept. Questions are launched in an interactive group setting where participants are free to talk with other group members. Each group should be composed of 6-8 people in order to allow for discussions under the control of the moderator.

In the case of MOBSERV, the addressed topic for discussion is centered on the service to be developed under the project, aiming at identifying perceptions, opinions, beliefs and attitudes of potential users regarding the new system. Thus, groups of potential users of two categories of age and both genders have been selected for the MOBSERV FG sessions according to previously defined criteria.

Being an innovative service to be available in the context of mobile banking operations, there are particular issues to be addressed: the users' willingness to adopt the service or any potential resistance to its use. This general attitude may be related to users' needs and expectations, their level of experience on the technology use, their level of confidence on the technology and the security of the services to be provided, as well as the costs related to the use of these services.

The previously identified users' needs represent the main factor to guarantee the adoption of the services, particularly if they comply with the users' expectations. In the case of innovative technological systems or services, the innovation is usually beyond the common user's ability to feel its need or to develop any expectations regarding its use. Then, people require some information or even education to accept and adopt the innovation. However, experienced users in innovative technologies are supposed to adopt the innovation if they find it reliable and secure. Finally, the cost of the service and its comparison with the same service provided through Internet or directly will make the difference. Another point in favor of the adoption of the service could be the easiness for personalization that these technologies can offer. Therefore, FG sessions are supposed to arouse discussions on the factors surrounding the potential adoption or resistance to the services use.

2.1 The objectives of the MOBSERV Focus Group

The MOBSERV FG sessions aimed at identifying:
1. Potential users' attitudes, believes, opinions and feelings regarding the use of m-banking services;
2. Potential factors of adoption of m-banking services;
3. Potential factors of resistance regarding the use of m-banking services (lack of confidence, low security, high cost, other);
4. Particular situations in favor of adoption or leading to resistance;
5. The value that potential users attribute to the service;
6. The support that should be provided in order to favor the services adoption;

7. Users' needs for information or training on the use of the services;
8. Users' feelings regarding security issues in m-banking services.

3 FOCUS GROUPS METHODOLOGY

3.1 Participants' Selection

In order to conduct the FG sessions and achieve the objectives of this task, one of the first concerns was to define the characteristics of the participants that should take part on the FG. The main aim of this first stage was considered essential to allow the definition of the independent variables. Two major independent variables were considered important to analyze how users' attitudes, believes, opinions and feelings regarding the use of m-banking services: Age (as being part of the individual differences capable of influencing behavior) and User's Experience in terms of mobile phone multi-use. Therefore, age and experience with new technologies were defined as the main criteria to help characterizing the groups of participants required for the FG sessions.

*   Age: defined in two categories (Table 1)
    – Participants between 18 and 23 years old
    – Participants between 30 and 45 years old

Table 1 – Focus Groups Sessions

	Mobile communications	
	Age category	
Participants	18-23 years old	30-45 years old

Due to practical aspects, a limited number of FG sessions were planned: four sessions for each category of age. Considering that each FG will have from 6 to 8 participants, the sample will involve between 24 and 32 participants for each age group.

Aiming at recruiting participants, a selection questionnaire has been developed in order to be filled in by the researcher during a previous phone interview. This questionnaire consists of general information regarding age, gender and experience with mobile phones.

3.2 FG Protocol

The FG Sessions have been prepared to last around 1h30m, being the first 10 minutes dedicated to the presentation of the project, the provision of information about the m-banking services to be developed and the procedure to be followed. Each FG session has been conducted in a comfortable room prepared uniquely for that purpose. The characteristics of the rooms are of special importance, as they should offer to participants relaxed and friendly feelings (Newman, 2005). For each FG session one moderator and one assistant have been present. Following a short description of the project and the m-banking services to be developed, the moderators clarified some participants' doubts concerning the study. The moderator launched the questions and encouraged the discussion during the session.

An assistant to the moderator has been present in the room during the entire FG sessions, performing the following tasks: marking down when a question started in order to ease the transcription process; making a scheme of the participants' distribution on the room (information very useful for the analysis), and taking notes regarding important details occurred during each FG session. The assistant should also write the main ideas that participants expressed aiming at easing the further transcription process.

Several materials and tools have been used during the sessions:

- Consent form that has been signed by each participant as his/her authorization to participate in the study;
- A short document with general introductory information about the study and the session agenda that has been distributed to each participant;
- Paper plaques (tags) with the name of each participant;
- Sheets of paper and a pen to each participant;
- PPT presentation about the targeted m-banking services;
- Video with sound recording to register the comments of participants during the sessions;
- Discussion guide with the questions to be launched during the session and important notes;
- Assistant guide to be used by the moderator's assistant during the FG session;
- Stopwatch to manage the session time;
- Refreshments and snacks to be served before the session while waiting for the arrival of all participants and the established time to start.

3.3   The Pre-Test

A pre-test has been carried out with the aim of assessing the methodology and allowing moderators and assistants to train the sequence of questions. The pre-test session was intended to highlight any inconsistencies that the methodology might have. This is an important aspect as it allowed moderators to anticipate problems and handle the discussion more easily during the actual sessions.

The pre-test session has been entirely conducted until the end. Then, the researchers made the transcription of the session in order to allow analyzing its contents and foreseeing any problems regarding the analysis process. No inconsistencies were found, so the actual FG sessions were considered ready to start.

4   PRELIMINARY RESULTS AND DISCUSSION

FG sessions have started but are still running; so, the qualitative results presented and discussed below are preliminary. So far, every participant from both age categories uses Internet for baking operations, as well as most of the available functions in their mobile phones, although avoiding or being cautious on the use of commercial functions. Just one of them is using a bank phone line open to clients for very simple operations: consult, transfer and small payment. Participants owning the most recent and developed mobile phones consider the artifact rather as a computer allowing communicating by phone than as a mobile phone. Therefore, they rely on the artifact the same way they rely on their own computer or laptop, which leads them to explore and use the available functions.

Due to the fact that participants own different types of mobile phones and different technologies, they do not have the same level of experience with new functions but they use the available functions in their own devices. These participants expressed interest in using new functions although being limited by the technology available in their own mobile phones. When discussing potential factors leading to avoid the use of m-banking functions, most participants and the more experienced ones in using mobile access to Internet referred the cost as the main factor, once it is still very high. Furthermore, they consider that the information regarding the use of Internet services on mobile phone, as well as the related rules and costs, is not enough clear neither well disseminated, which explains some resistance to the use of this function. The second factor referred by participants is the system security, being very important to initiate and increase the adoption of the service, together with an efficient dissemination of the required information. Participants prefer to have the access to the system provided by each bank than having an open service to be used on their communication with any bank where they have an account. They consider this open service insecure and unreliable. The use of the service is based on the user's trust on the system, as well as on the system usability, which was the third factor referred by participants to have a potential to generate a resistance to the use of the service or to favor its adoption.

The following list of m-banking services was presented to participants in order to get their opinions, feelings and attitudes regarding their adoption and to identify any potential resistance to their use:

- Transferring between different banks, using the banking identification number (NIB);
- Services pay orders, similar to procedure on Automated Teller Machine (ATM);
- Pay orders to the Portuguese state, similar to procedure on ATM;
- Real-time account detailed balance;
- Viewing the NIB, IBAN, SWIFT information;
- Viewing monthly account balance in PDF format;
- Pre-paid mobile phones recharge.

Participants have shown interest in having access to these services on their mobile phones but they raised observations falling into usability issues. Thus, some examples have been used to highlight the importance of following the same logics in the navigation when using either the mobile phone or an ATM or even Internet for the same operation. Different navigation procedures will favor the increase of errors and thus, some resistance to adopt the available service. Furthermore, they suggested security procedures to prevent frauds, including the situations in which the user spends more than a pre-defined time without any action.

Regarding diffusion models to promote the new m-banking services, participants suggested that the bank should provide clear and useful information on the service access and use. This information should be personalized according to the client age and experience with new technologies. A participant suggested the offer of the mobile device with the application to new bank clients instead of other usual types of offers. This could be subject to a customer loyalty period.

In brief, participants evidenced interest on the service under development and willingness to use it if the related costs are acceptable, if good and effective security is provided together with relevant information and if the system will have a good usability.

5 NEXT STEPS

At this stage, the advertising procedure regarding the composition of the sample is ready to start. Meanwhile, the copies of the FG tools and materials will be produced and the selection procedure will start as soon as the potential candidates will send their applications. A total of 8 FG sessions (4 of each age category) are previewed to start next February following a pre-test FG session to test the discussion guide and the corresponding timing. As each session will be video recorded and the transcription process will be supported by MAXQDA software, the final report with recommendations for the design of the targeted services will be finalized in due time

In order to quantitatively measure the adoption of m-banking services a set of mobile software prototypes are actually being designed and developed to be tested by participants in the frame of usability evaluations.

The goal of developing software prototypes is to compare the perceptions, opinions, attitudes and beliefs of potential users regarding m-banking services with real banking applications experience. Using mobile software applications the users are able to directly validate their perceptions without the interference of a natural language or other participant's opinion. In the end of each experiment the participants will be asked to fill a survey regarding their personal experience valuation.

Concerning this purpose of developing mobile software prototypes and evaluating them, a constructional perspective design founded in design science research (DSR) (Hevner et al., 2004) (Winter, 2008) is being considered. Piirainen et al. (Piirainen et al., 2010) studies the DSR developments proposing the following steps to implement an IT artifact in an organization: (i) identify a IT organizational problem, the aim is to identify and solve real problems that exist in the daily life of the organizations; (ii) demonstrate that no solution exists; (iii) develop IT artifact that addresses this problem; (iv) rigorously evaluate the artifact; (v) articulate the contribution to the IT knowledge base and community; (vi) explain the implications of IT management and also to the reality of the enterprise.

Moreover, due to the serious and continuous change that are identified in the nature of functional requirements for the m-banking services a split is performed between the user interface, e.g., the buttons and forms that changes for each mobile device, and the core functionalities, e.g., servers, database and web services that are required to operate the applications. The core functionalities follow the specification obtained from a Portuguese bank institution. A decoupling of aspects between interface and core functionalities will be considered during the software architecture design phase. We expected that this approach would result in a consequent faster adaptability to the requirements imposed by the participants during the FG.

6 REFERENCES

Anacom (2008). Autoridade Nacional de Comunicações, Serviço Nacional Móvel: relatório do 2º trimestre de 2008.

Blackmana, C. Forgeb, S.; Bohline, E.& Clements, B. (2007). Forecasting user demand for wireless services: A socio-economic approach for Europe,Telematics and Informatics, 24 (3), pp. 206-216.

Cooper, L., & Baber,C., (2005). Focus Groups (Chapter 32). In N. Stanton, A. Hedge, K. Brookhuis, E. Salas, H. Hendrick (Eds) Handbook of Human Factors and Ergonomics, (pp. 32.1-32.7) CRC Press

Cruz, P. & Laukkanen, T.; Muñoz, P. (2009). Exploring the Factors Behind the Resistance to Mobile Banking in Portugal. International Journal of E-Services and Mobile Applications, Volume 1, Issue 4

Cruz, P.; Neto, L.; Muñoz, P.; Laukkanen, T. (2010). Mobile banking rollout in emerging markets: evidence from Brazil. International Journal of Bank Marketing, Volume 28 issue 5, pp. 342 – 371.

Dahlberg, T.; Mallat, N.; Ondrus, J. & Zmijewska, A. (2008). Past, present and future of mobile payments research: A literature review, Electronic Commerce Research and Applications, 7 (2), 165-181.

Eurostat (2007). Science and Technology/Information Society Statistics: Individuals. Computer Devices.

Hevner, A., March, S., Park, J., and Ram, S. "Design Science in Information Systems Research," MIS Quarterly (28:1) 2004, pp. 75-105.

Krueger, R. & Casey, M. A. (2009). Focus Groups. A practical guide for applied research. SAGE

Kleijnen, M.; Ruyter, K. & Wetzels, M. (2007). An assessment of value creation in mobile service delivery and the moderating role of time consciousness. Journal of Retailing. 83 (1), 33-46.

Newman, L. (2005). Focus Groups (Chapter 78). In N. Stanton, A. Hedge, K. Brookhuis, E. Salas, H. Hendrick (Eds) Handbook of Human Factors and Ergonomics, (pp. 78.1-78.5) CRC Press.

Piirainen, K., Gonzalez, R., and Kolfschoten, G. (2010). Quo vadis, design science? a survey of literature. In Winter, R., Zhao, J. & Aier, S., editors, Global Perspectives on Design Science Research, volume 6105 of Lecture Notes in Computer Science, pages 93-108. Springer Berlin / Heidelberg.

Rogers, E. (2003). Diffucion of Innovations. 5th edition. Free Press, New York.

Winter, R. (2008). Design science research in Europe. European Journal of Information Systems (2008) 17, 470–475.

Section III

Societal Factors

Human Centered System Integration (HCSI): Case study of Elderly Care Service Design

Santosh Basapur, Keiichi Sato

Institute of Design
Illinois Institute of Technology
Chicago, USA
{Basapur, Sato}@id.iit.edu

ABSTRACT

This paper presents a part of attempt to develop applied human system integration (HSI) methodology for service design particularly focusing on educational aspects in graduate level education. In this Design School Human-Centered System Integration (HCSI) course, we had our students participate in a course-long project. They had to use HSI principles taught in the class for designing a service for elderly care. By making use of known methods of Human Factors (HF) and Human Computer Interaction (HCI), we sought to bridge the gaps in traditional system architecture development. The application of known techniques from HF and HCI to allow students to analyze users, usages and use contexts helped development and articulation of an applied HCSI method. The goal of this research is to further develop HCSI methodology to make it scalable, adaptive, reliable and repeatable in design education, research and practice, especially in complex system development such as healthcare and assisted living environments.

Keywords: Human System Integration, Service Design, User centered service architecture, computer aided service design, Design Information Framework

1 INTRODUCTION

In a 2010 talk to the National Research Council, Norman critiqued Human-System Integration as a failure. Reasons cited by Norman included "...We all try hard to be scientists. A second problem is the lack of good, solid practical, handbook-like knowledge. The third is the lack of design skills..." Norman emphasizes the need for applied methods to make HCSI useful as well as effective. Traditional System Integration has always accounted well for the core tasks of planning and design process to compose the system structure for complex and large-scale system development based on goals and requirements of users. But, the critique has always been that the user-centricity is superficial and methods have fallen short in identifying difficulties and issues. Hence, there is an unmet need for full-scale system integration methodology with human centricity.

The basis of Human-Centered System Integration (HCSI) is to develop and apply extensive information about users and use context for developing system architecture and interactive features in order to enhance human values addressed from various perspectives. When traditional subjects of design have transformed from single products to networked systems with advanced technologies such as embedded technologies and dynamic network technologies, design needs to encompass new dimensions such as business models, logistics, service systems and system lifecycles. This requires integration of many different types of information from different disciplines and sources such as user studies, social and cultural studies, task analysis, organizational requirements, ergonomic data, and different types of system models. Even a simple product manifests complex design problems when it is placed in contexts such as users' daily life, social environments, services, cultural practices, and environmental concerns. There is a major gap between information inquiry, analysis and use of the information in solving problems and developing a comprehensive (aka complex) service system.

This paper presents a case study of HCSI methodology application to service design through a graduate level design course (8 week lecture/seminar, three hour class time once a week). The goal of the case study was to uncover and identify insufficient or missing elements of HCSI approach through applications in educational setting. The course goal was to introduce the basic philosophy, principles and methodology of HCSI while keeping the approach as pragmatic as possible. In the class students were also engaged in a course-long project along with lectures and class discussions. The exercise project was to apply HCSI principles and methods introduced in the class for designing a service system for elderly care environments.

By using known methodologies of Human Factors and Human Computer Interaction, we can bridge the gap between traditional system architecture and HCSI. Also, different available ways of organizing complex design information and later, concept generation for solutions was taught. Information management from the needs inquiry phase and then the mechanisms of transformation of the information into systems concepts from multiple viewpoints such as system operation, interaction, service, and user experience was introduced and discussed.

2   RELATED WORK

This paper is influenced by research in the areas of Human System Integration and Senior Wellness.

2.1   Human System Integration (HSI)

Manpower and Personnel Integration (MANPRINT) management and technical program designed an effective integration of human factors into the main stream of system definition, development and deployment (Booher, 2003). Following are the seven topical areas where the impact of new technological systems must be considered: Manpower, Personnel, Training, Human factors engineering, System safety, Health hazards, and Survivability with specific emphasis on methods and technologies that can be utilized to apply the HSI concept to system integration. While this has been the research mainly undertaken with and for the armed forces of USA, researchers have taken the HSI concept and applied it to civil projects like the Rail Road system of USA (Reinach and Jones, 2007), Civil Aviation (Hewitt, 2010), and HealthCare Industry (National Research Council, 2011). Application of HSI has been quite appropriate because all these systems are large and obviously complex socio-technical systems.

2.2   Senior Wellness

Wellness of seniors as a research topic has become more important than ever before in our rapidly aging society. Census estimates in 2009 show that the world's 65-and-older population has already increased by 23 percent since 2000 to 516 million. Moreover, the biggest shift is expected soon when the post-war baby boomers will be of this age, tripling the world's senior population by 2050 (AARP, 2009 and US Census Bureau, 2009). As a result, the United States and other developed nations face enormous challenges in supporting health care for this population. In response to this challenge, research in senior wellness and fitness programs, have received increased attention. The work in wellness is seen as a means of lifting the overall health and well-being of seniors while reducing overall health care costs (AARP, 2009). We believe, Wellness is much more than merely physical health, exercise or nutrition (National Wellness Institute, USA, 2010). It inherently involves social, occupational, spiritual, intellectual, and emotional well-being. Despite the multi-dimensionality of wellness, we observe an unbalanced focus on the physical aspects of wellness in technology development. We believe that there is a growing need to understand how services (people and technology), both inside and outside the home, can improve seniors' wellness beyond the physical aspect.

3   HCSI AS AN ACADEMIC COURSE AT DESIGN SCHOOL

The concept of human centered design emphasizes the importance of users' needs and contexts of use in the design of products or services. But, the keeping of human perspective throughout the design cycle becomes particularly hard when it comes to complex services. HCSI particularly focuses on the quality of system performance from users viewpoints throughout its lifecycle, based on goals and requirements identified through the information inquiry phase. The system integration brings together discrete subsystems such as physical subsystems, software, physical and information infrastructure, operation management, and business mechanisms in order to create desirable services to users. This HCSI course introduces methodologies that allow different ways of organizing complex design information from the inquiry phase to the conceptualization phase. The mechanisms of transformation of the information into systems concepts from multiple viewpoints such as system operation, interaction, service, and user experience is also covered. Contents discussed in the class constitute three main topic areas:

3.1   Users, use context and user experience

This part of the class includes discussion of user centered design methods. Emphasis is on user, context of use and any cognitive, behavioral and cultural patterns that the design and designer should consider. The coursework covers methods of investigation and data collection to understand user activities, tasks and functions mapping of the products or services in current use. Following which students are taught how to analyze and document problems, user goals, needs/requirements, and finally ecological constraints and criteria. Students are exposed to wide variety of user research methods qualitative, quantitative and mixed.

3.2   Design Information Framework (DIF)

DIF is a means of representing and bridging a wide range of viewpoints, information concepts, representation frameworks, and design activities required for human-centered design (Lim and Sato, 2006). It provides mechanisms to define concepts, frameworks of design information by combining a set of very basic types of information elements such as entities, attributes, states, actions and time. DIF is based on General Design Theory and provides the foundation to theoretically define these different information concepts and their relationships (Sato, 2009). One of the tools under DIF is Modular Script Scenarios (MSS) method. MSS utilizes the terminologies and conventions of scenario descriptions and breaks down an event description into a spreadsheet format. The modularized format of MSS with clear ontology provides a project information management platform for systematically performing analysis and solving problems but also functions as a platform for developing new methods.

3.3   Viewpoints, Structured Planning and Architectures

Human-Centered System Integration requires integration of many different types of information from different disciplines and sources such as user studies, social and cultural studies, task analysis, organizational requirements, ergonomic data, and system models. Even a simple product manifests complex design problems when it is placed in contexts such as users' daily life, societal environments, services, cultural practice, and environmental concerns. DIF allows designers to take different viewpoints of analysis and see the system concepts in different lights. DIF allows structured planning such that designers can account for all factors and conceptualize solutions that are innovative and not obvious in the traditional design methods. Students learn in this class how human centered architecture for products and services should account for both Internal and External Viewpoints. The internal viewpoint represents qualities such as technical integrity, manufacturing ease and service-ability and the external viewpoints represents such factors as business organization, user and use context, values and ecological sustainability etc. In order to achieve over all system integrity and human centric qualities, both viewpoints need to be well integrated in the system models and conceptual solutions throughout the development process.

4   CASE STUDY

There were four registered students for the HCSI course out of which three completed the full project. This paper presents meta-analysis of three class projects done during the course of eight weeks. Similarities and differences are discussed later. This HCSI class was a lecture class with a course-long project which student did on their own. The project being part of a 9-week class was a very compact exercise rather than a full development process. The explicit purpose of the case study was to see how the students transform their own viewpoints and the project work as the HCSI course unfolded over weeks of class. The mechanism of meta-analysis of the projects was simple in that we compared project progress over weeks. After each week, projects were compared and contrasted with each other such that similarities and differences could be ascertained.

4.1   An Integrated Approach to Preventing Falls

This project was by a first year PhD Student who we refer to as Student G. She has a Masters degree in Social Sciences and so very little background in traditional HF/HCI/HSI. Student G was very interested in prevention of falls among the elderly. About one third elderly populace living in community-dwelling falls each year and she wanted to provide a solution to prevent these falls. In the first week when she proposed her project, the solutions mentioned were "a fall preventing shoe", "a walking aid" and "an alerting device to call for help." As the weeks progressed and she learned the basic philosophy of HCSI and her project evolved.

The project solution turned away from being a product to a systematic service solution for fall prevention. Rather than a fall preventing shoe device the new solution had product, care taker and information system all working in harmony to prevent falls and keeping the elderly person informed and in the loop.

The integrated approach to fall prevention included:

1. Device (MotoActv/FitBit) that monitored participant activity
2. Group therapy class to have the element of sociality
3. Family Involvement to help adoption and sustenance of participation
4. Professional's involvement like a Physio-Therapist and
5. Electronic Medical Record management integration

In eight weeks student G presented a comprehensive service concept for fall prevention. She incorporated Personas to empathize with users, POEM (people, objects, environments, and messages) analysis, UML diagrams of system, Modular Script Scenarios, Information flows and System Architecture diagrams.

4.2 An Integrated Approach to Healthy Snacking for the Elderly

This project was by a first year Master's Student, who we refer to as Student A. He has a Bachelor's degree in Industrial Design, and so, very little background in HF/HCI/HSI. Student A was very interested in snacking habits of elderly and particularly in how to encourage healthy snacking among seniors. In the first week his project included solutions such as "an app to remind snack time", and "an alert system for unhealthy snacking walking aid." Although his initial approach addressed user problems, the scope was quite limited. As he learned the basic philosophy of HCSI, his project evolved from being an stand alone application on phone to a systematic solution to the problem of snacking/ unhealthy eating habits. Student A created a new solution which had a soft application, a monitoring system, interaction with care provider/dietician/grocer and refrigerator working in harmony to prevent unhealthy snacking and to encourage healthy eating.

The integrated approach to snacking included:

1. Devices (Intelligent Refrigerator, Smart Grocery Shopping Application and SmartPhone)
2. Snacking advice from dietician and monitoring from doctor's office
3. Social circle involvement to help adoption and sustenance of participation
4. Information system to support the above co-ordination and delivery of persuasive solution to snacking.

In eight weeks student A presented a full conceptual solution to snacking. He presented the solution with Personas, Information flows, Modular Script Scenarios and analysis, Fishbone diagrams of snacking activity, and system architecture.

4.3   An Integrated Approach to Healthy TV entertainment and interactivity for the Elderly

This project was by another first year Master's Student, who we refer to as Student B. He too had a Bachelor's degree in Industrial Design, and so, very little background in traditional HF/HCI/HSI. Student B's interest was in improving sedentary lifestyle of senior citizens in assisted living. He observed that a lot of seniors just watch TV all day in assisted living facilities and he wanted to design such that entertainment and interactivity are built into the TV. In the first week when he proposed his project, the solutions mentioned were mostly information based applications like "weather app on TV" and "a video communication app" to keep in touch with family.

As the weeks progressed and he learned the basic philosophy of HCSI, his project evolved. The project solution turned away from being a simplistic application on TV to a somewhat systematic solution that encouraged activity via TV. Student B created an iTV solution that integrated exercise activity, food-ordering service internal to the assisted living facility. He thought of inter-room communications and mini-social-community formations for activities.

The integrated approach to active life at assisted living included:

1. Devices (interactive TV services, SmartPhone etc…)
2. Exercise application for encouraging and tracking activity via TV
3. Social circle involvement to help adoption and sustenance of participation
4. Information system to support the active life

In eight weeks student B presented a full conceptual iTV system that encouraged more activity in sedentary senior citizens via TV. His solution included Personas, FishBone Analysis of daily activities, Information flows, Modular Script Scenarios and analysis, iTV feature description, hierarchical diagram and service architecture.

5   ANALYSIS OF PROJECTS

In this section, comparisons have been drawn to show the difference in how students were thinking at the early stage of the class and at the later stage of the class. This comparison highlights: 1) the types of information elements inquired and used in the process, 2) factors or aspects of the system considered important for design of solutions, 3) types of subsystems incorporated in the system concepts, and 4) the methods used to analyze, represent and explore solutions during the project.

5.1 Use of different information elements

Here is a comparison of initial state of the projects and the factors considered:

Table 1 Comparison of initial state of the projects and the factors considered

Factor considered	Student G	Student A	Student B
User need/problem	Only a statement of interest	Statement of personal interest with a Mind Map of influences	Only a statement of interest
Analysis	Lit review but no analysis	Some non-peer-reviewed articles without analysis	None
Project Proposal	Proposals were "a fall preventing shoe", "a walking aid" and "an alerting device to call for help."	Proposals were "an app to remind snack time", and "an alert system for unhealthy snacking walking aid."	Proposals were mostly information based applications: "weather app on TV" & "video chat app"

Here is a comparison of final state of the projects and the factors considered:

Table 2 Comparison of final state of the projects and the factors considered

Factor considered	Student G	Student A	Student B
User need/problem	Personas used	1. Personas used 2. Mindmap of factors	Personas used
Task and user action analysis	POEM – People, objects, environments and messages analysis	Snacking behavior deconstruction into actors, function, information and environment/spaces	Analysis of daily activities using fishbone structure
User movement and spatiality	Used UML for interaction diagram of person, devices and spaces	Used Information Flowchart to describe user activities of interaction with devices, other people including professionals and spaces	Spatial map of assisted living facility and hierarchical structure of organization was used
Analysis and problem identification	Modular Script Scenarios (MSS)	Modular Script Scenarios (MSS)	Modular Script Scenarios (MSS)
New concept representation	System Architecture of new solution	System Architecture of new solution	System Architecture of new solution

Benefits representation	1. New solution strategies discussed 2. System requirements identified include user, health professionals and Electronic medical records	1.Multiple Solution Strategies shown 2.Prevention of unhealthy snacking and 3. Encouragement of healthy food habits highlighted	1. 2X2 matrix of proposed solution 2. Incorporation of Food services, iTV and social networking 3. New flow diagrams and fishbone to show off benefits

5.2   Discussion

We know that these student projects are limited to conceptual explorations–with a quick information inquiry process at the beginning. Although real design projects will be significantly different in scale, rigor and depth development process, the class projects assimilate critical characteristics of projects in practice. HCSI concepts and methodology were introduced together with common design methods and tools such as information flow modeling, personas, and task analysis. Students can better grasp the HCSI thinking and integrate solutions at a human level which is what we designers want (or need). The three cases of student projects illustrate what HCSI approach brings to projects. In short time students learned the value of different viewpoints of different stakeholders and the value of synthesizing solutions with a broad scope of the system that addresses basic human needs.

The patterns of the three projects were significantly similar. All three students used Personas, to depict the user profiles including demographics, behavioral patterns and some personal values, needs and the different contexts of use. All three students used information flow diagrams to show the information flow patterns between different sub-systems to enable a delivery of service. They all used information flows to clearly describe the different touch points and interactions betweens systems and users. Some variations of causality analysis such as fishbone analysis were applied to identify problems and their causes relations. One student used UML diagrams for describing the system. All three students were able to effectively use Modular Script Scenarios (MSS) for developing structured scenario description of use cases. MSS was proved to be very helpful in allowing students to break down narratives (user interviews/scenarios of use) into analyzable elements of data. It also allowed the students to see the interaction patterns and potential related problems from different viewpoints. MSS allowed easier data management and reconstruction of future scenarios when that phase of project came about. Last but not the least, the students were definitely overwhelmed with this much material for a short class of eight weeks. Their feedback was that this class is desirable to be semester long – 16 weeks and there needs to be lot more theoretical foundation covered in the lecture as well as off-class reading. The open-ended nature of the project in elderly care was felt as too much of uncertainty at the beginning. The ambiguity of the project introduction was exactly the course intention to allow students to learn about complex systems and about decisions to make under that uncertainty.

6   CONCLUSION AND FUTURE WORK

At the beginning of the class, each student had shown apprehension about coming up with system architecture on their own. They were saying we are designers we don't do system architecture. How can we be software architects or business model creators? But at the end, they were talking the "external and internal" viewpoints and discussing the business model for the services they were envisioning. This HCSI course structured the complex process of understanding and designing socio-technical systems and also took the designers from thinking about products or services to service concepts that are truly integrated and human centered. In conclusion, this class was illustrative of how students understood HCSI and the applied methods for it. However, many issues remain for improvement. HCSI application till requires extensive theoretical and methodological research and tool development. Further empirical verification and improvement cycles will also help HCSI become an effective and accessible methodology for practice.

ACKNOWLEDGMENTS

The authors would like to acknowledge the students in the HCSI class who provided valuable discussion and their project engagement.

REFERENCES

AARP International (2009) World: World's 65 And Older Population to Triple By 2050 http://www.aarpinternational.org/news/news_show.htm?doc_id=953328

Booher, H. (2003). Handbook of human systems integration. Hoboken, NJ: Wiley & Sons.

Lim, Youn-Kyung and Sato, Keiichi. "Describing Multiple Aspects of Use Situation: Application of Design Information Framework (DIF) to Scenario Development," Design Studies, Vol.27-1, 57-76, 2006.

National Research Council. (2011). *Health Care Comes Home: The Human Factors.* Committee on the Role of Human Factors in Home Health Care, Board on Human-Systems Integration, Division of Behavioral and Social Sciences and Education. Washington, DC: The National Academies Press.

Norman, Don. (2010) Why Human Systems Integration Fails (And Why the University Is the Problem). Invited talk for the 30th anniversary of the Human-Systems Integration Board of the National Research Council, the National Academies. Washington, DC. December 2, 2010.

Reinach, Stephen and Jones, Michael (2007) An Introduction to Human Systems Integration (HSI) in the U.S. Railroad Industry. A Report for Federal Railroad Administration.

Sato, Keiichi (2009) Perspectives on Design Research. Chapter in the book Design Integrations, Edited by Poggenpohl and Sato. The University of Chicago Press, USA

Sato, Keiichi, Yong Chen, Deborah Cracchiolo, Xiaoshan He, Eui-Chul Jung, Tom MacTavish. "DIF Knowledge Management System: Bridging Viewpoints for Interactive System Design". Proceedings of the 11th Human Computer Interaction International, Las Vegas, Nevada, USA (July 2005).

U.S. Census Bureau (2009) International Data Base (IDB) http://www.census.gov/ipc/www/idb/index.php

Integrating Healthcare Service with Ambient Interactive Systems: Conceptual Framework

Jihyun Sun, Santosh Basapur, Keiichi Sato

Illinois Institute of Technology
Chicago, Illinois USA
{jihyuns, basapur, sato}@id.iit.edu

ABSTRACT

Many issues of supporting elderly living urgently need rethinking from the perspective of overall service systems including technological systems, human roles of care giving, infrastructure, and business models. Service system plays a major role for integrating components and subsystems for achieving better human-qualities of overall service performance. This research proposes the concept and architecture of Ambient Interactive Systems (AIS) that provides general platform for deploying various services for creating desirable senior living experience for individual life styles in different living environments such as independent living, assisted living and hospital environments. AIS intends to integrate physical space and cyber space in order to enhance synchronizations of overall activities performed by the system constituents in the space for enhancing the quality of work and life.

Keywords: Ambient Interactive System, Spatiality, senior care, context-sensitivity, cyber-physical space, service system architecture, Human-Centered System Integration

1 INTRODUCTION

New technologies such as dynamic networking and embedded systems provide technological environments that seamlessly support our work and daily life. These new technologies can pervade into our activity spaces and deliver functional and informational services where and when they are needed. For effective and satisfactory deployment and use of a system with such technologies, the quality of interaction between people and the system needs to be finely tuned to users' needs and the context of use. Since system functions are interwoven into people's activities and their environments, understanding contexts of use particularly becomes critical for assessing how a system could be accepted, understood, incorporated, used, and valued in people's activities and minds.

In this research, the term "Ambient Interactive System (AIS)" is introduced in order to represent a system composed of subsystems and components including people involved in the activities of concerns, physical space, embedded technology subsystems, independent objects, cyber space created by computing and network subsystems. Examples of AIS can be implemented in spaces for various human activities such as home, office, public spaces, health care facilities and transportation systems. AIS intends to integrate and coordinate physical space with objects and functions and cyber space in order to enhance synchronizations of overall activities performed by system constituents in the space for the quality of work and life.

The performance of interactive systems particularly AIS can be only determined in relation to the context in which the system performs its intended roles. While the context dynamically changes, systems are usually designed to remain the same and to be operated within a limited range of contexts. In order to maximize the system performance particularly from users' viewpoints, the system needs to be sensitive to the change of the contexts. In order to ensure the quality of users experience with the system, contexts of use from users' viewpoints need to be well understood and incorporated into design consideration as the core information for human-centered design practice (Sato 2004).

In our rapidly aging society, elderly care and health care areas are casting major social and economic problems. Recent research efforts in areas such as ergonomics, architecture and design have been making significant improvements of individual design solutions for elderly population. Yet many issues of elderly living support have not been addressed and urgently need rethinking from the perspective of overall service systems including technological systems, human roles of care giving, infrastructure, and business models. In such environments, service design gains a major role for integrating different components and subsystems for achieving better human-qualities of overall service performance. In this research, AIS provides general platform for deploying various services for creating desirable elderly living experience for individual life styles in different living environments such as independent living, assisted living and hospital environments.

2 CONCEPT OF AMBIENT INTERACTIVE SYSTEMS (AIS)

2.1 AIS model: Spatiality and System Structure

Space embodies social, cultural, and physical elements that all together compose the environment as a stage setting for the user-system interaction. Conceptual models of spatial interaction, AIS with *spatiality*, is developed to provide a basic framework that enables system designers to develop a system that can enrich users' experience in the space (Figure 1). It is introduced to represent this complex and emergent mechanism in a particular situation that dynamically stages users activities. The proposed AIS, therefore, manifests its emergent mechanism composed of the relations among various factors of the user, the technical sub-systems, and the environment. The user interprets a states of *spatiality* generated by continuous interactions with the systems in an environment and constructs the models based on existing knowledge of the system. The models of the space developed by users represent different aspects including interaction, systems, space, relations to space, and images of himself/herself. An environment is the stage for user actions where the space is configured with objects that represent social, cultural, and physical elements. An environment therefore affects users' activities by providing a situation created by its components and configuration. The user assigns meanings to the entities and relations between the entities in the space, constructs models of the space or modifies existing models through interactions with the system.

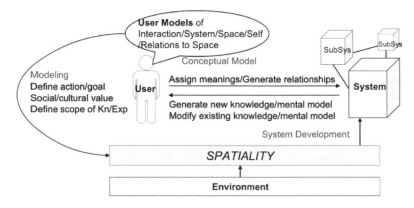

Figure 1 Conceptual AIS Model with SPATIALITY

2.2 Conceptual Structure and Boundaries of Space

The physical layout of the space greatly affects users' activities and behaviors. Physical spaces are structured according to uses and needs for interaction (Harrison and Dourish 1996). Systems are embedded in the physical structure of the space that reflects users' value system or cognitive information processing for effective interaction. This research distinguishes three different types of spaces: *attentive, peripheral, and latent spaces*. At first, *attentive space* is the primary working space where actions take place, more specifically where attention is paid in relation to body's positions and intended tasks. The second layer of the space, *peripheral space*, is a space that needs to be accessible and visible for users to take the next step of actions or inquire information while a certain task is executed in *attentive space*. This space accommodates actions and operations necessary for concurrent auxiliary tasks or a sub-goal achievement process. The third space is *latent space*. For the user it is out of immediate concern since the entities in this area are irrelevant to the current task or the current task engagement that imposes a limit to the cognitive and physical capacity to access the entities in this area. This space needs to be readily prepared for users' and system's actions with system components such as sensors and functional devices for other expected tasks and events.

2.4 The Qualities of User-System Interaction in Space

Space characteristics can greatly influence on user experience. System designers attempt to prescribe and achieve qualities of space. There are several kinds of variables that may affect the quality of the users' experience within a space as shown in Table 1.

Table 1. Qualities of User Experience in the Space

Quality	Definition
Availability	Well-identified and adequately distributed knowledge source for acquiring relevant knowledge with less cognitive load
Accessibility	Well-structured and represented information for users to easily activate and retrieve knowledge from the long-term memory
Connectivity	A number of subsystems distributed across the places
Controllability	A user can manipulate and transfer knowledge in controlling multiple systems and those functions in system interaction.
Predictability	Knowing the past history of interaction with the system a user can anticipate the result of the future interaction.
Adaptability	A user can assess the system interaction comparing with the previous interactions to formulate a model of the behavior of the system for future operations.
Locality	Specified information representation that is tailored to user's certain situation

2.5 Context Sensitivity

As stated in the introduction, context-sensitivity of the AIS is the key for the satisfactory applications to individual user cases. Since the term "context" has been used in many different meanings and concepts but yet there is no clear definition in the area of system design. Therefore this section introduces the definition of the term "context" used in this research.

Some aspects of context take significant roles in forming situations for the current action; some aspects become irrelevant to the current action. We call the former manifesting aspects of context and the later latent aspects of context. As a result of the action, some of the aspects of context change and evolve over time. Context is developed internally in the cognitive system as a model or pattern of knowledge that is associated by recognizing a current situation. Some contexts emerge as personal knowledge that is hard to communicate or not sharable with others. Some contexts emerge as socially shared knowledge through common experience among people in the community. In this research context is defined as a set of mental models activated by triggering elements in the situation. In order to achieve context-sensitivity in AIS design, the system needs to have a mechanism to monitor situational conditions and estimate user contexts for providing required services.

2.6 AIS Architecture and System Design Methods

AIS consists of sensor network subsystems, coordinating subsystem and service action subsystems as shown in Figure 2. The embedded sensor network monitors the states, behavior and interaction of AIS constituents including the users, the system itself, objects, and environments.

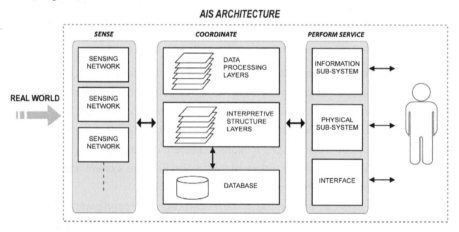

Figure 2 Architecture of Ambient Intelligent System (AIS)

AIS requires a wide range of metrics to capture user behavior, needs, and contexts for performing effective interactive services. Ontology of each AIS consists of application independent ontology and application dependent ontology. Application dependent ontology needs to be determined through extensive studies of each application domain in addition to the core AIS ontology. Sensor network composition and topology must be determined based on variables that constitute application domain models and application specific parts of the AIS itself. Table 2 shows categories of information that can be collected by the embedded sensor network. Sensor network composition and topology must be determined based on variables that constitute application domain models and application specific parts of the AIS itself (Ren, Yu et al. 2007)

Table 2 Categories of Information

User(s): *Usr*	Object(s): *Obj*	Environment: *Envt*
Action (*Act*)	Critical Objects (Crt)	Floor plan (*FP*)
State (*St*)	Functions (*Fnt*)	Lighting (*Lit*)
Relationship/role (*Rel*)	Objects' relations (*Rel*)	Sound (*Sd*)
Location (*Loc*)	Arrangement (*Arng*)	Temperature (*Tmp*)
Profile (*Pf*)	Distance (*Dis*)	Humidity (*Hmd*)
Knowledge: *Knl*	Location (*Loc*)	Scent (*Sct*)
Task Goal (Gl)		Weather (*Wth*)
SubGoal (sGl)		

3 AIS DESIGN CASE IN HEALTH CARE AND ELDERLY CARE

An example scenario of taking medication as follows depicts the situation that an AIS for senior living support typically faces.

"Mary, 85 year-old, had lunch with her daughter, Jess. Mary didn't have a good appetite. After lunch she decides to go to bed to rest instead of watching TV on the sofa in the living room. Jess gets ready to go out for errands. Before Jess is out she reminds Mary to take her cardio-vascular medication in two hours and drink at least a glass of water or juice. Jess gets a text message from the home system on her cell phone reporting that Mary has not taken the medication or drink. The status monitor indicates Mary is in bed and is awake after one hour nap."

Their actions primarily take place within the house. They have relationships, mother and daughter, and caretaker and caregiver. Jess's goal is to take care of her mother. Mary's tasks for the afternoon are to take medication and to drink a glass of water.

In order to identify application specific ontology, Modular Script Scenario (MSS) based on Design Information Framework (DIF) concept provides information structuring platform for representing application domain studies and deriving an ontology that effectively form an information platform for AIS both development and system operation to bridge the real world situation, user contexts, user needs and AIS service actions (Lim and Sato 2006). MSS can accommodate both qualitative and quantitative data as well as multi-media data. Figure 3 shows an example of a domain study data representation in MSS format.

Seq #	User (Usr)				Object (Obj)			
	PROFILE: Mary is 85 year old cardio vascular patient who is taken care of her daughter, Jess.				HOME CARE SYSTEM			
	TASK GOAL	Mary's TASK is to take medication and water while Jess is out.						
		Act	St	Rel	Loc	sGl	Fnt	Crt
D03	Mary	go to bed	tired	caretaker	bedroom A	rest		bed
D04	Jess	gets ready to go out	busy	care giver	dressing room	do errand		mirror
D05	Jess	Tell Mary to take medication & water		care giver	bedroom A			
D06	Mary	Agree to take them		caretaker				
D07	Jess	set reminder for Mary		care giver	living room		home system	cell phone
D08	Jess	say goodbye to Mary	ready to go out	care giver	living room			
D09	Mary	None	sleeping	caretaker	bedroom A		sensor in bed	bed
D10	Mary	wake up	resting	caretaker	bedroom A			bed
D11	Mary	read a book	forget to take medication	caretaker	bedroom A		monitor medication bottle	bed med bottle
D12	Mary	read a book	forget to drink water	caretaker	bedroom A		monitor water bottle	bed water bottle
D13	Jess	receive a text message about status		care giver	post office		send out text	cell phone
D14	Mary	receive reminder		caretaker	bedroom A			
D15	Mary	take medication & water		caretaker	bedroom A			
D16	Jess	receive a text message about status		care giver	grocery		send out text	cell phone

Figure 3 Scenario Example Applied Using Modular Script Scenario (MSS) format

4 AIS SERVICE STRUCTURE FOR ELDERLY CARE

Some service design models have been proposed with specific focuses such as IBM's Service Oriented Architecture (SOA) as a framework for IT services and Service Oriented Modeling and Architecture (SOMA) for system analysis and design in the SOA development.

The example case of section 3.1 indicates a wide range of potentially problematic situations to be addressed. The daughter, Jess, could simply call her mother, Mary, for reminding, but Mary could be on the phone, or is not picking up the phone for some reason. If so, where is she?, what is her condition?, and what is she doing? In order to assist Mary in this situation, the system needs to determine a strategy for series of actions to understand Mary's states, to remind about the medication and further confirm with Jess if Mary had medication. If the system detects that Mary's condition requires immediate attention and assistance, it must inform her daughter, medical agency for immediate assistance, or provide some direct assistance to Mary if possible.

In this particular instance, AIS should detect Mary's location, states, activity and the objects in use and deliver the most appropriate service in the right service format to the most appropriate device for the situation. For instance, alternative contact channels include wearable devices and peripherally located devices such as a bracelet, connected photo frame, TV or even an ambient display designed specifically to benefit from a system like AIS. The reminder could be displayed as a message scrolling at the bottom of a TV screen or as flashing of the daughter's picture on the connected photo frame or even as a cuddly toy bear waving and talking for the mother to see and get reminded of her medication.

A comprehensive service structure accounting for user goals, context and appropriate function allocation to different devices and people within the system is needed for such a user experience to unfold seamlessly. Services are complex technical systems with people and technology playing equally important roles. Service Design is to create desirable experience with people through many different touch-points that unfold over time. AIS service structure for senior care attempts to provide a physical and informational platform and deploying such tangible and intangible touch points that enables the service to achieve high qualities user experience for both caregivers and caretakers. As a service, AIS, provides a platform for deploying various stages of service delivery at the right time, at the right location through right device with the right quality of information and/or physical functions such that interaction is meaningful, intuitive and pleasant. With the increased availability of cloud based platforms it has become possible to deploy comprehensive systems as envisioned by AIS architecture (Schaper 2010). Figure 5 shows and example embodiment of the AIS service mechanism. AIS consists of backend of systems, which will monitor the persons and the space and interpret goals by the interaction users initiate with devices and environments.

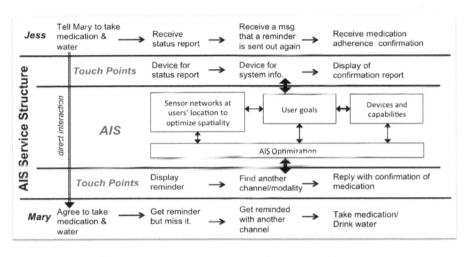

Figure 5 AIS Service Structure and an example of its application to senior care

4.1 AIS Development and Evaluation

The specification of service functions and qualities of the AIS has to be composed based on operational and user requirements, and evaluated in terms of both effectiveness and experiential satisfaction. Experiential satisfaction can be defined at two levels: the interaction with tangible touch points and the overall service experience with the AIS over time. The effectiveness needs to be evaluated from many different viewpoints including ones specific to application areas and ones generic in system development. Examples include relevance of functions to the system goal, efficiency of function performance, and adaptability to varying conditions, reliability and safety. The experiential satisfaction is hard to assess since it highly depends on individuals' values and subjective reflection. There are some framework and guidelines introduced for understanding and evaluating qualities of experience, particularly at the interaction level. Norman refers to three types of emotional responses to interaction experience with a product Visceral, Behavioral and the Reflective emotional response (Norman 2004). Similarly, Forlizzi and Battarbee propose three types of experiences with a product or service, namely Experience, An Experience and a Co-Experience. An Experience (Forlizzi and Battarbee 2004) has a start and an end point and inspires a behavioral and emotional change when the experience is occurring and/or when it is being retold. Co-Experience on the other hand occurs in a social context where two or more people might be interacting with a product or service and co-constructing meaning and thus emotional experience. Desmet (Desmet 2004; 2005) and Hassenzahl (2001; 2001; 2006) on the other hand, suggest ways of designing products and service to affect emotional responses and how to measure those emotional responses. Both Desmet's PrEmo and Hassenzahl's AttrakDiff intend to quantifiably assess service's user experience. On the other hand, mechanisms and evaluation criteria of the overall experience quality over time have not been well explored. Some qualities of this category are universally applicable to any AIS applications as introduced in section 2.4: Availability, Accessibility, Connectivity, Controllability, Predictability, Adaptability, and Locality.

No single framework of user experience evaluation can cover all aspects of AIS service qualities. Each framework has its viewpoint focusing on certain qualities. Further research and empirical studies needs to elaborate on the utility and suitability of frameworks for understanding user experience of such ambient services with tangible and intangible touch points.

One single aspect model cannot resolve a problem (Lim and Sato 2006). Thus, in order to achieve high qualities and integrity of AIS, many aspect models representing different perspectives of the system such as information flow, operation sequence, user cognitive structure, and social and cultural norms need to be incorporated in the design foundation for AIS design. These aspect models also will become a part of the AIS knowledge base that will be used for its operational decision-making as shown in Figure 6.

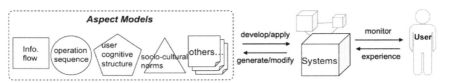

Figure 6 Aspect Models for AIS

5 DISCUSSION

This paper introduced a general framework for developing AIS as a platform for service systems for senior care environments. The proposed general AIS architecture is modularized to accommodate different application specific sensor networks, physical and information subsystems, and a database. In AIS applications, intensive and dynamic sensor networks effectively capture critical information for understanding user needs and contexts, and determining appropriate system actions in order to provide context-sensitive services.

AIS as a platform attempts to address these issues. We also deliberate on evaluating services deployed with AIS as the basis. We believe that no one framework of evaluation covers all aspects of different services. Future research needs to explore more integrated forms of frameworks built on the existing ones that particularly enable evaluation of AIS's service qualities over time. The conceptual framework of AIS introduced in this paper also provides a clear scope for further research and empirical studies that elaborates the utility and suitability of frameworks for understanding user experience of such ambient services with tangible and intangible touch points.

REFERENCES

Desmet, P. (2005). "Measuring emotion: development and application of an instrument to measure emotional responses to products." Funology: 111-123.

Desmet, P. M. A. (2004). From disgust to desire: how products elicit emotions. Design and emotion: the experience of everyday things. London, Taylor & Francis: 8-12.

Forlizzi, J. and K. Battarbee (2004). Understanding experience in interactive systems. DIS '04 Proceedings of the 5th conference on Designing interactive systems: processes, practices, methods, and techniques, ACM.

Harrison, S. and P. Dourish (1996). Re-place-ing space: the roles of place and space in collaborative systems. Proceedings of the 1996 ACM conference on Computer supported cooperative work, ACM.

Hassenzahl, M. (2001). "The effect of perceived hedonic quality on product appealingness." International Journal of Human-Computer Interaction 13(4): 481-499.

Hassenzahl, M., A. Beu, et al. (2001). "Engineering joy." Software, IEEE 18(1): 70-76.

Hassenzahl, M. and N. Tractinsky (2006). "User experience-a research agenda." Behaviour & Information Technology 25(2): 91-97.

Lim, Y. K. and K. Sato (2006). "Describing multiple aspects of use situation: applications of Design Information Framework (DIF) to scenario development." Design Studies 27(1): 57-76.

Norman, D. A. (2004). Emotional design: Why we love (or hate) everyday things, Basic Civitas Books.

Ren, S., Y. Yu, et al. (2007). "The role of roles in supporting reconfigurability and fault localizations for open distributed and embedded systems." ACM Transactions on Autonomous and Adaptive Systems (TAAS) 2(3): 10.

Sato, K. (2004). "Context-sensitive approach for interactive systems design: modular scenario-based methods for context representation." Journal of physiological anthropology and applied human science 23(6): 277-281.

Schaper, J. (2010). Cloud Services. 4th IEEE International Conference on Digital Ecosystems and Technologies (DEST), Germany, IEEE.

CHAPTER 13

Entrusting the Reply of Satisfaction or Physical Condition for Services to Unconscious Responses Reflecting Activities of Autonomic Nervous System

Hiroaki Okawai, Keiichi Kato, and Daiki Baya

Iwate University
Morioka, Japan
hokawai@iwate-u.ac.jp

ABSTRACT

The suppliers usually desire to know how much the service satisfied customers. However, the reply from customers obtained by questionnaire survey, for example, is not always accurate to express honest mental activity. This paper describes a detecting method of a more accurate reply by entrusting mental and physical activities for the services to unconscious responses reflecting activities of autonomic nervous system during sleep. Some experiments were performed by well known services for healing, for example, in order to demonstrate accuracy of the present method. As a result, the unconscious responses reflecting the above nervous system such as respiration and pulse rate shifts beyond several hours at night showed satisfaction instead of conscious responses reflecting mental activities.

Keywords: autonomic nervous system, unconscious response, body motion wave (BMW), dynamic air pressure sensor, satisfaction.

1   INTRODUCTION

Society as well as circumstances of nature gives human various stimuli. The stimuli of articles, information, energy and labor et al, produced by suppliers can be dealt with service. The suppliers usually desire to know how much the service satisfied customers. Questionnaire survey is often used to get replies of customers, however, the reply is usually modulated by reason, knowledge or consideration, so that it is not always honest, i.e. accurate, to express mental activity. Then, can we determine a scale in the reply? This paper describes a detecting method of an more accurate reply by entrusting the reply of satisfaction or physical condition for the services to unconscious responses reflecting activities of autonomic nervous system in charge of our living.

In order to study the response, and to demonstrate a method suggested here, healing music et al, widely recognized as heeling samples, and some beddings now on developing were adopted in concrete at the present study.

2   HYPOTHETIC FLOW OF A SERVICE FROM STIMULI TO RESPONSES FOR A HUMAN SYSTEM

A human is an elaborate system. Here, a usual idea, a system has a mechanism to produce an output modulated by its characteristics for an input, was applied. The services are input to the system both intentionally or unintentionally. Then, the system is influenced and makes a response, viz., an output. Such phenomena were shown in Fig.1 suggesting a mechanism of resulting a conscious response and an unconscious response for an input of a service to the human system.

Here, article, information, energy and labor were listed for example for input. The output of speech or action, for example, is a conscious response, on the other hand, the variation in physical condition is an unconscious response.

At first, the flow to conscious response is as follows. A biological signal, produced as sensing stimuli, i.e. service by, for example, sense of sight at channel 1, flows into two channels. The one, channel 2, processes it as mental activity, and the other, channels 3, does it as physical activity. Then, these mental and physical activities interact through channel 4. Mental activity, such as satisfaction or emotion, probably makes physical activity better. Also good physical activity will make mental activity better.

Thus, the mental activity will reply to the service by satisfying for a conscious response through channels 5 and 6. However, there would be possible to reply not satisfying, because a signal of mental activity was sometimes distorted by reason or consideration et al, as shown in the bracket between channels 5 and 6. This output is a conscious response also. Thus, the reply from here is not always accurate because of above distort.

On the contrary, there are two lines in the physical activity as a physical condition to express an unconscious response. The one is direct channel 3 and the other is channel 4 via a mental activity. The signals from such two pass ways are combined and then sent by channel 7 for output. This unconscious response might be an honest answer because it was not distorted by above reason.

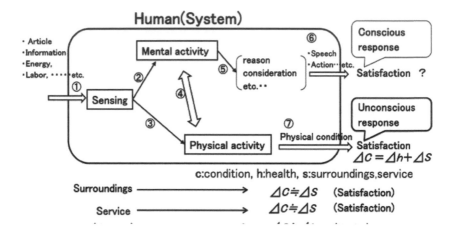

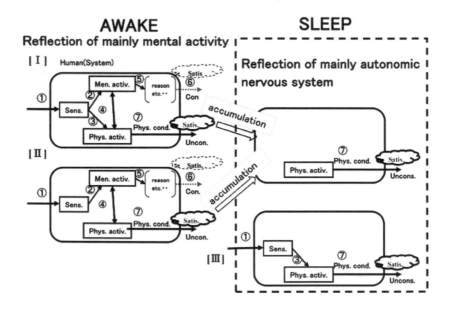

Figure 1 Hypothetic model revealing flow of a service for a human system from sensing to response

Figure 2. Hypothetic model revealing three types of flow of service: I, II, and III

Physical condition generating unconscious response as shown in Fig.1 was determined by a factor c. Then the factor c was assumed to be a product of health related factor h and service related one s. If these factors were shown by delta as they varied in a small amount, and were simplified, then, they were given by,

$$c + \Delta c = (h + \Delta h)(s + \Delta s) = (1 + \Delta h)(1 + \Delta s) .$$

Then the variation Δc is given by,

$$\Delta c = \Delta h + \Delta s .$$

As shown in Fig.1, as a new sensitive surroundings or service is input under health maintained, the output will mainly occurs as $\Delta c = \Delta s$. When Δc varied with no sensitive matter, the fraction of health condition, Δh, must have changed.

Daily life is roughly classified into two situations of in awake and in sleep as shown in Fig. 2. In the former, mental activities produced in cerebrum will generate a conscious response. On the contrary, in the latter, because of unconscious state, autonomic nervous system activity is superior so that physical activities will generate an unconscious response. This is why the answer is put into unconscious response.

Then, service input can be classified into three types as shown also in Fig. 2. Articles and food are examples of type I having channels 2 and 3. Information is an example input of type II not having channel 3. Unconscious responses would be accumulated in the body and, therefore, would appear during sleep. During sleep we are in unconscious state, therefore, the output through several hours is free from reason, consideration, et al. Some bedding materials, input to human system without awareness during sleep, are examples of type III which having neither channels 2 nor 4.

3   METHOD FOR TAKING DATA OF UNCONSCIOUS RESPONSES

3.1   Method

The unconscious response can be expressed in such as respiration and pulse because of reflecting autonomic nervous system activities. A pressure sensor, named "dynamic air pressure sensor" (M.I.Labs), was adopted in order to fabricate a non-restraint, non-attachment measurement system. As shown in Fig. 3, it was set on a bed to measure dynamic air pressure arisen between the sensor and a subject's body at lying. The pressure variation detected with the sensor was converted to electric signals, sampled at the rate 400 Hz, 16 bit and stored in a personal computer. The signal was processed with Chart v4.2.2 (AD Instrument) and programmable software VEE Pro ver6.0 (Agilent Technologies).

Here, in subject's body at sleep some continuous motions are generated resulting in respiration, pulse and other unconscious action et al, so that all of these motions was detected through the pressure wave. Thus obtained pressure wave here is neither electrocardiogram nor respiration gas flow. All of above wave components expresses body motions. Therefore, we named this wave "BMW: body motion wave". This wave was filtered to two wave components at this paper in the right part in Fig.3. The one is "R-BMW: respiration-origin BMW" and the other is "P-BMW: pulse-origin BMW".

This method would be applicable not only for checking effects of services by detecting physical condition but also for checking health state in daily life and evaluating welfare living surroundings as shown in section 2.

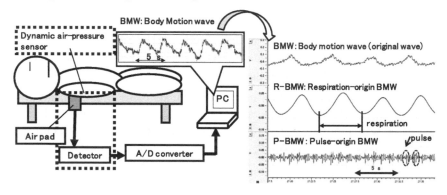

Figure 3   Measurement system for body motion waves (BMW) and filtered components of respiration-origin and pulse-origin BMWs (R-BMW and P-BMW)

3.2   Experimental Samples to Demonstrate Reliability of the Method and Human Charicteristics in Service Sience

(1) Reproducibility and Development of Accuracy in the Present Method

In order to check the method, experimental data were taken for two consecutive nights for a subject under the condition: a) two days in awake under approximately same state, and b) the same environment during sleep.

As a result, it was found as follows in Fig.4,

1) Both respiration and pulse rate shifts showed approximately the same patterns between upper and lower figures, respectively. The former shows approximately 16-18 before 3.5 h and a little large rate after 3.5 h. The latter does also approximately 50-60 in both the upper and the lower.   2) The pattern showed periodicity having increase and decrease whose period was 60 -120 minutes depending upon individuals. 3) For the bad physical condition from a few days lack of sleep before experiment, not shown here, the pattern showed a distinctly different manner for the same subject.

The above findings obtained from experiments showed reproducibility of the present method and of physical condition in a normal subject. The total number of subjects we obtained is already more than 300 for 20s normal subject, and several tens of the elderly.

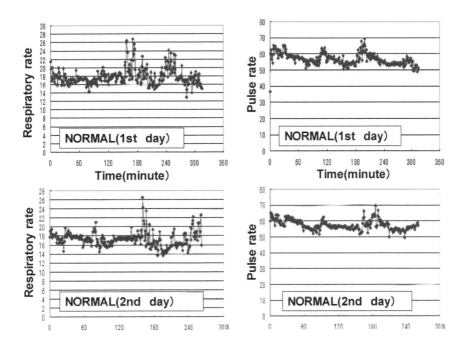

Figure 4   Example of respiratory and pulse rates shift for two consecutive nights to demonstrate roproducibility both of the present method and of the physiological state detected for a normal subject. "Normal" in the figure means usual sleep surroundings.

(2) Service Samples and Data Acuision

Consecutive four days were set. For the first and second days, normal day (usual day), physical reproducibility for a subject was checked. For third and fourth days, a stimuli i.e. service was tried. At the present study, the service of healing, relaxing or improving circumstances were adopted as follows.

1. (a) aroma, (b) healing music, (c) moss plant : a subject feel satisfaction at awake. The first two have been particularly well known to be effective to heal and the moss has been well appreciated particularly in temples in Japan.

For aroma, a subject does not use aroma daily. The subject spent time with any aroma chosen by himself/herself for 1-2 h before going bed. The aroma was available for approximately 3h after going bed. For healing music, a subject does not hear healing music daily. The subject hear any music chosen by himself/herself for approximately 30 min (minutes) to 1 h(hour) in the time of 2 h to 0.5 h before going to bed. For moss, a subject stayed for 30 min at the third and fourth days in the moss plant area.

2. (d) carbon bedding: a subject cannot recognize by senses of sight nor touch. As this material itself is often adopted in the citizens unaware of what makes sleep well, authors are now on developing. A subject cannot recognize its difference from the normal bedding by senses of sight nor touch. The normal bedding is an article on the market; major materials are polyester and cotton.

In the experiment for above a to c, the normal bedding was used. Subjects, 22 -30 normal male and female, were asked to spend four days approximately in the same state as usual and stayed at the health care experiment house in University campus for four nights. Thus, consecutive four days test was performed between Monday to Thursday, or Tuesday to Friday.

4 RESULT

For 4 to 20 subjects for each service trial, both or either of respiration rate and pulse rate showed decrease clearly as follows.

(a) Aroma

As shown in fig.5, respiration rate showed 12-14 in both the normal and the aroma, that is, it did not decrease. On the contrary, pulse rate started with 65 in both and shifted with approximately 60 in the normal and 50-55 in the aroma. It decreased by approximately 7. Four subjects of five showed approximately the same response, that is, it did not decrease in respiration rate and decreased in pulse rate.

(b) Healing music

As shown in fig.6, the attention was put, first of all, in the first 180 min, then it was found that respiration rate showed 14-18 /min at the normal, and 13-15 /min at the music night. Thus, it showed decrease by 2-3/min. Pulse rate started with 78 and became less to 65 at the normal. However, it showed 62-65 at music night. Thus, it showed decrease by approximately 10/min. Four subjects of four showed decrease approximately at the same.

(c) Moss Plant

Though data is not shown because of lack of space, the respiration rate did not vary, while pulse rate decreased by 7/min.

(d) Carbon bedding

Though the data is not shown here, the respiration rate showed 17-18 in the normal and 15-18 in the carbon bedding. It decreased by approximately 2. A great deal of decrease was seen in 4 h. The pulse rate showed 55-60 in the normal and decreased by keeping less than 55 in the carbon.

5 DISCUSSION

At the present study, the physical condition was influenced by healing services. In concrete, the decreases in respiratory and/or pulse rates at sleep i.e., unconscious responses, were confirmed with reproducibility.

(1) Validity in the Unconscious Response for the Service

A subject in awake chose one or two aromas or songs he/she liked. This means undoubtedly the one replied by mental or emotional activity of the subject's satis-

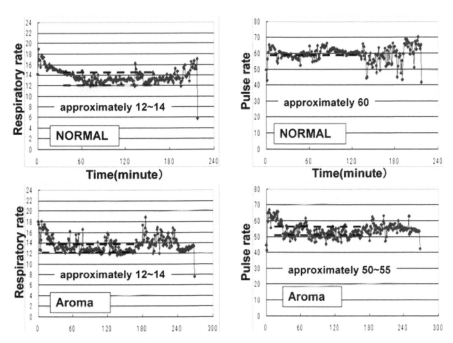

Figure 5   Respiratory and pulse rate shifts measured for a subject enjoyed aroma.

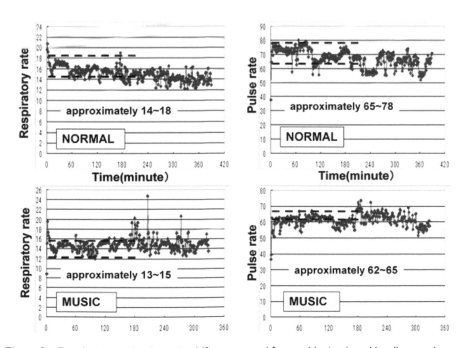

Figure 6   Respiratory and pulse rate shifts measured for a subject enjoyed healing music

faction, i.e. a conscious response. Judging from this experiment circumstances, this reply was honest one. Then, the other reply by unconscious response, the decreases in respiratory and/or pulse rates at sleep physiologically explained to be relaxed reflecting activities autonomic nervous system, would correspond to satisfaction. For the moss plant, though subjects felt nothing particularly with awareness, the physical activity replied. Therefore, the subject was sure to accept something through both mental and physical activity during stay in moss area.

On the other hand for the bedding, a subject cannot recognize satisfaction through mental or emotional activity in awake and also in sleep, however the physical activity in sleep, unconscious response, revealed the decreases as same as aroma did. Therefore, it would be explained that new bedding tried here produced satisfaction physically.

As above discussion was evolved, it would be possible to explain that the unconscious response appeared honestly regardless of the conscious response.

(2) Flows of service input to response classified into three types

Firstly, the service of aroma, because of chemical material, input and flowed into channels 2 and 3, simultaneously. Thus, this service can be classified into type I. Secondly, the service of music, because of not material but only information, flowed into channel 2. Thus, this can be classified into type II. In final, for the bedding used at the present study, the service was input only at sleep. Therefore, as mental activity cannot afford to work. This service, because of not material nor information but energy of mechanics, flowed through channel 3 into physical activity. Thus, this flow yielding only unconscious responses, can be classified into type III.

Thus, for the services to heal or relax etc. mentally and/or physiologically, experimental studies were performed to verify entrusting the reply to unconscious responses at the present study. Attention was paid to variation, i.e. decrease at the present experiment, in the respiratory or pulse rates between before and after a stimulus, not to rate level itself of how large or small. The rates themselves depend originally upon individual. So far the experiments for exciting or anger have not yet been tried because of ethical restriction.

6   CONCLUSIONS

A method to obtain customer's honest reply by detecting an unconscious response in sleep instead of a conscious response in awake for services suppliers prepared. Some basic experiments was performed to successfully demonstrate validity of the method, entrusting the reply of satisfaction or physical condition for services to unconscious responses reflecting activities of autonomic nervous system. In concrete, some well known services with aroma, music, plant, bedding were adopted in order to demonstrate accuracy of the present method. As a result, unconscious responses were detected and explained when mental activities, conscious response in awake, did or did not work. However, these results were only for healing or relaxing stimuli in this paper. Therefore, further more stimuli, i.e. services, and responses including excitement etc, would be worth studying.

ACKNOWLEDGMENTS

The authors would like to acknowledge Research Institute of Science and Technology for Science, Japan for NEXER project, 2009.

REFERENCE

H. Okawai, S. Ichisawa, K. and Numata 2011.Detection of influence of stimuli or sevices on the physical condition and satisfaction with unconscious response reflecting activities of autonomic nervous system. *N.A.Abu Osman et al.(Eds.);Biomed2011,IFMBE Proceedings35*,420-423.

CHAPTER 14

The Intelligent Space
for the elderly
— including activity detection

Somchanok Tivatansakul[1], Sikana Tanupaprungsun[2], Katchaguy Areekijseree[2], Tiranee Achalakul[2], Kazuki Hirasawa[1,] Shinya Sawada[1,] Atsushi Saitoh[3] and Michiko Ohkura[3]

[1]Graduate School of Engineering, Shibaura Institute of Technology
[2]Department of Computer Engineering, King Mongkut's University of Technology Thonburi
[3]College of Engineering, Shibaura Institute of Technology

ABSTRACT

Nowadays the ratio of the elderly to the whole population has been consistently increasing. Many elderly have to live alone. It is often found that they might suffer from depression and injuries that may lead to physically disabilities together with bad mental health. Therefore, assistance for old people to improve their quality of life and to cope with stressful events from aging is promptly required. Intelligent space is a concept designed to support smart interaction between humans and the space that can make humans in the space more comfortable. This idea can be adopted to help the elderly. Thus, this study designs and builds a prototype of an intelligent space for the elderly. From survey research, we learned that our intelligent space should include the following services: Emergency, Medical Consultation and Long Distance Social Interaction. These services help provide assistance so the elderly can feel safer and more comfortable about living independently. Depression might also be reduced. Thus, the quality of life of senior citizens will be improved including their physical and mental health.

Keywords: Intelligent space, the elderly

1 INTRODUCTION

1.1 Background

Due to the advanced research and development of technologies in medicine prolong human life, nowadays the ratio of the elderly to the whole population has been consistently increasing. Most elderly have to live alone by themselves. Without family members or any caregivers to assist and support them physically and mentally to maintain their health, old people may deteriorate faster than usual. It is often found that old people might suffer from depression and injuries that may lead to physically disabilities together with bad mental health. Therefore, assistance for old people in order to improve their quality of life and to cope with stressful events from ageing is promptly required. Next section describes the concept of the intelligent space.

1.2 The intelligent space

Intelligent space is the concept designed to support smart interaction between human and the space that can make human in the space more comfortable. Intelligent space (Morioka, 2004) was first conceived in 1995 by Hashimoto Laboratory of the University of Tokyo. Its goal was to assist individuals residing within it. Initially the space consisted of only two CCTV cameras and a computer equipped with 3D tracking software. Later they also installed a large video projector called the Actuator (Morioka, 2004 and Podrzaj, 2008) as well as some additional CCTV cameras as depicted in figure 1.

The intelligent devices can sense, process and connect other devices via network. These are called Distributed Intelligent Network Device or DIND. DIND concept is purposed by Hashimoto Laboratory of the University of Tokyo. DIND is composed of three main components: sensor, processor/computer and communication hardware. The sensors monitor the environment and feed the information into the processor that communicates with other DINDs through the network to share information as shown in figure 2.

Several technologies are integrated into the space, including information processing, speech recognition, and face recognition (Helal, 2005) that will make the space more intelligent. Therefore, the intelligent space idea can be adopted to help the elderly and improve their quality of life. Thus the intelligent space for the elderly is proposed. This study designs and builds the prototype of the intelligent space for the elderly to provide assistance so that they could safely and comfortably live by themselves.

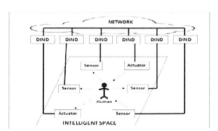

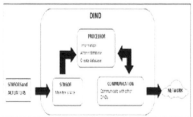

Figure 1 Intelligent Space.
(From K. Morioka. 2004.)

Figure 2 DIND. (From K. Morioka. 2004.)

138

2   RELATED RESEARCH

Smart House (Helal, 2005) is proposed as an intelligent space to assist the handicapped and the elderly. The house is filled with various technologies to maximize comfort and safety for the occupants. The features are as shown below.

- Social-Distant Dining which utilizes existing audio/video equipment in the dining room to provide a simultaneous dining experience with someone far away.
- Cognitive assistance which provide reminders for such as medications and appointments via auditory and visual cues.
- Emergency call for help which is a system to track potential emergencies, check the occupants. If the occupants have problems, it will call outside for requesting help.

Lim S. and research group proposed an interactive Cyber-Physical System (CPS) for People with Disability and Frail Elderly People (Lim, 2011). The functional design of an interactive CPS resolve inter space interaction issues caused by the users change their location and environment.   Thus, the interactive CPS helps provide the services anytime and anywhere the users need. For example, when the users stay at their home, there is a networked assistive robot called "Home Agent" to provide the users with interactive and intelligent services anytime such as

- Wake the users up with their favorite song.
- Remind the users to take medicine by sending voice notice via the speakers near the users.
- Recognize the activities that users want to do and notice them.

When the users want to go out from their home, the "Home Agent" reminds the users to put RFID tag on their cloth to detect their new location. Moreover the agent sends termination signal to the remote pervasive computing system to record users' service information. The new agent from anywhere can get the user's location from RFID reader and get their context from pervasive computing system in order to provide services.

From prior researches regarding intelligent space including mentioned above, it can be concluded that intelligent space can increase comfort of the residents. To make it comfortable for the elderly as well as their family to keep in touch, the intelligent space for the elderly is proposed.

3   THE INTELLIGENT SPACE FOR THE ELDERLY

The proposed intelligent space for the elderly can provide three services: Emergency service, Medical Consultation service and Long Distance Social Interaction service (Tivatansakul, 2012). Our proposed intelligent space for the elderly is shown in figure 3.

It consists of many I/O devices to sense the environment, feed the information to intelligent space system and display the information to human in the space e.g. the elderly, caregivers and family members.   To support every service in the intelligent space, we designed a framework and describe in next section.

4 THE DESIGN FRAMEWORK

The design framework of the intelligent space for the elderly is illustrated in figure 4. Our proposed design framework consists of twelve I/O devices, three services, three applications, server and database (Tivatansakul, 2012).

4.1 I/O Devices

The twelve I/O devices are ECG sensor (heart rate sensor), acceleration sensor, thermistor, humidity sensor, gas sensors, cdS photoresistor, microphone, speaker, button, touch screen, display and webcam. The input devices monitor the environment and feed the information for analysis and processing to provide appropriate services to the elderly. The output devices are used for display the important information to the elderly. Figure 5 shows the touch screen that user can use for requesting services.

4.2 Services

Our intelligent space can support the elderly by providing three services; Emergency service, Medical consultation service and Long-distance social interaction service. Each service is described in services design section.

4.3 Applications

There are three applications in our intelligent space; Analysis application, Alarm application and Communication application. Each application will send and receive information via Red5 Flash Server or Internet Information Server (IIS) (Tivatansakul, 2012).

To improve the reliability of the intelligent space for the elderly, we designed the database to collect information for each service.

5 SERVICES DESIGN

As we have mentioned above, our proposed intelligent space for the elderly can provide three services as follow.

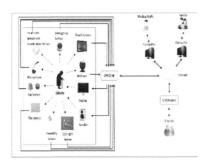

Figure 3 Proposed intelligent space for elderly.

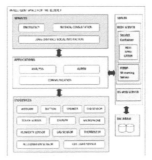

Figure 4 The design framework of the intelligent space.

5.1   Emergency service

Physical ability in the elderly always decreases (Tivatansakul, 2012). In case of emergencies, the elderly often cannot help themselves effectively and no one can help them. So we propose an emergency service to detect the falls down situation, heart attack situation and activity of the elderly using heart rate sensor, acceleration sensor, thermistor, humidity sensor, gas sensors and cdS photoresistor. Moreover if the elderly sense a health problem, they can put the emergency button to request emergency service. This service automatically calls an ambulance, warns the hospital staffs (doctor, nurse, etc.) to preliminarily diagnose and help the senior citizen at home. Furthermore, this service also reports the situation of the elderly to their family by sending emergency email.

There are seven functions in this service: Emergency button detection, heart rate detection, fall detection, activity detection, automatically call an ambulance, automatically send email to mobile phone and video conference call for preliminary diagnosis.

5.2   Medical consultation service

Senior citizens may have difficulty to meet their personal doctor at hospital because of their lack of physical ability (Tivatansakul, 2012).   We provide this service for the elderly to contact and consult with their personal doctor at home when they have problems.

There are two functions to support this service. First is the function to detect signal from touch screen that is shown in figure 5.   Second is a Video Conference Call for medical consultation.

5.3   Long distance social interaction service

Since senior citizens have to live alone and far away from their family (Tivatansakul, 2012), may have the mental health problems and feel alone. In order to improve their mental health, we provide this service for the elderly to easily contact their family members and friends.

There are two functions to support this service. First is the function to detect signal from touch screen that is shown in figure 5. Second is a Video Conference Call for social interaction.

Figure 5 The touch screen for requesting our services.

6   THE PROTOTYPE IMPLEMENTATION

To design and build the prototype of the intelligent space for providing three services, we have implemented three applications.

6.1   Analysis application

This application analyzes the information from input devices and select appropriate services for the elderly. This application has four functions.

6.1.1   Touch screen detection

Figure 5 shows touch screen for requesting our services. The request from the touch screen is sent to the intelligent space system via database (Tivatansakul, 2012).

6.1.2   Heart rate detection

This function detects heart attack situation by analyzing heart rate of the elderly in order to request the emergency service (Tivatansakul, 2012). This function receives the heart rate signal (millivolt, mV) from heart rate sensor or ECG sensor. To find QRS interval, we use the basic structure of heart rhythm. Then we find R-R interval and calculate the heart rate in beat per minute (BPM). If the measured heart rate is less or more than the normal range, heart attack situation will be detected. The emergency service is selected to provide next appropriate function to help the elderly.

6.1.3   Fall detection

Falls is the major issues for the elderly that lead to injury situation (Jantaraprim, 2010 and Hitachi Metals Ltd., 2012). So we designed and implemented fall detection function to detect falling down situation of the elderly by using 3D acceleration sensor in order to request the emergency service. This function uses tri-axial accelerometer with threshold based algorithm. After this function gets input data from 3D acceleration sensor, the norm value is calculated from input data as shown below.

$$= \sqrt{(A_x)^2 + (A_y)^2 + (A_z)^2}$$

Ax, Ay and Az is accelerations (g) along the x, y and z axes. In addition, we find the maximum and minimum norm value in every 2.5 seconds. This function uses three thresholds: Maximum Acceleration Threshold, Minimum Acceleration Threshold and Free Fall Threshold that are set from falling down experiment. If three threshold values that we have mentioned above are reached and the free fall situation occurs before the maximum norm value occurs, this means that the elderly might fall down. The emergency service is requested to provide the next appropriate function to the elderly. Figure 6 shows graph of 3d acceleration when human fall down.

6.1.4   Activity detection

Normally, the activities of the elderly in daily life are relatively simple. The elderly usually do similar and a few activities in everyday such as taking a bath, cooking, eating, going out and sleeping. We design and implement the monitoring function. Several monitoring systems in other previous research use camera to detect unusual situation but it might violate the user's privacy. Our multi-sensor unit (Hirasawa, 2009) can reduce the violations of privacy as much as possible as follows.

The multi-sensor unit can measure environment change in gas concentration, temperature, humidity and brightness at user's home using two types of gas sensors, thermistor, humidity and CdS light sensor. The detail of each sensor is shown in table 1.

Table 1 Sensors Detail (Hirasawa, 2009)

Sensor	Model	Remarks
1. Gas	TGS880	High sensitivity for alcohol and fragrance
	TGS800	High sensitivity for hydrogen, ethanol, isobutane, carbon monoxide. Used to detect the dirt of air.
2. Thermistor	103ET	Thermal time constant τ=3s
3. Humidity	HS-15P	Range of humidity : 10%RH~ 90%RH
4. Light sensor	CdS	Resistance under dark condition :1MΩ

In the experiment, the sampling period is one second. The frequency information of each sensor is measured. The measured data is shown in figure 7. From the result, in figure 7(i) the frequency of humidity sensor rapidly increases because a resident was taking a bath using hot water and then he came to living room with some wet state. So this can indicate the activity of taking a bath. Moreover, the rapid response of humidity sensor can indicate for cooking activity as is shown in figure 7(ii). This is caused by steam from rice cooker or by some moisture and gas from boiling or baking. The humidity sensor and the gas sensor rapidly respond, and their frequencies are converging to stable state because of ventilators. The other stable states indicate no activity or the activity when the resident was sleeping or going outside. So this can be summarized that the frequency change of each sensor depends on each daily activity. However, the measured frequency data might have noise. Therefore, the moving average processing as shown below, is used to reduce that noise.

The moving average processing:

$$A_n = \frac{1}{N}\sum_{k=0}^{N-1} x_{n-k} \tag{1}$$

A_n is the result of the moving average processing, N is the number of data for the moving average processing and x_{n-k} is measured data at time $(n-k)$. Figure 8 shows the result of the moving average processing to figure 7 when N is equal to 301. In this figure, it is shown that noise is reduced and graph of result is smoother than figure 7.

Furthermore, to detect the beginning of activities and to reduce the difference of output level from each sensor, we used the differential processing as shown below.

The differential processing:

$$D_n = A_n - A_{n-1} \tag{2}$$

From equation (1) we get:

$$D_n = \frac{1}{N}\left(x_n - x_{n-N}\right)$$

D_n is the result of the differential processing. Figure 9 shows the result of the differential processing to figure 8. In this figure, we can indicate the beginning point of activities and the difference of output level between sensors and the influence of slow environmental change from daytime to nighttime other than daily activities are reduced.

Finally, we can detect the activities of the elderly and find unusual situation. For example, if the measured data shows that the elderly have only sleeping activity, and neither have eating and nor going out activity; this means that there is something strange about the elderly. The emergency service is requested to provide the next appropriate function to the elderly.

6.2   Alarm application

This application warns and reports other people such as the family members, caregivers and hospital staffs about particular situation involving the elderly person. There are two functions to support this application. First is automatically call to telephone with recorded voice. We developed this function by using a Skype application programming interface (API). Second is automatically send a mobile phone email. We developed this function by using Email API for .NET application that support within System.Net.Mail namespace. This function sends email via the simple mail transfer protocol (STMP).

6.3   Communication application

We integrated the communication application into our intelligent space for the elderly in order to make a video conference call between the elderly and the other people such as family members, friends, personal doctor and hospital emergency staffs from different location (Tivatansakul, 2012). This application is designed and implemented based on Modular Web-based Collaboration

(MWC) concept (Intapong, 2010). This communication application sends audio/video data via Red5 Flash Server.

7 CONCLUSIONS

The proposed intelligent space for the elderly can provide three services: Emergency, Medical consultation and Long distance social interaction services. Emergency service helps detect heart attack, falling down and unusual situations of the elderly in order to automatically call an ambulance, warn their family and caregiver about situation and allow the hospital staff to do preliminary diagnosis. Medical consultation service allows the elderly to consult with their personal doctor at their home. Long distance social interaction service allows the elderly to contact with their family and friends. Thus the elderly will feel safe and comfortable to live alone by themselves in our society. Moreover, the quality of life of the elderly in both physically and mentally will be improved.

REFERENCES

Helal, S., W. Man., H. El-Zabadani , J. King, Y. Kaddoura, and E. Jansen. 2005. The Gator Tech Smart House: A Programmable Pervasive Space. *Computer* 28: 50-60.

Hirasawa, K., and A. Saitoh. 2009. Study on multi-sensor unit for monitoring of living space environment and classification of daily activities. *IEEE/SICE International Symposium*. 101-106.

Hitachi Metals, Ltd. "Free-Fall Detection utilizing H48C." Accessed March 1, 2012. http://www.parallax.com/Portals/0/Downloads/docs/prod/acc/H48CPrinciplesofFree-FallDetection.pdf.

Intapong, P.,S. Settapat, B. Kaewkamnerdpong, and T. Achalakul. 2010. Modular Web-Based Collaboration Platform. *International Journal of Advanced Science and Technology* 22: 37-48.

Jantaraprim, P., P. Phukpattaranont, C. Limsakul, and B. Wongkittisuksa. 2010. Improving the accuracy of a fall detection algorithm using free fall characteristics. *International Conference on Electrical Engineering/Electronics Computer Telecommunications and Information Technology*. 501-504.

Lim, S., L. Chung, O. Han, and J. Kim. 2011. An Interactive Cyber-Physical System (CPS) for People with Disability and Frail Elderly People. *Proceedings of the 5th International Conference on Ubiquitous Information Management and Communication*.

Morioka, K., J.H Lee, and H. Hashimoto. 2004. Human-Following Mobile Robot in a Distributed Intelligent Sensor Network. *IEEE Transactions on industrial electronics* 51: 229-237.

Podrzaj, P. and H. Hashimoto. 2008. Intelligent Space as a Framework for Fire Detection and Evacuation. *Fire Technology* 44: 65–76.

Tivatansakul, S., S. Tanupaprungsun, K. Areekijseree, T. Achalakul, and M. Ohkura, 2012. The Intelligent Space for the Elderly. *Proceedings of the 10th Asia Pacific Conference on Computer Human Interaction (APCHI)*. (in submission).

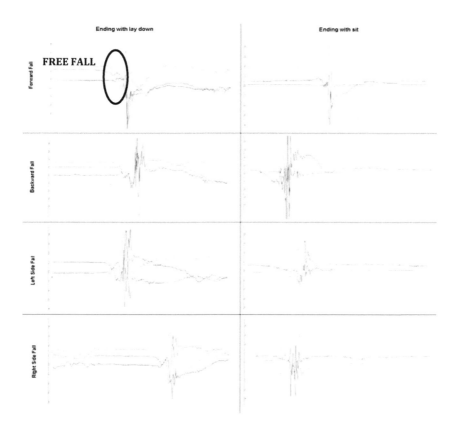

Figure 6 Graph of 3d acceleration when human fell down.

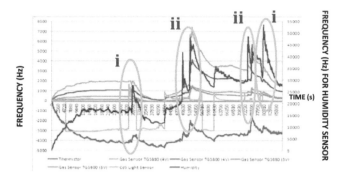

Figure 7 Measured Data. (From K. Hirasawa. 2009.)

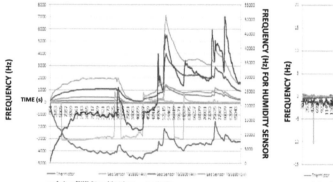

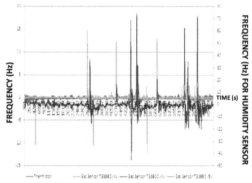

Figure 8 Result of moving average processing. (Form K. Hirasawa. 2009.)

Figure 9 Result of differential processing. (From K. Hirasawa. 2009.)

CHAPTER 15

Patient Engagement – Implications to Service Engineering

Iiris Riippa, Karita Reijonsaari

Aalto University
Espoo, Finland
iiris.riippa@aalto.fi

ABSTRACT

The importance of patients' engagement in the management of chronic conditions is broadly acknowledged in medical research. By engaging patients and supporting their self-management of health, health care providers can both improve the quality as well as contain the costs of chronic care. However, limited academic attention has been paid to the conceptualization and operationalization of patient engagement for service engineering and management purposes.

In this systematic literature review, we investigate how patient engagement has been operationalized in medical research. We categorize the approaches to the operationalization of patient engagement and, based on these approaches, we suggest a preliminary definition of patient engagement for health service engineering and management purposes. We draw on the service research insights into the role of the customer as a co-creator of value. Patient engagement is thus understood as patient's participation in the co-creative process of joint management of health with the care provider.

The definition of patient engagement emerging from medical research adds to the understanding of the essentials of an engaging health service. In addition, the operationalizations of patient engagement identified in the literature serve as a basis for measuring patient engagement. Conceptualization and operationalization of patient engagement aid in the engineering and provision of engaging services to the patients in different levels of engagement.

Keywords: patient engagement, patient activation, patient-centered care, customer engagement

1   INTRODUCTION

The importance of patients' engagement in the management of chronic conditions is evident. Regardless of the type of the chronic condition, the patient is responsible for the day-to-day management of his/her health (Bodenheimer et al. 2002; Lorig & Holman 2003). Self-management of health is inevitable, and it is demonstrated, that by supporting the patient in this task, health care providers can improve patients' health outcomes as well as contain the costs of chronic health care (Lorig et al. 1999; Hibbard et al. 2004; Hibbard et al. 2007). The assumption of cost-effectiveness is not solely based on transferring some of the tasks previously performed by a health care professional to the patient. It is suggested that an empowered patient is both proactive and reactive in the management of his/her condition, preventing and responding to the health-related challenges, and thus requiring less input from the health care professionals (Connor et al. 2010).

The objective of engaging patients in their own care has lead to the emergence of new patient-centric care models in the medical research and practice. Patient-centric care models, such as Chronic Care Model introduced by Wagner et al. (1996), have shifted the focus of chronic disease care from tackling acute conditions to a more longitudinal and holistic view of care (Barlow et al. 2002). This shift to a patient-centric longitudinal care sets a demand for engineering and management of engaging health care services. Health care systems designed for acute care often do not meet the needs for effective clinical management, psychological support, and information of the chronic patients (e.g., Wagner et al. 2001).

Patient-centered care of a chronic condition should aim at reaching the patients' health-related goals (Wagner & Majchrzak 2006). The perspective of provider supporting the customers in their value creating process is broadly embraced in service research (see e.g., Vargo & Lusch 2004; Grönroos 2000). The customer's value creation process is defined as a series of activities performed by the customer to achieve a particular goal, and the service provider's competence is defined by its ability to assist the customer in this process (Payne et al. 2007). To approach the customer's role in the co-creation of value, the concept of customer engagement has been presented in service research. According to Brodie et al. (2011) customer engagement "reflects customers' interactive, co-creative experiences with other stakeholders in specific service relationships". The insights on co-creation of value from service research have been suggested to be applied in the field of health care (Bitner et al. 1997; Prahalad & Ramaswamy 2002).

Even though the importance of patients' engagement in the management of chronic conditions is broadly acknowledged in the medical literature, limited academic attention has been paid to the characterization of an engaging service. In the medical research, a vast number of different self-management interventions have been evaluated, but no single intervention has proved to meet the needs of all patients at all points in time (Hibbard et al. 2007; Barlow et al. 2002). In order to better understand the aspects of an engaging service in different given contexts, an operationalization of patient engagement is needed.

This study aims to investigate how patient engagement in self-management of health has been operationalized in the medical literature. In addition, a preliminary definition of patient engagement is suggested based on the identified approaches to operationalization. We draw on the service research insights into the role of the customer as a co-creator of value. Patient engagement is thus understood as patient's participation in the co-creative process of joint management of health with the care provider.

2 METHOD

We conducted a systematic literature review in order to categorize the approaches to the operationalization of patient engagement in medical literature (Figure 1.). In order to find relevant articles discussing patient engagement in the self-management of health, we searched the Pubmed database with a combination of search terms. We searched for articles with 1) at least one of the terms "patient-centered care", "chronic care" and "primary care" in the title or the abstract of the article and 2) at least one of the terms "self-management of health", "patient engagement", "patient activation", "patient empowerment" and "patient self-efficacy" in the title or the abstract of the article. Through the preliminary search 109 articles written in English were found.

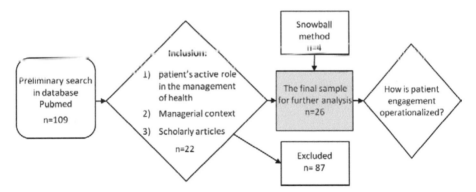

Figure 1. The inclusion of the articles in the systematic literature review.

The term "patient engagement" has been used in its most restricted sense to refer to patient use of certain health care related services (Lindenmeyer et al. 2010; Schillinger et al. 2008). In this study, we aimed at a broader view of patient's engagement in the co-creative process of joint health management, and thus only chose articles that discussed patient engagement in the sense of patient's active role in the management of his/her health through lifestyle-choices and management of possible chronic conditions. Further, we only chose articles in which patient engagement was discussed from a managerial point of view i.e. in its relation to other managerial aims, mainly cost-effectiveness.

From the preliminary search, 22 scholarly articles meeting the inclusion criteria were chosen for further evaluation. By using the snowball method, we found 4 additional articles meeting the inclusion criteria. The content of the articles in the final sample was analyzed in order to categorize the approaches to operationalization of patient engagement in the medical literature.

3 FINDINGS

3.1 The definition of patient engagement

In the literature review, three approaches to the operationalization of patient engagement in self-management of health were identified. (Table 1.) The approaches were 1) not to operationalize 2) operationalization through patient behaviors, and 3) operationalization through patient's emotional/psychological state.

Table 1. Three different approaches to patient engagement in the reviewed literature

Approach	Articles
Black box	Eaton et al. 2011 ; Cooper et al. 2011; Lu et al. 2011; Musacchio et al. 2011; Thiboutot et al. 2010
Operationalization through health behaviors	Cho et al. 2012; Parchman et al. 2010; Stafford & Berra 2007); Meglic et al. 2010; Frei et al. 2011)
Operationalization through psychological/emotional state	Ledford 2012; Greene & Hibbard 2011; Lee et al. 2011; Maindal et al. 2011; Marshall et al. 2011; Donald et al. 2011; Bann et al. 2010; Friedman et al. 2009; Rohrer et al. 2008; Seligman et al. 2005; Glasgow et al. 2005; Fernandez et al. 2004; Heisler et al. 2003; Wolever et al. 2010; Brazier et al. 2008; Iverson et al. 2008

In the first category, patient engagement was not operationalized, but was instead used as an explaining factor between a health intervention and the desired health outcome related to the intervention. In these articles, patient engagement was treated rather as a black box. The explaining characteristic of patient engagement was grounded in previous research on the interventions used in the studies. In two of the articles (Eaton et al. 2011; Musacchio et al. 2011) the interventions tested were based on the chronic care model refined by Wagner (2001). According to Wagner (2001), the chronic care model summarizes the evidence of effective system changes that improve chronic care.

In the second approach patient engagement was operationalized in health behaviors performed by the patient. Some of the behaviors were condition specific and the interventions were designed to address these particular behaviors. Behaviors that were considered at least partial operationalization of patient engagement were changes in diet (Cho et al. 2012), adherence to medication (Meglic et al. 2010; Parchman et al. 2010), physical activity (Stafford & Berra 2007) and smoking (Frei et al. 2011).

The third approach to patient engagement was to operationalize it through the patient's emotional or psychological state. Often a previously validated psychometric measure or several measures were used but in some studies the patients were approached with questions designed for the particular study that were drawn from psychological theories related to patient engagement. The main three theories referred to in the articles discussing patient engagement, were health locus of control (see e.g., Wallston & Wallston 1981) , self-efficacy theory (see e.g. (Bandura 1982; Lorig et al. 1999) and self-determination theory of motivation (see e.g., Ryan & Deci 2000a; Ryan & Deci 2000b). These theories discuss an individual's perceived control over one's health (health locus of control), confidence to carry out a behavior necessary to reach a desired goal (self-efficacy) and the source of motivation to adopt certain behaviors (self-determination theory of motivation).

The most frequently used psychometric measure in the reviewed articles was Patient Activation Measure (PAM) developed by Judith Hibbard (Hibbard et al. 2004; Hibbard et al. 2005). PAM is a probabilistic Guttman-like scale that reflects a developmental model of patient's activation. In order to include a broad range of elements involved in patient activation, PAM was built on the existent psychological theories of health locus of control and self-efficacy in self-managing behaviors. In addition, the model of the stages of readiness to change health-related behaviors by Prochaska (1994) was applied to form a developmental measure indicating patient's knowledge, skill, and confidence needed in the self-management of chronic conditions.(Hibbard et al. 2004)

According to the analyzed articles, patient engagement has been approached as a psychological state which is manifested through different behaviors adopted by the patient that result in desired health outcomes. Based on the analysis of the included articles we suggest a preliminary definition for patient engagement emerging from medical research: "Patient engagement is a psycho-social and emotional state of engagement in self-management of health, which is manifested through different actions taken by the patient that result in desired health outcomes."

To further specify the definition of desired health outcomes we draw on medical research and service research. In both fields of research, it is stated that the service provider's goal is to support the customer/patient in reaching their goals. In health services patient often has limited knowledge of his/her condition and needs the support from the health care provider in the very setting of health-related goals (Ouschan et al. 2006). Thus, we specify the "desired health outcomes" further to "the health and wellness targets jointly agreed on by the patient and the care giver".

The definition for patient engagement suggested in this study is the following:

> *Patient engagement is a psychosocial and emotional state of engagement in self-management of health, which is manifested through different actions taken by the patient that result in the health and wellness targets jointly agreed on by the patient and the care giver.*

3.2 Implications of patient engagement for service engineering

Service research as well as service engineering and management have recognized customer engagement as an important aspect. However, in the healthcare context the recognition of patient engagement has not been fully "engineered" into the new self-management services. The definition of patient engagement emerging from medical and service research adds to the understanding of the essentials of an engaging health service. Further, in order to efficiently manage the engineering of engaging services, measures for patient engagement are needed. The two operationalizations of patient engagement identified in this systematic literature review (i.e. health behaviors and psychometric measures) serve as a basis for measuring patient engagement.

It is suggested that no one approach to self-management interventions will meet the needs of all patients at all points of time (Barlow et al. 2002; Stone et al. 2005). Different types of services with varying intensity should thus be offered depending on the engagement level of the patient. The level of an individual's engagement to self-management of health could be an important basis for segmentation of customers. Efficacy of segmentation based on engagement remains an important question for future multidisciplinary research.

Due to resource constraints, healthcare organizations are increasingly seeking to provide more of their care online. Although remote communications offer a possibility to cost-effective encounters between a patient and a health care professional, they also diminish the physician-patient communication and psychosocial factors become harder to detect. This shift in the use of different service channels poses a further challenge to service engineering to plan for ways to facilitate engagement and find ways for measuring it.

4 Discussion and conclusions

In the cost-effective management of chronic conditions, patient's role is essential. Based on the systematic literature review conducted, we suggest a concept of patient engagement that should be used in the engineering and management of engaging health services. In addition, two operationalizations of patient engagement, health behaviors and psychometric measures, were identified.

Patients are currently treated without regard to their level of engagement into their own care. An engaged person is likely to react very differently to services offered to them than a non-engaged person. As customer engagement is crucial for chronic disease management, healthcare should offer services that can be targeted to right people and customized based on the level of engagement. A service engineering challenge is to design systems that drive increased engagement with lower cost. This will be a multibillion-dollar challenge as chronic diseases increase exponentially and care must be offered through lower cost channels instead of the physician office.

Psychometric measures related to patient engagement have been evidenced to predict health outcomes. The findings of this literature review support Ledford's (2012) suggestion of linking psychometric measures, such as PAM, to the medical records. The other operationalization of patient engagement, i.e. health behavior, is to date more often used in the clinical setting. Patient health behaviors such as physical activity and smoking are often included in different condition specific risk factors (e.g. Frei et al. 2011; Stafford & Berra 2007).

By combining the insights on customer engagement in service research and the extensive body of knowledge on patient self-management interventions in medical research, a better understanding of patient engagement for managerial purposes can be attained. Through the operationalization of patient engagement, the essence of an engaging service can be better grasped. We would like to encourage research on conceptualization and operationalization of patient engagement in order to better understand the aspects of an engaging service as well as to better manage the engineering and provision of engaging services to the patients in different levels of engagement. Further, we would like to encourage the development and pragmatic use of predictive patient engagement measures so that health care providers would be able to support the patients in their self-management of health even in the early or preceding stages of life-style related conditions.

REFERENCES

Bandura, A., 1982. Self-efficacy mechanism in human agency. American psychologist, 37(2), p.122.

Bann, C.M., Sirois, F.M. & Walsh, E.G., 2010. Provider support in complementary and alternative medicine: Exploring the role of patient empowerment. The Journal of Alternative and Complementary Medicine, 16(7), pp.745–752.

Barlow, J. et al., 2002. Self-management approaches for people with chronic conditions: a review. Patient Education and Counseling, 48(2), pp.177–187.

Bitner, M.J. et al., 1997. Customer contributions and roles in service delivery. International Journal of Service Industry Management, 8(3), pp.193–205.

Bodenheimer, T. et al., 2002. Patient self-management of chronic disease in primary care. JAMA: the journal of the American Medical Association, 288(19), p.2469.

Brazier, A., Cooke, K. & Moravan, V., 2008. Using mixed methods for evaluating an integrative approach to cancer care: a case study. Integrative Cancer Therapies, 7(1), pp.5–17.

Brodie, R.J. et al., 2011. Customer Engagement: Conceptual Domain, Fundamental Propositions, and Implications for Research. Journal of Service Research. Available at:

http://jsr.sagepub.com/content/early/2011/06/21/1094670511411703.abstract [Accessed February 10, 2012].

Cho, A.H. et al., 2012. Effect of genetic testing for risk of type 2 diabetes mellitus on health behaviors and outcomes: Study rationale, development, and design. BMC Health Services Research, 12(1), p.16.

Connor, A., Mortimer, F. & Tomson, C., 2010. Clinical Transformation: The Key to Green Nephrology. Nephron Clinical Practice, 116(3), p.c200–c206.

Cooper, L.A. et al., 2011. A randomized trial to improve patient-centered care and hypertension control in underserved primary care patients. Journal of General Internal Medicine, 26(11), pp.1297–1304.

Donald, M. et al., 2011. The role of patient activation in frequent attendance at primary care: a population-based study of people with chronic disease. Patient Education and Counseling, 83(2), pp.217–221.

Eaton, C.B. et al., 2011. Translating Cholesterol Guidelines Into Primary Care Practice: A Multimodal Cluster Randomized Trial. The Annals of Family Medicine, 9(6), pp.528–537.

Fernandez, A. et al., 2004. Physician Language Ability and Cultural Competence. Journal of General Internal Medicine, 19(2), pp.167–174.

Frei, A. et al., 2011. The chronic care for age-related macular degeneration study (CHARMED): study protocol for a randomized controlled trial. Trials, 12, p.221.

Friedman, B. et al., 2009. Patient Satisfaction, Empowerment, and Health and Disability Status Effects of a Disease Management–Health Promotion Nurse Intervention Among Medicare Beneficiaries With Disabilities. The Gerontologist, 49(6), p.778.

Glasgow, R.E. et al., 2005. Development and validation of the patient assessment of chronic illness care (PACIC). Medical Care, 43(5), p.436.

Greene, J. & Hibbard, Judith H, 2011. Why Does Patient Activation Matter? An Examination of the Relationships Between Patient Activation and Health-Related Outcomes. Journal of General Internal Medicine. Available at:

http://www.ncbi.nlm.nih.gov/pubmed/22127797 [Accessed February 14, 2012].

Grönroos, C., 2000. Service management and marketing. European Journal of Marketing, 15(2), pp.3–31.

Heisler, M. et al., 2003. When Do Patients and Their Physicians Agree on Diabetes Treatment Goals and Strategies, and What Difference Does It Make? Journal of General Internal Medicine, 18(11), pp.893–902.

Hibbard, J. H et al., 2005. Development and testing of a short form of the patient activation measure. Health Services Research, 40(6 Pt 1), p.1918.

Hibbard, J. H et al., 2004. Development of the Patient Activation Measure (PAM): conceptualizing and measuring activation in patients and consumers. Health Services Research, 39(4 Pt 1), p.1005.

Hibbard, Judith H et al., 2007. Do Increases in Patient Activation Result in Improved Self-Management Behaviors? Health Services Research, 42(4), pp.1443–1463.

Iverson, S.A., Howard, K.B. & Penney, B.K., 2008. Impact of internet use on health-related behaviors and the patient-physician relationship: a survey-based study and review. The Journal of the American Osteopathic Association, 108(12), pp.699–711.

Ledford, C.J.W., 2012. Exploring the Interaction of Patient Activation and Message Design Variables: Message Frame and Presentation Mode Influence on the Walking Behavior of Patients With Type 2 Diabetes. Journal of Health Psychology. Available at: http://hpq.sagepub.com/content/early/2012/01/17/1359105311429204.full.pdf+html?frame=sidebar [Accessed February 13, 2012].

Lee, A. et al., 2011. General practice and social service partnership for better clinical outcomes, patient self efficacy and lifestyle behaviours of diabetic care: randomised control trial of a chronic care model. Postgraduate Medical Journal, 87(1032), pp.688 – 693.

Lindenmeyer, A. et al., 2010. Patient engagement with a diabetes self-management intervention. Chronic Illness, 6(4), pp.306–316.

Lorig, K. R et al., 1999. Evidence suggesting that a chronic disease self-management program can improve health status while reducing hospitalization: a randomized trial. Medical care, 37(1), p.5.

Lorig, Kate R. & Holman, H.R., 2003. Self-management education: History, definition, outcomes, and mechanisms. Annals of Behavioral Medicine, 26(1), pp.1–7.

Lu, W.-H. et al., 2011. Activating Community Health Center Patients in Developing Question-Formulation Skills: A Qualitative Study. Health Education & Behavior. Available at: http://heb.sagepub.com/content/early/2011/05/06/1090198110393337.abstract?rss=1&p atientinform-links=yes&legid=spheb;1090198110393337v1 [Accessed February 14, 2012].

Maindal, H.T. et al., 2011. Effect on motivation, perceived competence, and activation after participation in the "'Ready to Act"'programme for people with screen-detected dysglycaemia: A 1-year randomised controlled trial, Addition-DK. Scandinavian journal of public health, 39(3), p.262.

Marshall, B., Floyd, S. & Forrest, R., 2011. Clinical outcomes and patients' perceptions of nurse-led healthy lifestyle clinics. Journal of Primary Health Care, 3(1), pp.48–52.

Meglic, M. et al., 2010. Feasibility of an eHealth service to support collaborative depression care: results of a pilot study. Journal of Medical Internet Research, 12(5), p.e63

Musacchio, N. et al., 2011. Impact of a chronic care model based on patient empowerment on the management of Type 2 diabetes: effects of the SINERGIA programme. Diabetic Medicine: A Journal of the British Diabetic Association, 28(6), pp.724–730.

Ouschan, R., Sweeney, J. & Johnson, L., 2006. Customer Empowerment and Relationship Outcomes in Healthcare Consultations. European Journal of Marketing, 40(9/10), pp.1068–1086.

Parchman, M.L., Zeber, J.E. & Palmer, R.F., 2010. Participatory Decision Making, Patient Activation, Medication Adherence, and Intermediate Clinical Outcomes in Type 2 Diabetes: A STARNet Study. Annals of Family Medicine, 8(5), pp.410–417.

Payne, A.F., Storbacka, K. & Frow, P., 2007. Managing the co-creation of value. Journal of the Academy of Marketing Science, 36(1), pp.83–96.

Prahalad, C.K. & Ramaswamy, V., 2002. The co-creation connection. Strategy and Business, pp.50–61.

Prochaska, J.O. et al., 1994. Stages of change and decisional balance for 12 problem behaviors. Health psychology, 13(1), p.39.

Rohrer, J.E. et al., 2008. Patient-centredness, self-rated health, and patient empowerment: should providers spend more time communicating with their patients? Journal of Evaluation in Clinical Practice, 14(4), pp.548–551.

Ryan, R.M. & Deci, E.L., 2000a. Intrinsic and extrinsic motivations: Classic definitions and new directions. Contemporary educational psychology, 25(1), pp.54–67.

Ryan, R.M. & Deci, E.L., 2000b. Self-determination theory and the facilitation of intrinsic motivation, social development, and well-being. The American Psychologist, 55(1), pp.68–78.

Schillinger, D. et al., 2008. Seeing in 3-D: examining the reach of diabetes self-management support strategies in a public health care system. Health Education & Behavior: The Official Publication of the Society for Public Health Education, 35(5), pp.664–682.

Seligman, H.K. et al., 2005. Physician notification of their diabetes patients' limited health literacy. A randomized, controlled trial. Journal of General Internal Medicine, 20(11), pp.1001–1007.

Stafford, R.S. & Berra, K., 2007. Critical factors in case management: practical lessons from a cardiac case management program. Disease Management: DM, 10(4), pp.197–207.

Stone, M. et al., 2005. Empowering patients with diabetes: a qualitative primary care study focusing on South Asians in Leicester, UK. Family Practice, 22(6), pp.647–652.

Thiboutot, J. et al., 2010. A web-based patient activation intervention to improve hypertension care: study design and baseline characteristics in the web hypertension study. Contemporary Clinical Trials, 31(6), pp.634–646.

Wagner, C. & Majchrzak, A., 2006. Enabling Customer-Centricity Using Wikis and the Wiki Way. Journal of Management Information Systems, 23(3), pp.17–43.

Wagner, E.H. et al., 2001. Improving Chronic Illness Care: Translating Evidence Into Action. Health Affairs, 20(6), pp.64–78.

Wagner, E.H., Austin, B.T. & Michael Von Korff, 1996. Organizing Care for Patients with Chronic Illness. The Milbank Quarterly, 74(4), pp.511–544.

Wallston, K.A. & Wallston, B.S., 1981. Health locus of control scales. Research with the locus of control construct, 1, pp.189–243.

Vargo, S.L. & Lusch, R.F., 2004. Evolving to a New Dominant Logic for Marketing. The Journal of Marketing, 68(1), pp.1–17.

Wolever, R.Q. et al., 2010. Integrative health coaching for patients with type 2 diabetes: a randomized clinical trial. The Diabetes Educator, 36(4), pp.629–639.

Section IV

Service System Frameworks

A Distributed-Cognition Based Method for Finding Social Feature Opportunities in Business Services

Kelly Lyons

University of Toronto
Toronto, Canada
kelly.lyons@utoronto.ca

Stephen Marks

Scholars Portal
Toronto, Canada
stephen@scholarsportal.info

ABSTRACT

Social media and social networking sites are increasingly being used in organizational settings to connect and engage customers and employees. In this paper we show how a distributed-cognition based method that analyzes representational activity in terms of the movement of information across media can be extended to find opportunities to integrate social mechanisms in business services. We describe the results of applying the extended method on an academic library service.

Keywords: social media; distributed cognition; information systems

1. INTRODUCTION

In the fast-moving world of today, organizations must continually innovate as they adapt and react to business opportunities, changing environments, systems, and customers. In order to do so, organizations must first understand and articulate their current business services and then identify existing information systems and/or implement new ones that can implement those services (Demirkan, Kauffman, Vayghan, Fill, Karagiannis, & Maglio, 2008). Most organizations have extensive information technology (IT) systems in operation; however, in some cases, current business services may still be executed with little or no IT support. Often, these existing offline business services have been in use for many years and have been well-honed to provide specific needs to the organization and its customers.

At the same time, in an effort to continually innovate through the introduction of new technology, organizations are increasingly incorporating social computing techniques within their business services such that their customers and employees may interact with the organization and with each other effectively online (Abram, 2005; Farkas, 2007; Stephens, 2007). Social computing has also been referred to as social media, social networking, and social technologies and has been described in different ways by different authors (Boyd and Ellison, 2007; Kim, et al, 2010; Parameswaran and Whinston, 2007; Thelwall, 2007). We use as a definition of social computing technology, the keys social features identified in (Lyons and Lessard, 2012). They define social features as those features that distinguish a socially-oriented technological system or application from one that is not. The five key social features identified in (Lyons and Lessard, 2012) are: personal profiles; articulated networks; communities or groups that can be created and joined; user-generated content (UGC) that can be created and shared; and, comments and information that can be added to existing content. Efforts to adopt social features are often being undertaken with little analysis of existing business services but instead by looking at how other similar organizations have incorporated social features in their business services.

Carr (2004) examines the theory that the more available computing is, the less it becomes a competitive advantage. He states that in order to gain an edge over rivals (in order to be truly innovative), one has to do something or have something that the others cannot do or do not have. It is not sufficient to do what others are doing but to understand what makes your organization or service unique. In this paper, we show how extending a distributed-cognition based method can help organizations analyze their current business services in order to determine where social features can support their existing unique business activities and service offerings. Our method is an extension of a method presented in (Flor & Maglio, 2004) that views businesses as distributed cognitive systems and proposes a step-by-step mechanism to determine which aspects of an offline customer service activity can be enhanced if modified to use online technology. We describe how our extended method was used to diagnose technology and social feature opportunities in a library service.

2. ANALYZING A BUSINESS AS A COGNITIVE SYSTEM

Distributed cognition extends the notion of what is cognitive beyond individuals to include interactions among people and with other parts of their environment (Hollan, Hutchins, & Kirsh, 2000). Hollan et al (2000) have analyzed the bridge of a ship and an airline cockpit as distributed cognitive systems. Flor and Maglio (2004) view a business as a distributed cognitive system. Based on Simon's (1969) argument that the computer and the human brain are kinds of physical symbols systems and using the fact that a physical symbol system is a mechanism (necessary and sufficient) for cognition, Flor and Maglio (2004), state that, "a business is a cognitive system because it is a physical symbol system, and we can more precisely define a business as a coalition of individuals *and* technologies that store, manipulate, and distribute representations in the service of tasks..." (page 41). Distributed cognition is a joint activity that is distributed across the individuals, workgroups, and technologies in this coalition (Lintern, 2005).

Flor and Maglio (2004) propose a method for modeling a business's symbolic activities based on this notion that a business is made up of people and technologies that create, store, manipulate, distribute, and exchange symbols. They use media constellation diagrams to represent: the agents (people, technologies); the symbols that are created and exchanged; and, the media across which the exchange is made. Media constellation diagrams are made up of nodes and labelled arrows connecting nodes. Nodes are either a medium that contains a symbolic state (such as paper) or a physical symbol that can store, process, or send symbols (such as people or technologies). Arrows show the movement of symbols between nodes and are labelled with the symbol's content and the channel (or media) across which it is conveyed. An example of a simple media constellation diagram given in (Flor and Maglio, 2004) represents two people who meet face to face: one person (P1) says 'happy birthday' to the other person (P2) who responds 'thank you'. A media constellation diagram for this activity has two nodes, P1 and P2, and an arrow from P1 to P2 labelled with 'a:happy birthday' and an arrow from P2 to P1 labelled with 'a:thank you'. The prefix 'a:' indicates the audio channel for this exchange. Figure 1 shows two media constellation diagrams for the submission and return of an assignment in a university course.

Flor and Maglio (2004) argue that modeling symbols and the channels that propagate them enables business owners to get a better sense of where to infuse computational technology than simply mapping symbol movement alone. They present a three step process that uses media constellation diagrams to investigate opportunities to introduce technology into a business's activities. Specifically, each node (agent) in a media constellation diagram represents an opportunity to substitute in technology and each arrow (exchange) represents an opportunity to mediate that exchange with technology.

In this paper, we extend their method to also find opportunities to introduce social features within existing business services. It makes intuitive sense that media constellation diagrams which are representations based on distributed cognition theory can be used to effectively diagnose social technology opportunities for the following reason. One of the three kinds of distribution of cognitive processes observed by Hollan et al (2000) is that of cognitive processes distributed across members of a social group.

2.1 Method for Diagnosing Social Feature Opportunities in Distributed Cognitive Systems

In this section, we describe an extension to a distributed-cognition based method presented in (Flor & Maglio, 2004). We follow their underlying assumption for injecting technology into offline processes - that the best place to look for opportunities to insert social features is to look at an existing, well-honed, well-used business service rather than trying to mimic how others have used social features in their business services. Our extension retains the systematic aspect of their method. Our systematic method can help people efficiently identify how to bring business services on line while enhancing the service with social features and maintaining alignment with well-honed processes and practices.

The three steps presented in (Flor & Maglio, 2004) are:

Step 1: Identify Products: The products of a business service are defined as anything that the customer leaves the business with that they did not have upon entering. This includes knowledge or information acquired as a result of interacting with people, technology, or entities in the business. In the hair salon example, the products are identified as the hair cut, any purchased hair products, and a reminder card for the client's return visit.

Step 2: Model Representations and Representational Activity: Once the products are identified, the next step is to model the representations and representational activities that make the creation and delivery of the products possible, specifically looking at those that are relevant to tasks within the business service. Representational activities are defined as "any work in which individuals or groups store, process, or distribute information" (Flor & Maglio, 2004, page 40). Representational activities are modeled as media constellation diagrams which depict people, computational technologies, and information as agents (or nodes) and model items that are moved, communicated, or exchanged on media channels between agents as arrows (or directed edges). Media channels identify the way in which the exchange takes place such as information communicated verbally, by mail systems, over email, on paper, etc.

For example, if we consider a university course offering as the business service, one product is the graded assignment that a student receives in the course. The media constellation diagram for this product that depicts submission and grading using a paper media channel is shown in Figure 1 (a). The media constellation diagram modeling submission and grading through a learning management system (LMS) is given in Figure 1 (b). In both cases, the submission moves from student to professor and the grade moves from the professor to the student. When paper is used, the media channel in both cases is paper. When the LMS is used, the submission moves from student to the LMS using a keyboard channel and from the LMS to the professor using a screen channel and the grade moves from the professor to the LMS (on the keyboard channel) and from the LMS to the student (on the screen channel).

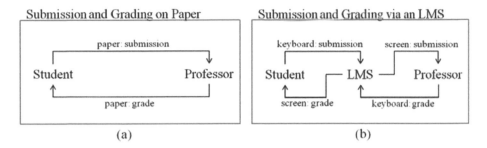

Figure 1: Media constellation diagrams depicting assignment grading (a) using paper submission, and (b) using a learning management system (LMS).

Flor and Maglio (2004) also identify functional organs within the media constellation diagrams which represent a grouping of agents that work together to carry out a business-related task. These functional organs are like business processes (from the user's or customer's view point) that make up the business service that is being modeled. Functional organs are represented as dashed lines encircling the group of agents involved in the business-related task.

Step 3: Diagnose Computational Opportunities: The final step involves looking at the models of how products are created (the media constellation diagrams developed in Step 2) and identifying places where technology can be used to augment, substitute for, or give a new way of facilitating the creation of those products. Flor and Maglio (2004) investigate two places in their models to incorporate technology: as substitutions for agents in the models; and, as mediators in exchanges between agents. In order to keep existing practices intact, they also pay careful attention to maintaining the functional organs and patterns of exchange. For example, in Figure 1 (b), the LMS is added to mediate the exchanges between student and professor represented in Figure 1 (a).

Our extension to their method adds a fourth step that diagnoses opportunities to introduce social features in the business service, where social features are those defined in (Lyons and Lessard, 2012): personal profiles; articulated networks; communities or groups that can be created and joined; user-generated content (UGC) that can be created and shared; and, comments and information that can be added to existing content. In order to diagnose social feature opportunities, we examine: people agents; agents that represent electronically stored information (which we call "data"); technology-mediated exchanges between people; exchanges of data between two agents; and, technology-mediated exchanges between people and data. We do this through four sub-steps:

Step 4.1: Examine each person agent: Within each functional organ, if there is an agent that represents multiple people, we consider whether those people could be part of a community and if so, we look for opportunities to introduce user-profiles, articulated networks, and community formation. For example, the student agent in Figure 1 typically represents multiple students in a class and, together, they could be considered part of a community of people. This is one place we may consider introducing user profiles, articulated networks, or communities in the assignment grading process.

Step 4.2: Examine each technology-mediated exchange between two people agents: A technology-mediated exchange between two people agents represents an opportunity to introduce profiles and community formation between those two people agents. For example in Figure 1 (b), there is an opportunity to enable connections and community formation between students and professors. We note that not all representations examined will result in the introduction that specific social feature just as not all representations examined may result in technology substitutions in (Flor and Maglio, 2004). The goal is to help business owners identify opportunities where social features may be introduced into an existing business service. Other factors may come into play when determining which of the identified opportunities should be implemented in the resulting service.

Step 4.3: Examine each agent that represents data and each exchange involving data: Each data agent or exchange involving data is an opportunity to introduce user-generated content. For example, in Figure 1 (b), the assignment being submitted might be a candidate for tagging, rating, or commenting. There are no data agents in the Figure 1 (b).

Step 4.4: Examine each technology-mediated exchange between people and data: The exchange between a person agent and data agent is also an opportunity to introduce user-generated content. For example, in Figure 1 (b), the student agent exchange of assignments with the LMS agent represents electronically available content. This exchange is a candidate for introducing user-generated content. Again we note that in most cases it would not be appropriate to enable students to tag, rank, or comment on each other student's assignments but there may be situations where this could be desirable.

3. EVALUATING THE EXTENDED METHOD ON A LIBRARY SERVICE

We applied the method with our extension to an existing offline library service called the Current Contents (CC) service. The CC library service provides current tables of contents (TOCs) of periodicals to users and performs photocopying and mailing of articles selected by users from the tables of contents. Technology is currently not used to implement any part of the library service. We selected this library service to demonstrate our extended method in part because it is a business service that has been offered successfully without technology for over 20 years and in part, because we felt it was an ideal candidate for the introduction of technology and social features.

In order to understand how this well-honed business service is currently offered, we conducted face-to-face interviews with eight users or potential users of the service. We also interviewed four library staff with an administrative role in the CC service. Of the users who were interviewed, the largest group of respondents had been a member of the Faculty for four years or less, while two had been faculty members for more than twenty years. The amount of time that users had subscribed to the CC service generally corresponded to the length of time they had been at the Faculty, though some of the newer hires did not use the service at all. Half of the people interviewed stated that they had published at least one article in a journal covered by the CC service. We analyzed the interviews to understand the ways in which the CC service is currently used. We also used the interviews to establish a set of requirements for the technology and social feature enhancements as a way of comparing opportunities found by the extended method.

3.1 The CC Library Service

The CC service makes use of existing library IT services (cataloguing and circulation) but is itself an offline process. The basis for the CC service is a subscription list that matches periodicals with faculty members such that, for each periodical, there is a list of faculty members who receive a copy of that periodical's table of contents (TOC) each time a new issue arrives. This list contains only periodicals for which the library has a subscription to a physical copy. New periodicals are added to the subscription list in two ways: at the discretion of the librarian in charge of collection development or at the request of a faculty member. The subscription list is maintained manually on a printed copy of a spreadsheet. Individuals are added to the subscription list when they are first hired by the Faculty and asked to indicate the titles to which they would like to subscribe. The list is updated annually but new subscriptions that involve an existing periodical and an existing individual can be added at any time by contacting the library staff.

Upon receiving a new copy of a periodical, a library staff member adds the issue to the library's holdings and checks the subscription list to see how many copies of the TOCs need to be made. After the appropriate number of photocopies is made, the master subscription list is used to write the name of individuals on the top of the photocopies. The photocopies are sent to the individuals via campus mail.

When an individual receives copies of TOCs, they have the option requesting articles by contacting the library by phone, email, or by mailing back the TOC copy with the appropriate articles marked. Many faculty members reported forwarding TOCs to graduate students or other colleagues with notes about articles of interest or filing TOCs with course preparation documents for future reference.

Upon receipt of the individual's request, a technician or student assistant in the library will make a photocopy of the article so requested, which is then sent to the requestor, again through campus mail.

3.2 Results

We identified four products in the CC service: TOCs; articles; subscription list; and, list of periodicals. For each of the products, we identified agents and exchanges used in the creation of the products and modeled them as MCDs. Details of applying the extended method to the CC service are presented in (Lyons, Lessard, and Marks, 2011). In Table 1, we compare the technological and social feature opportunities that resulted from the application of the extended method (right hand column) with the requirements gathered through interviews (left hand column). We see that most of the requirements obtained through the interviews were identified as opportunities through the extended method.

The issue of allowing users to control their own privacy settings was not identified by Step 4 of our extended systematic method but came up during our interviews. The other two requirements identified by users involved adding more third-party information into the system (statistics from third-party providers and information about periodicals). Each of the requirements would be easy to include in the resulting implementation; however, since this kind of activity is not carried out in the original CC service, it was not uncovered by the systematic method. This enforces the importance of gathering user feedback on items identified through the systematic method.

Table 1 Comparing Opportunities Found through the Extended Method with Requirements Gathering

Requirements Gathered through Interviews	Technology and Social Features Identified through Extended Method
A. Avoid having paper copies that are often thrown away	1. Substitute paper copies of the TOCs online versions and allow faculty members to electronically request articles from the TOCs 2. Use an electronic version of the master subscription list 3. Distribute online versions of the requested articles to faculty members electronically
B. Create an interface to let faculty easily update their subscriptions	4. Enable faculty members to subscribe to TOCs electronically
C. Include periodicals on the subscription list published by all providers including freely available ones	5. Enable faculty members and staff to update the list of available periodicals electronically

D. Allow the creation of communities of practice E. Provide visualizations of connections between people and their subscriptions F. Let faculty members annotate and evaluate articles they have read	6. Include user profiles of faculty members and allow faculty members to connect and join communities 7. Enable faculty members to comment on, and make suggestions about TOCs 8. Enable faculty members to comment on, rank, tag articles. 9. Enable faculty members to tag, rank, and comment on their subscriptions and that of their colleagues
G. Faculty members should be able to choose privacy levels H. Provide information to help faculty members choose periodicals and articles I. Add statistics from third-party providers to articles	10. Not identified by the extended systematic method

4. CONCLUSION AND FUTURE WORK

In this paper, we proposed an extension to Flor and Maglio's (2004) method that diagnoses opportunities for social features in service businesses. The method makes use of media constellation diagrams to represent service activities as distributed cognitive systems. We described the results of applying the extended method to a library service.

Additional research into the design of the extended method includes understanding how and where to include checks for privacy, security, and addressing social and cultural issues. In particular, designers who use this method should be aware of concerns with replacing people agents with technology. It is also the case that, as with other kinds of models, media constellation diagrams may be created differently by different designers. While this may be an issue, we feel that the process of studying the business service, the identification of products, and the agents and activities used in the creation of products is an important exercise. It would be interesting to understand how different models affect the resulting social features and technologies identified. Finally, the extended systematic method should be evaluated on other kinds of business services.

ACKNOWLEDGEMENTS

The authors would like to thank Lysanne Lessard and Sarah Flynn for contributions to this project. Funding for this research was provided by an NSERC Discovery Grant and a University of Toronto Connaught Research Grant.

REFERENCES

Abram, S. (2005, September). Web 2.0 – huh?! Library 2.0. Library 2.0 Information Outlook, 9(2): 44-6.

boyd, d.m. and N.B. Ellison. (2007). Social network sites: Definition, history, and scholarship. Journal of Computer-Mediated Communication, 13(1): 210-230.

Carr, N. G. (2004). Does IT matter? Information technology and the corrosion of competitive advantage. Boston(Massachusetts): Harvard Business School Press.

Demirkan, H., Kauffman, R. J., Vayghan, J. A., Fill, H.-G., Karagiannis, D., & Maglio, P. P. (2008). Service-oriented technology and management: Perspectives on research and practice for the coming decade. Electronic Commerce Research and Applications, 7: 356-76.

Farkas, M. G. (2007). Social Software in Libraries: Building Collaboration, Communication, and Community Online. Medford(New Jersey): Information Today.

Flor, N. V. & Maglio, P. P. (2004). Modeling business representational activity online: A case study of a customer-centered business. Knowledge-Based Systems, 17: 39-56.

Hollan, J. Hutchins, E., and Kirsh, D. (2000). Distributed cognition: Toward a new foundation for human-computer interaction research, ACM Transactions on Computer-Human Interaction, 7(2): 174–196.

Kim, W., O.-R. Jeong, and Lee, S.-W. (2010). On social websites. Information Systems, 35(2): 215–236.

Lintern, G. (2007). What is a cognitive system? Proceedings of the Fourteenth International Symposium on Aviation Psychology, April 18-21, 2005, Dayton, OH, 398-402.

Lyons, K., Lessard, L., and Marks, S. (2011). Integrating social features in service systems: the case of a library service. AMCIS 2011 Proceedings - All Submissions. Paper244.

Lyons, K. and Lessard, L. (2012). S-FIT: A technique for integrating social features in existing information systems, in Proceedings of the iConference 2012, February 7-10, 2012, Toronto, Ontario. 263-270.

Parameswaran, M. and Whinston, A.B. (2007). Research issues in social computing, Journal of the Association for Information Systems, 8(6): 336-350.

Simon, H. (1969). The Sciences of the Artificial, MIT Press, Cambridge, MA.

Stephens, M. (2007). Best practices for social software in libraries. Library Technology Reports, 43(5): 67-75.

Thelwall, M. (2007) Social network sites: Users and uses. Advances in Computers, 76: 19-73.

CHAPTER 17

Distributed Cognition, Service Science & 3D Multi-User Virtual Environments

Nick V. Flor

University of New Mexico
Albuquerque NM, USA
nickflor@unm.edu

ABSTRACT

The goal of an information systems designer is to create useful applications (apps) for individuals and groups by combining existing technologies. This can be a problem since new technologies are emerging constantly. This article presents a step towards solving one facet of this complex problem: how to create both useful and desirable apps when the unit of analysis is the individual-using-a-toolset. Specifically, I present a framework for guiding the creation of useful & desirable apps derived from basic principles of distributed cognition and of service science, in particular the principles of expert systems and value co-creation, respectively. The framework consists of four steps: (1) find a task requiring individual expertise, which has value—either economic, social, or cognitive; (2) discover the skills needed to perform the task using current technologies; (3) design a new distribution of skills across new technologies, that allows users to take advantage of the basic cognitive skills that all people possess; (4) execute the realization of the app. I apply this framework to the task of representing the human figure. The resulting app allows even non-artists to represent a human figure in an imagined pose, using technology originally intended for building multi-user virtual environments. I present the results from students using this app, and end with a discussion of how to generalize the framework to larger units of analyses.

Keywords: distributed cognition, service science, multi-user virtual environments, virtual worlds, digital ecosystem, digital bionics

1   THE CHALLENGE

How can distributed cognition and service science inform the design of applications using technology intended for multi-user virtual environments? The basic goal of an information-systems designer is to create applications using existing technology—along with some customization—that allows individuals or groups in organizations either to do existing activities better or to perform entirely new activities altogether, where the notion of "better" is along dimensions such as cost, time, or quality. Discovering such applications (hereafter, "apps") can be problematic since new technology is constantly emerging.

This paper attempts to provide a solution to part of this problem. Specifically, I provide a framework for designing useful *and* desirable apps, for a unit of analysis consisting of an individual using a toolset. The framework combines basic principles from distributed cognition and service science, specifically the principle of expert systems and the principle of value co-creation, respectively.

2   A BRIEF OVERVIEW OF DISTRIBUTED COGNITION, SERVICE SCIENCE, AND 3-D VIRTUAL WORLDS

2.1   Distributed Cognition

The concept of distributed cognition (Hutchins, 1995; Norman, 1993; Salomon, 1993) emerged from the realization that an individual without access to external resources was in many ways handicapped in terms of intelligent behavior—that the material, social, and cultural environment in which an individual was embedded played a significant role in that individual's intelligent behavior. One need only consider solving math problems in one's head, versus doing math on a calculator or spreadsheet, or solving math problems as part of a team, to understand the important role the environment can play in contributing to what is considered intelligent behavior. Thus, when studying intelligent activity all distributed cognition researchers are careful to analyze and attempt to characterize all the resources that are used by an individual or team during the performance of a task.

A key design principle in distributed cognition is the "expert systems principle" which is alluded to by Hutchins in the following: "… this should also give us a new meaning for the term 'expert system.' Clearly, a good deal of the expertise in the system is in the artifacts (both the external implements and the internal strategies)— not in the sense that the artifacts are themselves intelligent or expert agents, or because the act of getting into coordination with the artifacts constitutes an expert performance by the person; rather, the system of person-in-interaction-with-technology exhibits expertise." (Hutchins, 1995, p. 155). The system exhibits expertise because a well-designed tool allows the person using it to leverage what people do best, namely: pattern recognition, predicting future states of the world based on existing states, and manipulating things in the world so that they come to mean something (Rumelhart, Smolensky, McClelland, & Hinton, 1988).

2.2   Service Science

Service science is the study of service systems (Maglio & Spohrer, 2008), where the term "service" is defined as the application of competencies by a system for the benefit of others (Vargo & Lusch, 2004), e.g., banking, and where the "service system" consists of people, technology, organizations, and shared information. Moreover, the service system includes both service provider and service consumer. At the smallest unit of analysis, you have an individual and the resources that he or she uses, and larger units of analyses can include nations and alliances between nations. One of the key principles in service science is the co-creation of value—that the service provider together with the service consumer create value jointly.

2.3   3-D Multi-User Virtual Environments (Virtual Worlds)

A 3-D virtual world is a computer program that displays to users 3-D representations of people, places, and other things found in the "real world," and that allows users to flexibly interact with these representations via standard input devices like a keyboard and mouse, or more sophisticated input devices like accelerometers or motion capture controllers.

A *third-person* 3-D virtual world, hereafter "virtual world", also contains an avatar—a character under a user's control that the user can see and that is central to the user's feeling of immersiveness in the virtual world (see Figure 1). Avatars do not have to look anything like a user, nor do they have to look human, but human-like figures are the most common type of avatars in video games.

Figure 1. Virtual UNM Campus Developed by the Author on a Mobile Device, Avatar in the Center

The technology used to create virtual worlds include: (a) 3-d modeling and animation software—for creating the people, terrain, buildings, and other artifacts found in the virtual world, as well as to animate avatars and computer-controlled characters; (b) the software development environment—for displaying the virtual world and adding interactivity; and (c) the target system—used to host the virtual world, which can be either a desktop or a mobile device.

3 A FRAMEWORK FOR CREATING USEFUL & DESIRABLE APPS FOR INDIVIDUALS-USING-TOOLSETS

For a unit of analysis consisting of an *individual using a toolset*, I propose the following framework for creating useful & desirable apps, which consists of four steps: (1) find a task requiring individual expertise, which has value—either economic, social, or cognitive; (2) analyze the task to discover the skillset needed to perform it with one toolset; (3) design a new distribution of skills across a new toolset, that allows users to take advantage of the basic cognitive skills that all people possess; (4) execute the realization of the app.

By taking a valuable task involving an expert using a particular toolset, and distributing the task-relevant skills across a non-expert using a different toolset, which is based on the expert systems design principle from distribute cognition, the new activity system of user-with-tool co-creates value, which is a service science principle.

3.1 Step 1. Find a Task that Requires Individual Expertise

The first step in designing an app for an individual that is both useful and desirable is to identify an activity that requires expertise using a given toolset, where acquiring the expertise takes significant time, and where the product of the activity—information in some medium—provides benefits to others. While one could always perform a strategic market analysis (e.g., Aaker, 1995) to identify a useful app, this heuristic is simpler to apply and its result can always be verified with a strategic market analysis.

The notion of "others" includes individuals, groups, or organizations in domains such as business, education, and entertainment. The notion of "benefits" can be broad. In particular, a product could provide a benefit because it is used by others as a component of a larger product, or because it allows others to do existing activities with greater speed, or with lower cost, or with higher quality, or at a greater reach, among many others. A product could also benefit others because it allows them to do entirely new activities. Finally, the benefit could be to the producer him- or herself, in terms of entertainment or other forms of personal satisfaction, e.g., an expert musician playing music for enjoyment.

3.2.1 Example

An example of a task that fits the criteria mentioned is drawing the human figure. The product is a figure in some pose (information) on paper (medium). The toolset includes pencil, eraser, and other instruments used by the artist to draw the figure on paper. The ability to draw a human figure at a level that is useful to others requires considerable expertise acquired through years of practice. Lastly, the drawing can be used for a variety of purposes, including business and entertainment.

3.2   Step 2. Analyze the Task to Discover the Skillset

In the task identified in Step 1, an individual expert co-creates a valuable product with his or her toolset. What we want to do in the next couple of steps is to find a way for non-experts to create a similar product using a different toolset. If we are successful, then this different toolset along with whatever instructions are required for the non-expert to operate the toolset becomes a useful and desirable app for a larger number of individuals.

Finding a way for a non-expert to perform an expert task begins with discovering the representational skills employed by the expert during task performance, where I define a representational skill as an ability that results in information in some medium. There are a variety of methods one can use to analyze expert performance including protocol analysis (Ericsson & Simon, 1980), cognitive ethnography (Hutchins, 1995), and representational analysis (Flor & Maglio, 2004). Whatever the method employed, the outcome of the analysis is a description of task-relevant representations and the skills needed to produce them.

3.2.1   Example

Performing a representational analysis of existing instructional materials on figure drawing reveals a common three-stage sequence of representations. In stage 1, the artist draws a rough action sketch using mainly lines. In stage 2, the artist overlays basic geometric primitives over these lines such as cylinders, cubes, and spheres. Finally, in stage 3, the artists layers detail like muscles, clothing, and shadows over the basic geometric primitives (see Figure 2). See Flor (2012), for a more extensive analysis of these representations.

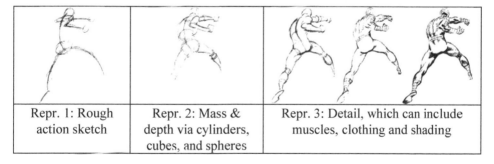

Repr. 1: Rough action sketch	Repr. 2: Mass & depth via cylinders, cubes, and spheres	Repr. 3: Detail, which can include muscles, clothing and shading

Figure 2. Figure Drawing as a Sequence of Layering Three Representations. See Text for Explanation; figures adapted from Lee & Buscema (1978)

The skillset needed for producing the representations found in this three-stage sequence are (Loomis, 1943): proportion, anatomy, perspective, value, color, impact, and media & materials. Briefly, an artist uses *proportion* skill to draw body parts in aesthetically pleasing sizes relative to one another. *Anatomy* skill is used to detail the surface of the body as influenced by the underlying skeletal and muscular structures. *Perspective* skill is used to draw the body parts from different viewpoints and with *foreshortening*, in which body parts closer to an observer appear bigger than those that are further away. Foreshortening is the single most difficult skill for drawing figures (Hogarth, 1996, p. back cover). *Value* skill is used to create the appearance of light and shading to the figure (see last image in Figure 2). *Impact* skill is used to pose the figure in a way that viewers will find exciting. Finally, *knowledge of mediums and materials* is an operational skill—the ability to use paper and pencil or other media and instruments for drawing. Figure 3 is a graphical depiction of these skills.

Figure 3. The Skills Required for Figure Drawing

3.3   Step 3. Design a New Distribution of Skills

Step 2 yields a list of the skills needed by an expert using a particular toolset to perform a given task. A useful and desirable app would allow non-experts using a different toolset—or the same physical toolset and a different mental "toolset"—to perform a similar task without having to undergo the years of training that experts go through. The distributed cognition principle of an expert *system* provides guidelines for how to achieve this. In particular, one can design a distribution of task-relevant skills where the difficult skills are performed by the toolset, and the users perform those skills that all people are good at doing. It should be noted that because a different toolset is used, there may be operational versus representational skills in the new system that a user may need to learn.

3.3.1 Example of Skill Redistribution

In figure drawing on paper, the skills that are difficult to learn and apply are perspective, value, and color. To recap, perspective is the ability to draw parts from an arbitrary viewpoint and with foreshortening; value is adding shading to the drawing, which varies according to the type, direction, and distance of a lighting source; and color is the application of hues to a figure, which also varies in brightness and saturation depending on the type, direction, and distance of the lighting source. If these difficult skills could be redistributed to the toolset, then a non-expert could depict human figures while just possessing the skills of proportion and impact, along with whatever added skills are required by the new toolset.

3-d modeling software is such a toolset that can perform the skills of perspective, value, and color, which are difficult for individuals to learn. Figure 4 is a graphical depiction of the skill redistribution; see Flor (2012) for a more extensive analysis of the redistribution   Although this software is intended for creating characters and other artifacts for virtual worlds, the models that one creates can be stored or printed out on paper and used just like an expert's hand-drawn figures.

Figure 4. A New Distribution of Skills. Animation, rigging, and in-betweens are new skills that must be learned as a consequence of the new toolset.

Finally, when using 3-d modeling software, the drawing skill of making marks on paper using a pencil is substituted with the modeling skill of first placing basic geometric primitives like cubes, spheres, & cylinders; and secondly, extruding different areas of these primitives. An information systems designer can create customized instructions that combine modeling skill with proportion skill as in

figures 5-9, for creating various body parts. The 3-d modeling software combined
with the instructions for creating the body parts, constitute the app.

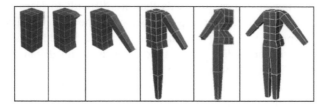

Figure 5. Modeling the Body

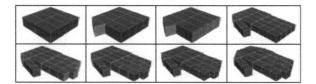

Figure 6. Modeling the Hand

Figure 7. Modeling the Foot

Figure 8. Modeling the Head (see text for explanation)

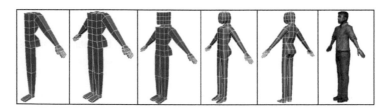

Figure 9. Attaching the Body Parts & Adding a Texture Map

3.4   Step 4. Execute the Realization of the App

The execution step includes all post-design activities such as implementation, testing, marketing, sales, and customer service. As these activities are well-defined and covered in depth by design and business researchers, I will not review them in this article. Instead, I present evidence of a successful realization of the example app, where non-experts perform a task with a new toolset that requires expertise using a current toolset.

3.4.1   Example of the Successful Realization of an App

The app consisted of a technology—3-d modeling software—and custom instruction (see Figures 5-9) for modeling human-like figures or androids. It was taught in three hours to 43 undergradate business students with no art experience as part of a management of information systems course. The students had one week to then turn in an android. Of these 43 students, 33 turned in an android, while 10 students did not turn in an android. Of the 10 students that did not turn in an android, 7 did not turn in any assignments at all during the semester. Of the 33 students that turned in an android, 3 did not save their files properly and only turned in an animated skeleton. However, when their work was checked in person, they indeed modeled an android correctly. If you remove the 7 students that did not show up to class, then 91.6% of the students successfully used the app to model an android. Figure 10 depicts the androids created by the students

Figure 10. Humanoids Figures Created by (non-artist) Management Students

5   SUMMARY AND CONCLUSION

I have presented a four-step framework based largely on principles from service science (step 1) and distributed cognition (steps 2 and 3). An information systems designer can use this framework to guide the design of apps that are useful and desirable for individuals. To recap, the steps are (1) find a task requiring individual expertise; (2) discover the skills needed by an expert to perform the task using current technologies; (3) design a new distribution of skills across new technologies; (4) realize the app. I applied this framework to the task of drawing human figures, resulting in an app that allowed non-artists to successfully model human figures in a short amount of time. The main limitation of this framework is that the unit of analysis is an individual using a toolset. To generalize this to more complex apps, one must extend the unit of analysis to include groups, where the individuals use both subtask-specific tools and shared tools, such as groupware or configuration management software. The steps remain the same, but designing a new distribution of skills (step 3) is more difficult because there are more technologies and more possible skill distributions in larger units of analyses.

6   REFERENCES

Aaker, D. A., 1995. *Strategic Marketing Management.* Fourth ed. New York: John Wiley & Sons.

Ericsson, K. A. & Simon, H. A., 1980. Verbal Reports as Data. *Psychological Review,* Volume 87, pp. 215-251.

Flor, N. & Maglio, P., 2004. Modeling business representational activity online: A case study of a customer-centered business. *Knowledge-Based Systems,* Volume 17, pp. 39-56.

Flor, N. V., 2012. Designing Instruction Outside Your Expertise: A Study of Distributed Cognition Applied to 3-D Figure Modeling. *Manuscript Submitted for Publication.*

Hogarth, B., 1996. *Dynamic Figure Drawing.* New York: Watson-Guptill Publications.

Hollan, J., Hutchins, E. & Kirsh, D., 2000. Distributed Cognition: Toward a New Foundation for Human-Computer Interaction Research. *ACM Transactions on Computer-Human Interaction,* Volume 7, pp. 174-196.

Hutchins, E., 1995. *Cognition in the Wild.* Cambridge: MIT Press.

Lee, S. & Buscema, J., 1978. *How to Draw Comics The Marvel Way.* New York: Simon & Schuster.

Loomis, A., 1943. *Figure Drawing for All It's Worth.* New York: Viking Press.

Maglio, P. P. & Spohrer, J., 2008. Fundamentals of Service Science. *Journal of the Academy of Marketing Science,* Volume 36, pp. 18-20.

Norman, D. A., 1993. *Things That Make Us Smart.* Reading(MA): Addison-Wesley Longman.

Rumelhart, D. E., Smolensky, P., McClelland, J. L. & Hinton, G. E., 1988. Schemata and Sequential Thought Processes in PDP Models. In: *Parallel Distributed Processing, Volume 2: Psychological and Biological Models.* Cambridge: MIT Press, pp. 7-57.

Salomon, G., 1993. *Distributed Cognitions: Psychological and Educational Considerations..* New York: Cambridge University Press.

Vargo, S. L. & Lusch, R. F., 2004. Evolving to a New Dominant Logic for Marketing. *Journal of Marketing,* Volume 68, pp. 1-17.

CHAPTER 18

The Role of Positive Emotions in the Creation of Positively Deviant Business-to-Business Services

Examination of employee and customer perceptions

Merja Fischer

Aalto University
Espoo, Finland
merja.fischer@aalto.fi

ABSTRACT

In this conceptual paper a novel theory is provided that explicates how positive emotions of four actors (supervisors, employees, peers and customers) in the service profit chain (SPC) can foster the creation of positively deviant service businesses. This paper suggests incorporating studies and theories of positive organizational scholarship and particularly studies on positive emotions to the services marketing literature. The suggestion of this study is that positively deviant performances create climate for positivity in the supplier-customer interaction and foster the co-creation of mutual value in service businesses.

Keywords: Positive emotions, service profit chain, positively deviant behaviors, positively deviant performances, climate for positivity, model of positively deviant service businesses and co-creation of mutual value in services

1   INTRODUCTION

An effective service organization understands that managing employees' perceptions regarding their own organization is important; what employees experience in their workplace is echoed in customer experiences (Pugh, Dietz, Wiley, & Brooks, 2002; Schneider & Bowen, 1993).

Recent studies on industrial marketing management emphasize the increasing importance of the role of the human element in service businesses. Ballantyne, Frow, Varey, & Payne (2011) argue that dialogical communication such as collaboration and learning together, embodies the potential to reveal new value creating possibilities for the industrial world. Likewise, Ramaswamy (2011) challenges the premises of services, claiming that service value is emerging from the human experience in interactions rather than from the efficiency of the service process. Further to this, Berry (2011) claims that the future competiveness of a service company is measured by the strength of its relationships and reciprocity commitments between its customers, employees, suppliers and other stakeholders such as communities.

This paper goes some way towards proposing a novel theory of the positive means embedded in the human experiences that can promote the profitability and productivity of service sector.

A fairly recent paradigm, Positive Organizational Scholarship (POS), has appeared that introduces positive behaviors as a means of building positive spirals in human interaction and sustainability in organizations (Dutton & Ragins, 2007; Fredrickson, 2003a; Fredrickson & Joiner, 2002; Walter & Bruch, 2008). The POS research focuses on the opportunities and strengths rather than weaknesses and pitfalls (Cameron, Caza, & Bright, 2002). POS deviates from traditional organizational studies in that it seeks to portray what is understood by the idea of what represents and approaches the best of the human condition in the workplace, rather than dwelling on the negative aspects or malfunctions (Cameron, Dutton, & Quinn, 2003:4). To concretize the opportunity of the human potential a summary of the relevant studies and theories in POS is provided in Table 1. The summary in Table 1 presents how positively deviant behaviors, such as appreciation, helping others and gratitude, of four actors in the SPC can cultivate the creation of positively deviant performances, as verified in the POS literature. According to Cameron (2008:2) positively deviant performance means outcomes that dramatically exceed common or expected performance.

Furthermore, this study takes the opportunity to integrate theories in positive organizational scholarship with service business and organizational climate research, and suggests how relationships in the workplace may contribute to improving overall business performance. Positive Organizational Scholarship (POS) contributes to the basis of this study by advocating that organizational outcomes can be positively influenced by adding elements of POS to existing service business research. The following research question is formulated to study this topic.

Research Question: Based on the existing POS theories and studies, how may the broaden-and-build processes of actors in the SPC enable them to foster the creation of positively deviant service businesses?

To answer this question, firstly, the broaden-and-build process is elucidated based on broaden-and-build theory of positive emotions by Fredrickson (Fredrickson, 2004) and a study provided by Folkman (1997).

Specifically a study by Fischer (2012) on the linkages between employee and customer perceptions in B-to-B services is elaborated. This study (Fischer, 2012) verified how employee perceptions of the behaviors and attitudes of their supervisors and peers impact on customers' perceptions of service quality in a business-to-business context.

As a conclusion, this paper provides a new model, model of positively deviant service businesses that contributes both to the existing services marketing literature as well as provides a novel application of the positive means embedded in the positive organizational scholarship studies and theories.

2   SERVICE PROFIT CHAIN AND LINKAGE RESEARCH

Heskett et al. (1994) introduced service profit chain thinking into the service literature. They describe the service profit chain (SPC) as a conceptual framework, which integrates leadership, suppliers' processes and procedures, employees' behaviors and customers' expectations, resulting in productivity and profitability. The purpose of linkage research is to identify the elements of the work environment–as described by employees–that correlate, or link, with critically important organizational outcomes, such as customer satisfaction and business performance (Wiley, 1996:330).

Several researchers have successfully tested some elements of the linkages in the business-to-consumer (B-to-C) context (Gelade & Young, 2005; Hallowell, 1996; Ryan, Schmit, & Johnson, 1996; Salanova, Agut, & Peiro, 2005; Schneider, Hanges, Smith, & Salvaggio, 2003; Schneider, White, & Paul, 1998). However, the first study to examine the linkages in a B-to-B context was published by Fischer (2012). Fischer could verify, using path modeling that employee perceptions of their organization impacted on customer perceptions of service quality.

Fischer's (2012) study included employee and customer survey data from 38 countries in a global company providing industrial services. In order to examine the boundary conditions, Fischer applied data from two employee groups: account managers and field service engineers. These two employee groups differ in the way they encounter with customers, one having a physical and psychological closeness with customers (field service engineers) and the other contacting customers mostly by email and phone (account managers). Her study showed that account managers' perceptions of their workplace climate impacted on customer perceptions of account managers' capability to express responsiveness and empathy and assurance. Further to this, her study showed that field service engineers' personal engagement predicted customers' perceptions of reliability and assurance provided by field service engineers. These results support the suggestion provided by Schneider and Bowen (1985:424) who have emphasized that the difference in boundary conditions in the SPC could be the physical and psychological closeness of employees and customers.

Research results provided by Fischer (2012) highlight the importance of the behaviors of four types of actors in the service profit chain: supervisors, employees, peers and customers. Thus, this study takes the opportunity to suggest integrating positive organizational scholarship studies and theories, and in particular, broaden-and-build theory of positive emotions to the service profit chain thinking.

3 THE IMPACT OF POSITIVE EMOTIONS IN THE SERVICE PROFIT CHAIN

The importance of the quality of interactions between actors in the SPC is increasing (Ballantyne et al., 2011; Berry, 2011; Ramaswamy, 2011). Positive organizational scholarship provides positive means of enhancing the quality of interactions by proposing positively deviant behaviors such as expressions of appreciation, gratitude, trustworthiness and helping others (Fredrickson, 2003). Fredrickson (2004) argues that positive emotions of members of an organization may, over time, have an influence on organizational functioning. Fredrickson's (1998; 2001) broaden-and-build theory of positive emotions is used in this study as the framework to indicate how individuals' positive emotions broaden-and-build their personal resources and elevate their health and well-being. According to Fredrickson (2004), experiences of positive emotions transform people and they may become more creative, knowledgeable, resilient, socially integrated and healthier individuals. On an organizational level, the increased personal resources of the members of an organization may foster positively deviant performances in the service profit chain, thus creating a strengthening upward spiral. Positively deviant behaviours, such as appreciation, gratitude, helping others, trustworthiness and unselfishness within the four types of actors in the SPC are likely to foster the creation of positively deviant performances in service businesses.

Table 1 presents examples of existing POS studies and theories and other related literature that may support each actor in the SPC in promoting his/her individual positively deviant behaviors. These behaviors may lead to the cultivation of positively deviant performances thus gaining benefits for the entire service business. The upper part of Table 1 provides examples of such positively deviant behaviors that may enhance positively deviant performances in the SPC. Further to this, Table 1 suggests, based on existing research (Folkman, 1997; Fredrickson, 2003a; Spreitzer & Sonenschein, 2003) that these positively deviant behaviors of the four types of actors: supervisors, employees, peers and customers can create positively deviant performances in SPC, both on an individual actor level and on the organizational level. Table 1 proposes the idea that supervisors applying leadership strategies such as: positive leadership strategy (Cameron, 2008), authentic leadership strategies (Avolio, Walumbwa, & Weber, 2009; George & Bennis, 2003; Shirey, 2006) and servant leaderships strategies (Liden, Wayne, Zhao, & Henderson, 2008; Sendjaya, Sarros, & Santora, 2008; Spears & Lawrence, 2004) may apply such positively deviant behaviors that create positive meaning and positive emotions among their subordinates. Furthermore, Table 1 proposes POS studies that recognize positively deviant behaviors such as personal engagement (Kahn, 1990), work engagement (Bakker, Demerouti, & Schaufeli, 2003; Demerouti, Nachreiner, Bakker, & Schaufeli, 2001; Hakanen, 2009; Hakanen, Perhoniemi, & Toppinen-Tanner, 2008; Salanova et al., 2005), and systems intelligence (Hämäläinen & Saarinen, 2004). Systems intelligence provides a potential framework for individuals to cultivate positively deviant performances through understanding of their own influence upon the system as well as the influence the system has upon him/her (Hämäläinen & Saarinen, 2004). Some examples of existing POS studies that include elements of how peers may cultivate the creation of positively deviant performances in an organization are: high performing teams as studied by Losada and Heaphy (2004), studies on high-quality connections (Dutton & Heaphy, 2003), positive organizational network analyses (Baker, Cross, & Wooten, 2003; Cross, Baker, & Parker, 2003) and interaction ritual theory by Collins (2004) customer relationship building that could create positively deviant performances. These theories could be seen as generally applicable in the supplier-customer interaction and they may help towards fostering such customer relationship building that could create positively deviant performances.

Table 1 . A summary of existing theories and studies that could foster the creation of positively deviant behaviors and performances of the four types of actors in the SPC. (Fischer, 2012:147)

The lower part of Table 1 provides examples of those POS theories and studies that have verified the impact of positive emotions and consequently positively deviant behaviors in the creation of positively deviant performances in various contexts. The lower part of Table 1 presents individual level positively deviant performances such as emotional well-being, flourishing, psychological resilience, personal resources and trust (Fredrickson, 2003a).

POS literature has studied several positively deviant performances that can be theorized as supporting the mutual value creation in the supplier-customer co-creation process. The relevant outcomes of positively deviant performances include: more creativeness (Rowe, Hirsh, & Anderson, 2007), better productivity (Wright & Cropanzano, 2004), high performing teams (Losada and Heaphy 2004), flourishing workplaces (Fredrickson & Losada, 2005b), better resilience (Cohn, Fredrickson, Brown, Mikels, & Conway, 2009; Fredrickson, Tugade, Waugh, & Larkin, 2003; Tugade, Fredrickson, & Feldman Barrett, 2004) and less absenteeism (Avey, Patera, & West, 2006). Additionally, positive emotions increase vitality, positive regard and mutuality (Dutton & Heaphy, 2003), increase productivity in supplier-customer interactions and provide a framework for better negotiation results (Kopelman, Rosette, & Thompson, 2006). Table 1 summarizes how the positively deviant behaviors of several actors: supervisors, employees, peers and customers may foster the creation of positively deviant performances in the SPC.

4   BROADEN-AND-BUILD PROCESS OF ACTORS IN THE SERVICE PROFIT CHAIN

Connecting the research and conceptual insights by Folkman (1997) and Fredrickson (1998; 2001; 2003b) with the results presented in the study by Fischer (2012), it is possible to suggest a novel theory of how an individual actor in the service profit chain can cultivate others' positive emotions by applying positively deviant behaviors and thus create positively deviant performance among themselves and with other actors in the SPC. Individual level positively deviant performance, such as trust in self and others, a feeling of oneness, creativity and seeing the bigger picture impact on the creation of organizational level positive deviance in the sense recognized by Cameron (2008).This novel theory of broaden-and-build process is presented in Figure 1.

The broaden-and-build process presented in Figure 1 expresses the importance of positively deviant behaviors, such as expressions of appreciation, love, gratitude, helping others, trustworthiness and unselfishness as the means of transforming ordinary events into positive events (Folkman, 1997; Fredrickson, 2003a). Positive emotions, in turn, emerge when positive events carry positive meaning (Folkman, 1997). According to Fredrickson (2003a:174), positive meaning at work can be triggered by experiences of competencies, involvement, significance and achievements and social connections.

Figure 1.The broaden-and-build process of four types of actors in the SPC: a reinforcing cycle of positively deviant behaviours and positive emotions in the creation of positively deviant performance.(Fischer, 2012:132)

Figure 1 suggests, based on existing research (Folkman, 1997; Fredrickson, 2003b; Spreitzer et al., 2003), how the broaden-and-build process can create individual level positively deviant performance. These individual level positively deviant performances, such as emotional well-being, trust in self and others (Dunn & Schweitzer, 2005), feeling of oneness (Waugh & Fredrickson, 2006), creativity (Amabile, Barsade, Mueller, & Staw, 2005; Rowe et al., 2007) and seeing the bigger picture (Fredrickson, 2003b) may foster the reinforcing cycle of positively deviant behaviors within those individuals (Fischer, 2012).

5   CLIMATE FOR POSITIVITY

This novel concept, a climate for positivity, is an organizational state in which individuals' positively deviant behaviors are cultivated, supported and rewarded (Fischer, 2012:153). A climate for positivity emerges from individuals' positive emotions and positively deviant performances that impact on and reinforce the climate for positivity. A climate for positivity is an organizational condition, where positive emotions are experienced and expressed and guiding individuals' interactions. Supervisors have a key role in establishing a framework in which a climate for positivity can occur through such means that cultivate, support and reward such behaviors which employees perceive as positively deviant behaviors. However, each individual is responsible for his/her own behavior in each interaction.

Each actor: supervisor, employee, peer and customer can cultivate the climate for positivity through their positively deviant behaviors. From this perspective, each interaction is seen as an opportunity to nurture the human potential in other members of an organization and its customers through the application of positively deviant behaviors. These interactions can transform ordinary events into positive events by creating positive meaning and positive emotions among participants. Moreover, the positive emotions of the members in an organization foster the creation of a climate for positivity through a reinforcing loop that strengthens the possibility for more positive events to take place.

7   The model of positively deviant service businesses

The new model presented in Figure 2 suggests that the positively deviant performances of the four groups of actors in the service profit chain create a climate for positivity that facilitates the creation of positively deviant service businesses. Climate for positivity foster the co-creation of mutual value in services, impacting on productivity and profitability of the entire service profit chain.

Figure 2 indicates how the positively deviant behaviors of the four types of actors in the SPC (i.e. supervisors, employees, peers and customers) can foster the creation of positively deviant performances that may consequently generate a climate for positivity which ultimately fosters the formation of positively deviant B-to-B service businesses.

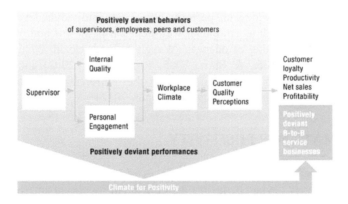

Figure 2. The model of positively deviant B-to-B service businesses. (Fischer, 2012:150)

To elucidate the model of positively deviant service businesses in Figure 2, a team member's positively deviant behavior such as: appreciation, gratitude and helping others transforms ordinary events involving other team members into positive events and by doing so, creates positive emotions in the team. Subsequently, positive emotions foster the creation of positive deviant performances resulting in outcomes such as flourishing workplaces (Fredrickson & Losada 2005b), high performing teams (Losada & Heaphy, 2004), more emotional energy in the team (Dutton & Heaphy, 2003), more trust (Dunn & Schweitzer, 2005) and the feeling of oneness (Waugh & Fredrickson, 2005).

These positively deviant performances, as proposed by the summary in Table 1 and explicated through the broaden-and-build process in Figure 1, foster the creation of a climate for positivity which provides a platform for positively deviant B-to-B service businesses to emerge.

8   DISCUSSION AND CONCLUSIONS

This paper has provided novel theory of the potential in releasing the human potential in services context. This paper provides a link between studies and theories of positive emotions and positive organizational scholarship with the existing services marketing literature. Research results provided by Fischer (2012) of the linkages between employee and customer perceptions are used as the bridge between these two disciplines.

In addition, this paper has contributed both to the services marketing literature as well as to the positive organizational scholarship studies and theories. Firstly, the new concept, climate for positivity, provides a novel approach how each individual can foster the creation of a positive organizational climate through their own positively deviant behaviors. Secondly, the new model, model of positively deviant service businesses, provides a new framework to explicate how the positively deviant behaviors of supervisors, employees, peers and customers can cultivate the co-creation of mutual value in services.

The model of positively deviant service businesses in Figure 2 suggests that the positively deviant behaviors embedded in the broaden-and-build theory of positive emotions and other POS theories and studies presented in Table 1 can build a unique competitive advantage for a service business. This competitive advantage is embedded in the minds, behaviors and interactions of both those working in and working for the suppliers' and customers' organizations. It seems safe to conclude that positive emotions and positively deviant behaviors when fostered appropriately, may lead towards positively deviant B-to-B service businesses that cannot be imitated or copied by the competitors.

As was indicated in the introductory chapter, service orientation is an increasing global trend among companies which creates the necessity for organizations to discover novel means of cultivating and uncovering the human potential. This paper has scrutinized the quality of interactions of four types of actors in service businesses (i.e. supervisors, employees, peers and customers) and has made a serious attempt to identify positively deviant behaviors which would improve performance in service businesses. Based on this paper, it is suggested that a service business, specifically its four types of actors, can impact on the performance by taking an intentionally and deliberately chosen stance towards positively deviant behaviors that furnish the upward spirals through individual emotional well-being and a climate for positivity. In the increasing service business markets it is vital that the perceived service quality is excellent and the performance keeps improving and customer loyalty is sustained.

To summarize, this paper indicates how the POS research can open new dimensions in the service profit chain research leading towards more positively deviant service businesses. This may happen through enhanced positively deviant performances, such as employee well-being, flourishing workplaces and higher customer quality impacting on the creation of a climate for positivity.

Ultimately, this paper suggests novel opportunities for development in both disciplines: SPC and POS. Implications to existing theories are several. Firstly, service profit chain and services research as a whole, can find numerous future research areas to study the impact of positive emotions on co-creation of value for both organizations and their employees.

REFERENCES

Amabile, T. M., Barsade, S. G., Mueller, J. S., & Staw, B. M. 2005. Affect and creativity at work. Administrative Science Quarterly, 50(3): 367-403..

Avey, J., Patera, J., & West, B. 2006. The Implications of Positive Psychological Capital on Employee Absenteeism. Journal of Leadership & Organizational Studies, 13(2): 42.

Avolio, B. J., Walumbwa, F. O., & Weber, T. J. 2009. Leadership: Current Theories, Research, and Future Directions. Annual Review of Psychology, 60(1): 421-449.

Baker, W., Cross, R., & Wooten, M. 2003. Positive Organizational Network Analysis and Energizing Relationships. In K. Cameron, J. Dutton, & R. Quinn (Eds.), Positive Organizational Scholarship: Foundations of a New Discip: 328-342. San Francisco: Berrett-Koehler Publishers.

Bakker, A. B., Demerouti, E., & Schaufeli, W. B. 2003. Dual processes at work in a call centre: An application of the job demands- resources model. European Journal of Work & Organizational Psychology, 12(4): 393-417.

Ballantyne, D., Frow, P., Varey, R. J., & Payne, A. 2011. Value propositions as communication practice: Taking a wider view. Industrial Marketing Management, 40(2): 202-210.

Berry, L. L. 2011. Lessons from high-performance service organizations. Industrial Marketing Management, 40(2): 188-189..

Cameron, K. 2008. Positive Leadership: Strategies for Extraordinary Performance. San Francisco: Berrett-Koehler Publisher, Inc.

Cameron, K., Caza, A., & Bright, D. 2002. Positive deviance, organizational virtuousness, and performance: Working paper, University of Michigan Business School.

Cameron, K., Dutton, J., & Quinn, R. 2003. Positive Organizational Scholarship: Foundations of a New Discipline. San Francisco: Berrett-Koehler Publisher Inc.

Carnevale, P. J. D., & Isen, A. M. 1986. The influence of positive affect and visual access on the discovery of integrative solutions in bilateral negotiation. Organizational Behavior and Human Decision Processes, 37(1): 1-13.

Cohn, M. A., Fredrickson, B. L., Brown, S. L., Mikels, J. A., & Conway, A. M. 2009. Happiness unpacked: Positive emotions increase life satisfaction by building resilience. Emotion, 9(3): 361-368.

Collins, R. 2004. Interaction ritual chains. Princeton, N.J.: Princeton University Press.

Cross, R., Baker, W., & Parker, A. 2003. What Creates Energy in Organizations? MIT Sloan Management Review, 44(4): 51-56.

Demerouti, E., Nachreiner, F., Bakker, A. B., & Schaufeli, W. B. 2001. The Job Demands-Resources Model of Burnout. Journal of Applied Psychology, 86(3): 499-512.

Dunn, J. R., & Schweitzer, M. E. 2005. Feeling and believing: The influence of emotion on trust. Journal of Personality and Social Psychology, 88(5): 736-748.

Dyer, J., H, & Hatch, N., W. 2006. Relation-specific capabilities and barriers to knowledge transfers: creating advantage through network relationships. Strategic Management Journal, 27(8): 701-719.

Fischer M.,H. 2012. Linkages between employee and customer perceptions in business-to-business services - Towards positively deviant performances. Doctoral, Thesis, Aalto University, School of Science, Espoo.

Folkman, S. 1997. Positive psychological states and coping with severe stress. Social Science & Medicine, 45(8): 1207-1221.

Fredrickson, B. L. 1998. What good are positive emotions? Review of General Psychology, 2: 300-319.

Fredrickson, B. L. 2001. The role of positive emotions in positive psychology: the broaden and build theory of positive emotions. American Psychologist, 56(3): 218–226.

Fredrickson, B. L. 2003a. Positive Emotions and Upward Spirals in Organizations. In K. Cameron, J. Dutton, & R. Quinn (Eds.), Positive Organizational Scholarship: Foundations of a New Discipline: 163-175. San Francisco: Berrett-Koehler Publisher Inc.

Fredrickson, B. L. 2003b. The Value of Positive Emotions The emerging science of positive psychology is coming to understand why it's good to feel good. American scientist, 91(4): 330-335.

Fredrickson, B. L. 2004. The broaden-and-build theory of positive emotions. Philosophical Transactions of the Royal Society B: Biological Sciences, 359(1449): 1367-1377.

Fredrickson, B. L., & Branigan, C. 2005a. Positive emotions broaden the scope of attention and thought of action repertoires. Cognition & Emotion, 19(3): 313-332.

Fredrickson, B. L., & Losada, M. F. 2005b. Positive Affect and the Complex Dynamics of Human Flourishing. American Psychologists, 60(7): 678–686.

Fredrickson, B. L., Tugade, M. M., Waugh, C. E., & Larkin, G. R. 2003. What good are positive emotions in crisis? A prospective study of resilience and emotions following the terrorist attacks on the United States on September 11th, 2001. Journal of Personality and Social Psychology, 84(2): 365-376.

Ganesan, S. 1994. Determinants of long-term orientation in buyer-seller relationships. the Journal of Marketing, 58(2): 1-19.

Gelade, G. A., & Young, S. 2005. Test of a service profit chain model in the retail banking sector. Journal of Occupational & Organizational Psychology, 78(1): 1-22.

George, B., & Bennis, W. 2003. Authentic leadership: Rediscovering the secrets to creating lasting value: Wiley India Pvt. Ltd.

Grönroos, C. 2004. The relationship marketing process: communication, interaction, dialogue, value. The Journal of Business and Industrial Marketing 19 (2).

Grönroos, C. 2008. Service logic revisited: who creates value? And who co-creates? European Business Review, 20(4): 298-314.

Grönroos, C., & Ojasalo, K. 2004. Service productivity: Towards a conceptualization of the transformation of inputs into economic results in services. Journal of Business Research, 57(4): 414-423.

Gummesson, E. 1998. Productivity, quality and relationship marketing in service operations. International Journal of Contemporary Hospitality Management, 10: 15.

Hakanen, J. J. 2009. Työn imua, tuottavuutta ja kukoistavia työpaikkoja? , Työsuojelurahasto, Tätä on tutkittu 2009. Helsinki.

Hakanen, J. J., Perhoniemi, R., & Toppinen-Tanner, S. 2008. Positive gain spirals at work: From job resources to work engagement, personal initiative and work-unit innovativeness. Journal of Vocational Behavior, 73(1): 78-91.

Hallowell, R. 1996. The relationships of customer satisfaction, customer loyalty, and profitability: an empirical study. International Journal of Service Industry Management, 7(4).

Hämäläinen, R. P., & Saarinen, E. 2007. The Way Forward with Systems Intelligence. In R. P. Hämäläinen, & E. Saarinen (Eds.), Systems Intelligence in Leadership and Everyday Life: 295-305. Espoo: Systems Analysis Laboratory, Helsinki University of Technology.

Hansen, E., & Bush, R. J. 1999. Understanding Customer Quality Requirements: Model and Application. Industrial Marketing Management, 28(2): 119-130.

Hennig-Thurau, T., Groth, M., Paul, M., & Gremler, D. D. 2006. Are all smiles created equal? How emotional contagion and emotional labor affect service relationships. Journal of Marketing, 70(3): 58-73.

Heskett, J. L., Jones, T. O., Loveman, G. W., Sasser Jr, W. E., & Schlesinger, L. A. 1994. Putting the Service-Profit Chain to Work. Harvard Business Review, 72(2): 164-170.

Isen, A. M. 2001. An influence of positive affect on decision making in complex situations: Theoretical issues with practical implications. Journal of Consumer Psychology, 11(2): 75-85.

Isen, A. M. 2009. A Role for Neuropsychology in Understanding the Facilitating Influence of Positive Affect on Social Behavior and Cognitive Processes. In C. R. Snyder, & S. J. Lopez (Eds.), Oxford handbook of positive psychology: 503-518. Oxford; New York: Oxford University Press.

Isen, A. M., Daubman, K. A., & Nowicki, G. P. 1987. Positive affect facilitates creative problem solving. Journal of Personality and Social Psychology, 52(6): 1122.

Isen, A. M., Shalker, T. E., Clark, M., & Karp, L. 1978. Affect, accessibility of material in memory, and behavior: A cognitive loop? Journal of Personality and Social Psychology, 36(1): 1-12.

Jackson, M. C. 2000. Systems Approaches to Management: Kluwer.

Kopelman, R., E., Brief, A., P., & Guzzo, R. A. 1990. The Role of Climate and Culture in Productivity. In B. Schneider (Ed.), Organizational Climate and Culture: 283-318. San Francisco: Jossey-Bass.

Kopelman, S., Rosette, A. S., & Thompson, L. 2006. The three faces of Eve: Strategic displays of positive, negative, and neutral emotions in negotiations. Organizational Behavior and Human Decision Processes, 99(1): 81-101.

Liden, R. C., Wayne, S. J., Zhao, H., & Henderson, D. 2008. Servant leadership: Development of a multidimensional measure and multi-level assessment. The Leadership Quarterly, 19(2): 161-177.

Losada, M. 1999. The Complex Dynamics of High Performance Teams. Mathematical and Computer Modelling 30(9-10): 179-192.

Luoma, J. 2009. Systems Intelligence in the Process of Systems Thinking. Helsinki University of Technology, Espoo.

Pugh, S. D. 2001. Service With a Smile: Emotional Contagion in The Service Encounter Academy of Management Journal, 44(5): 1018-1027.

Pugh, S. D., Dietz, J., Wiley, J. W., & Brooks, S. M. 2002. Driving service effectiveness through employee-customer linkages. Academy of Management Executive, 16(4): 73-84.

Ramaswamy, V. 2011. It's about human experiences... and beyond, to co-creation. Industrial Marketing Management, 40(2): 195-196.

Rousseau, D. M., Sitkin, S. B., Burt, R. S., & Camerer, C. 1998. Not so different after all: a cross-discipline view of trust Academy of Management Review, 23(3): 393-404.

Rowe, G., Hirsh, J. B., & Anderson, A. K. 2007. Positive affect increases the breadth of attentional selection. Proceedings of the National Academy of Sciences, 104(1): 383.

Ryan, A. M., Schmit, M. J., & Johnson, R. 1996. Attitudes and effectiveness: Examining relations at an organizational level Personnel Psychology, 49(4): 853-882.

Salanova, M., Agut, S., & Peiro, J. M. 2005. Linking Organizational Resources and Work Engagement to Employee Performance and Customer Loyalty: The Mediation of Service Climate. Journal of Applied Psychology, 90(6): 1217-1227.

Schneider, B., & Bowen, D. E. 1985. Employee and customer perceptions of service in banks: replication and extension. Journal of applied psychology 70(3): 423-433

Schneider, B., & Bowen, D. E. 1993. THe Service Organizations - Human Resources Management is Crucial Organizational Dynamics, 21(4): 39-52.

Schneider, B., Hanges, P. J., Smith, D. B., & Salvaggio, A. N. 2003. Which comes first: Employee attitudes or organizational financial and market performance? Journal of Applied Psychology, 88(5): 836-851.

Schneider, B., White, S. S., & Paul, M. C. 1998. Linking Service Climate and Customer Perceptions of Service Quality: Test of a Causal Model. Journal of Applied Psychology, 83(2): 150-163.

Selnes, F., & Sallis, J. 2003. Promoting Relationship Learning. Journal of Marketing, 67(3): 80-95.

Sendjaya, S., Sarros, J. C., & Santora, J. C. 2008. Defining and measuring servant leadership behaviour in organizations. Journal of Management Studies, 45(2): 402-424.

Senge, P. M. 1990. The fifth discipline: the art and practice of the learning organization. New York: Doubleday/Currency.

Senge, P. M. 2006. The fifth discipline, 2nd edition. London: Random House Business.

Shirey, M. R. 2006. Authentic leaders creating healthy work environments for nursing practice. American Journal of Critical Care, 15(3): 256-267.

Spears, L. C., & Lawrence, M. 2004. Practicing servant-leadership: succeeding through trust, bravery, and forgiveness. San Francisco, CA: Jossey-Bass.

Spreitzer, G. M., & Sonenschein, S. 2003. Positive Deviance and Extraordinary Organizing. In K. Cameron, J. Dutton, & R. Quinn (Eds.), Positive Organizational Scholarship: Foundations of a New Discipline. San Francisco: Berrett-Koehler Publisher Inc.

Spreitzer, G. M., & Sonenshein, S. 2004. Toward the Construct Definition of Positive Deviance. American Behavioral Scientist, 47(6): 828-847.

Sterman, J. D. 2000. Business dynamics: systems thinking and modeling for a complex world. Boston, MA McGraw-Hill

Tugade, M. M., Fredrickson, B. L., & Feldman Barrett, L. 2004. Psychological resilience and positive emotional granularity: Examining the benefits of positive emotions on coping and health. Journal of Personality, 72(6): 1161-1190.

Ulrich, W. 1993. Some difficulties of ecological thinking, considered from a critical systems perspective: A plea for critical holism. Systemic Practice and Action Research, 6(6): 583-611.

Waugh, C. E., & Fredrickson, B. L. 2006. Nice to know you: Positive emotions, self–other overlap, and complex understanding in the formation of a new relationship. The Journal of Positive Psychology, 1(2): 93-106.

Wiley, J., W. 1996. Linking Survey Results to Customer Satisfaction and Business Performance. In A. I. Kraut (Ed.), Organizational surveys : tools for assessment and change: 330-359. San Francisco: Jossey-Bass Publishers.

Wright, T. A., & Cropanzano, R. 2004. The Role of Psychological Well-Being in Job Performance:: A Fresh Look at an Age-Old Quest. Organizational Dynamics, 33(4): 338-351.

Yieh, K., Chiao, Y. C., & Chiu, Y. K. 2007. Understanding the antecedents to customer loyalty by applying structural equation modeling. Total Quality Management & Business Excellence, 18(3-4): 267-284.

CHAPTER 19

Change in Organization – Emerging Situations, Character and Praxis

Tapio Keränen D. Sc. (Tech)

Variantti Ltd/Fortum Power and Heat Oy

ABSTRACT

This study addresses local transformation and organizational character through analyzing events that are related to the development of service business in the context of a large energy company. Drawing on critical realism and assuming organizational change to be a practical and social performance, a mediating entity – organizational character as action dispositions – is postulated to account for relationship between structure, agency, and action.

The organizational character in context consists of prudence, and of the pursuit of technical excellence and pursuit of operational efficiency. Prudence is an action disposition in context when encountering new situations. Organizational character, in the context of service business development, overshadows creative and transformative capacities of human agents but does not prevent local changes to the context. Concerning the development of business to business services a suggestion is that attention must be paid to internalized action dispositions in which the organization, human agents, and social consciousness about context and action, are intertwined.

BACKGROUND AND INTRODUCTION

Throughout their existence many business organizations have been forced to change and renew, creating a paradox between the need for change and the assumption of stability that is tied to the concept of an organization. As an example of organizational renewal is the development of service business in the context of power and heat generating company in Finland. The dominant logic of the organization has been the construction and effective use of durable power generation assets. The period under investigation starts from the end of the construction era during 1980. The number of large construction projects was considerably reduced during 1980s and it became necessary to mobilize the released resources for other tasks: providing services to other industries. In this context the management started to develop service business based on expertise gained during construction era. This attempt to develop service business was a genuine and unique initiative especially because the overall objective was the growth of business through knowledge based services.

The development of service business entailed a considerable change in prevailing ways to think and act. The emphasis when providing services is on people, on their abilities, on the use enabling technologies, and on organizing and on related intangible elements such as organizational identity, culture and character. Furthermore complex and dynamic relations between customers and partners form complex social systems that *"are irreducible to other types of systems because they are made up of human beings who are purposeful and reflexive, and who make history as well are constrained by it"* (Demers 2007: 118).

Organizations embody stability and continuity and exist, through their reproduction, over time. Reproduction means that the consistency of organizational response in various situations results in continuity and also immutability. Immutability can be considered a drawback in a changing environment where emerging situations make new and novel actions necessary.

The immutability and organizational inertia is often explained by referring to organizational culture and identity that are slowly changing intangible structures. Culture entails internal models (e.g. patterns, exemplars) molded during the development of an organization. Culture means regularities in the behavior of members of a society, whereas identity is related to the mission of the organization. Identity is an inter-subjective structure that answers the question *"Who we are?"* (Hämäläinen 2007)

The question is "How does an organization act and behave when something occurs or is anticipated to occur?" An analogy to the character of individuals would be useful (Selznick 1957), but this is not often used in organizational studies. Character means deep and intrinsic aspects of action when the states of affairs are disturbed (Selznick 1957). Culture, identity, and character are three entities in which organization and human agents, and their consciousness about context and situation and how to act, are intertwined.

SURVEY OF PRIOR RESEARCH: THEORETICAL CONSIDERATIONS

The dialectical view on organizations (Benson 1977), and related views (Tsoukas and Chia 2002), emphasize that the social world is in a state of becoming. Critical realism postulates that the world has structure that predates action. Drawing on the assumptions of critical realism (Archer et al 1998, Fairclough 2005, Fleetwood 2005), and on the closely related theory of social becoming (Sztompka 1991), and on practice theories (Reckwitz 2002), I concluded that local transformation can be considered as practical and social performance.

The notion organization is considered as a social structure that is a network of persistent and regular relationships between components of social reality (Sztompka 1991). Each organizational outcome incorporates and implicates traces of its past – immanence (Chia 1999) – which both creates future potentialities and constrains them.

Concerning human agents and actors (Figure 1) they maintain the continuation of operations, and have the capacity to change the states of affairs. Agents are bodies and minds that carry and perform social practices (Reckwitz 2002) that maintain stability and order. The necessary prerequisites of the dynamics of organizations consist of creativity, learning potential, and the need for self-realization (Sztompka 1991). Furthermore, human beings have the capacity to exercise free will within the limits of pre-existing structures: to have done otherwise, to think and act creatively, and to do novel things.

Praxis entails both what is going on in an organization – practices – and what people actually do in emerging situations resulting in either reproduction or in the transformation of the organization. Established productive and social practices and their combinations are related to pre-existing organizational structures containing the traces of the organizational past. A genuine property of organization – character (Selznick 1957) (Birnholtz et al 2007) and internalized practices (Bourdieu 1977, Sewell 1992, Chia and McKay 2007, Jarzabkowski et al 2007) engender action and maintain consistency in emerging and evolving situations.

Based on literature the model of local transformation can be outlined as follows (Figure 1). Organization consists of social structures that have structuring properties; human agents as actors who accomplish social and productive practices; a postulated mediating entity – character – maintaining the consistency of action; the emerging and evolving situations that human agents have to cope with; and of praxis – what people really do in emerging situations.

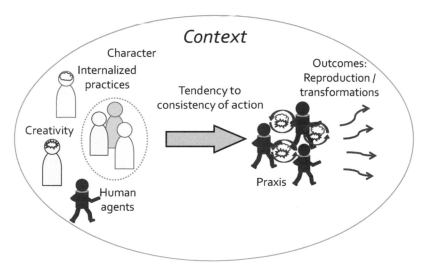

Figure 1 A model of local transformation

RESEARCH DESIGN

Within the organizational context I identified an established business unit that provided various maintenance services for various industries, henceforth referred to as *Alpha*. The origin of this business unit dates to 1988, when the firm's management made a decision to expand its maintenance business by a joint venture with other industries. The origin of the sub unit *Zeta* that is engaged in the operation and maintenance services for power plants is in the 1990s. *Zeta* has experienced first strong growth and later reduction abroad, and there have been ongoing attempts to continue the business abroad.

The research and development unit, referred to as *Gamma* came about during the 1980s and was an attempt to transform a research and testing laboratory into a product development organization. After several development stages, *Gamma* was dismantled, devolved and partly amalgamated into business sub-units by the end of 2001. The focus of the sub unit *Delta* was primarily on the design and construction, and later on the maintenance of power transmission lines, substations, and communication systems. Delta was broken up during strategic reorientations of the broader organizational context during 2001.

The development of *Beta* started to take shape in 1997. The basic idea was to develop a new business model for aggregate services: power, heat and facility management for industrial customers. After several development stages, Beta faded by 2001 without tangible results.

The sub units *Epsilon Alpha* and *Epsilon Gamma*, condition monitoring and condition management of equipments represent attempts to establish new service business activities that are based on the expertise of critical components of production processes. Epsilon Gamma emerged in the 1980s and Epsilon Alpha emerged in the 1990s. They have progressively developed into small businesses of their own.

Methodically I followed the idea of ethnographic interviews, which were repetitive, open and extensive. They provided the researcher with the opportunity to retrospectively document an account of events in an organization (Czarniawska 1998). The supplementary material I have used included annual reports, internal newsletters, internal unpublished memos, and personal notes and documents made by interviewees and obtained from them. I refer the written material henceforth as Text.

RESULTS OF THE RESEARCH

Traces of the past: organizational character

With regard to the traces of past, two consistent, immanent and related aspects emerge from the Text: pursuit of operational efficiency and pursuit of technical excellence. In particular the organization had been a trailblazer during the construction era. Plenty of resources were invested in solving technical problems that were encountered during the construction of power plants and transmission systems. The pursuit of operational efficiency concerns the utilization of production assets aiming to improve the reliability of power plant operations. The focus on the efficient use of human resources was also an embodiment of pursuit of operational efficiency.

The character of the organizational context is recognizably embodied in perceptions of action and the adoption of new issues and change. The capacity of actors to make judgments about the best course of action was characterized by prudence and ensuring, therefore the major dimension of organizational character is depicted using the notion of prudent.

In everyday language, prudence is understood as "the quality or fact of being prudent" (Webster's 1989: 1158) and prudent means ""judicious cautious in practical affairs" and "careful in providing for the future". Prudence was for a long period of time a consistent feature in the organizational context and its relatedness to decision making was prominent. Furthermore the prudence was related to ensuring; a seemingly internalized way of acting when coping with new and emerging situations. The prudence can be interpreted also as dominant organizational logics meaning "the underlying assumptions, deeply held, often unexamined, which form a framework within which reasoning takes place" (Ford and Ford: 1994: 758).

Praxis: what people really are doing

From managerial point of view often the change management seems to be simple. Just follow the Plan Do Check Act cycle. But we see in reality something else when attempting to develop new business.

The elements of the transformational mode of organizational praxis are related to the capacity of people to imagine future development paths (Emirbayer and Mische 1998). The embodiments of creative actions were; envisioning the future rather than doing exact plans, affiliations, experimentation, multiple progressions, and the redirection of ongoing actions and productive practices. Learning, creativity, identifying opportunities, and introducing new ideas, are examples of individual and group initiatives that precipitated the development and provided the impetus to action and interaction

Various contradictions, contingencies, and ruptures were extreme types of emerging situations. The development of the environment, including the development of technology and even an order from customer were emerging situations as well. The deceleration of activity, collapse of venture, segregation of new activity from ongoing operations, and adherence to the original mission resulted in unintended outcomes of the transformational mode of praxis. It was typical during the course of events that the creative action of human agents pushed the events forward.

Contradictions that had an effect on praxis were the discrepancies between ideas and intentions, asymmetry of power of actors and different identities among others. With regard to actions and interactions, there were also diverse concerns in the organization. Such diverse concerns resulted in fluid participations that disintegrated the actions and interactions. Concerning the development in the environment, changes in technology and the actions of competitors also resulted in contradictions that have had an effect on the outcomes of praxis. This mode of praxis can be called a contingent mode, including local improvisations.

SUMMARY AND CONCLUSION

The organizational character – prudence – would suggest that the development of service business would have been carefully planned and implemented. However the emerging situations during development of service business resulted in many unanticipated development paths embodying the creativeness of people. The service organization was "in a state of becoming".

The organizational character is related to immutability or to the resistance to change in organizations having effect of the development of business to business services as well. When people refer to the convention – *"resistance to change"* – they most likely refer to an opposition to change. In the management literature the resistance to change is conventionally treated as a psychological concept and resistance or support to change is seen to be within the individual (Dent and Goldberg 1999).

Organizational character as a genuine property of an organization tends to resist change. But, in fact, this is only one side of the coin. It is important to understand that this kind of resistance to change is immanent in the organization; it is not necessarily an opposition to any plan or intention. Organizational character, identity, and culture, are stabilizing elements in any organizational context. The question is not only about individuals and their personal traits, or psychology, but social practices as well.

REFERENCES

Archer Margaret, Roy Bhaskar, Andrew Collier, Tony Lawson and Alan Norrie (Editors) (1998) Critical realism: essential readings. Routledge.

Benson J. Kenneth (1977) Organizations: a dialectical view. *Administrative Science Quarterly* 22, 1-21.

Birnholtz Jeremy P., Michael D. Cohen and Susannah V. Hoch (2007) Organizational character: on the regeneration of camp poplar grove. *Organization Science*18 (2), 315-332.

Bourdieu Pierre (1977) Outline of a theory of practice. Cambridge University Press.

Chia Robert (1999) A 'rhizomic'model of organizational change and transformation: perspective from a metaphysics of change. *British Journal of Management* 10, 209-227.

Chia Robert and Brad McKay (2007) Post-processual challenges for emerging strategy-as-practice perspective: Discovering strategy in the logic of practice. *Human Relations* 60 (1), 217-242.

Czarniawska Barbara (1998) A narrative approach to organization studies. Sage Publications. United States of America.

Dent Eric B. and Susan Galloway Goldberg (1999) Challenging "resistance to change". *The Journal of Applied Behavioral Science* 35 (1), 25-41.

Demers Christiane (2007) Organizational change theories: a synthesis. Sage Publications.

Emirbayer Mustafa and Ann Mische (1998) What is agency? *The American Journal of Sociology* 103 (4), 962-1023.

Fairclough Norman (2005) Peripheral vision: discourse analysis in organization studies: the case for critical realism. *Organization Studies* 26 (6), 915-939.

Fleetwood Steve (2005) Ontology in organization and management studies: a critical realist perspective. *Organization* 12 (2), 197-222.

Ford Jeffrey D. and Laurie W. Ford (1994) Logics of identity, contradiction and attraction in change. *Academy of Management Review* 19 (4), 756-785.

Jarzabkowski Paula, Julia Balogun and David Seidl (2007) Strategizing: the challenges of a practice perspective. *Human Relations* 60 (1), 5-27.

Hämäläinen Virpi (2007) Struggle over "Who we are" – a discursive perspective on organizational identity change. Helsinki University of Technology. Yliopistopaino Helsinki 2007.

Reckwitz (2002) Toward a theory of social practices: a development in culturist theorizing. *European Journal of Social Theory* 5 (2), 245-265.

Selznick Philip (1957). Leadership in administration: a sociological interpretation. University of California Press.

Sewell William H. Jr (1992) A theory of structure: duality, agency, and transformation. *The American Journal of Sociology* 98 (1), 1-29.

Sztompka Piotr (1991) Society in Action: The theory of social becoming. The University of Chicago Press.

Tsoukas Haridimos and Robert Chia (2002) On organizational becoming: rethinking organizational change. *Organization Science* 13 (5), 567-582.

Webster's (1989) Webster's encyclopedic unabridged dictionary of the English language. Random House.

CHAPTER 20

Including Human Factors in Business Ecosystem Network to Expose & Predict Budgetary Risk

Neil Boyette, Haijing Fang

IBM Research Almaden
San Jose, USA
nboyette@us.ibm.com, hfang@us.ibm.com

ABSTRACT

During a project's budget life cycle there is always the risk of expected income not materializing, or expected expenses differing. Most large organizations have a complexity of departments, where one department consumes another's product, or multiple departments compete with one another by providing a similar product using different technologies. In such an interlocked environment, budgeting risks may propagate across related projects and impact many seemingly unrelated departments. Being able to predict the risks and expose them to all related parties is crucial.

Some risks are truly unpredictable and therefore beyond the scope of our discussion, but some are due to deliberate, and predictable, decisions made by people. Organizations typically use ERP systems to manage their budgets, however these systems are not able to manage this kind of risks. This paper seeks to show an approach to manage such risks by creating a business ecosystem network model to present the complex relationships among an organization from a budget management perspective. By incorporating human factors the network will allow analytics on decision making patterns for key contributors and thus predict budgetary risk propagation on both the individual and aggregate level.

Keywords: business ecosystem, budget allocation optimization, project management

1   INTRODUCTION

When Steve Jobs passed away in October 2011, while the millions of Apple fans around the world mourned their loss, AAPL investors as well as financial analysts worried whether Apple would still be able to keep up the same strong performance as it had during the period when it was under Jobs' management (History of Apple Inc. on Wiki).

In the pyramid of management hierarchy, doubtlessly the CEO is the soul of a company (a great one with vision, such as Jobs, can lead a company from cliff of bankruptcy to the 35th of the Fortune 500 (Fortune 500 Ranking)), on the other hand, every other employee, even the manager or team lead at the bottom tier, plays an important role as well. Individually or collectively, their decisions will affect the whole company's business results.

In large scale organizations, especially globally diverse corporations, there are many departments, each with different responsibilities. Developing a product usually takes multiple departments working together. The product can't (or shouldn't) be released until all its components have been completed. Sometimes the impact of a component is obvious, but other times it is not. For example department A could implement a piece of software multiple ways while providing the same immediate functionality. Suppose approach X takes more effort but enables easier extension for the future reuse, and approach Y, takes less effort but with no potential for reuse. Department B who consumes the software and sponsors department A does not need to reuse it, however, another department, C, may need something similar in the near future. If the technical lead of department A has a long term vision and decides to take approach X, their product will be used by both department B and C and realize more value for the company. In addition, department C will save costs by extending the product department A created for department B instead of creating a new one from scratch. In this case, risks to the budgets of departments A, B and C will depend on the key personnel in departments A and B.

Most ERP systems don't have the capability to manage these types of risks, because they only allow users to track their (biggest) customers on an isolated level, missing the fact that departments are related with each other in terms of budgeting changes. They do not track how different individuals can reach entirely different decisions under the same circumstances and therefore can not track how individual decisions impact a budget (either directly or indirectly), on an individual project, or department basis.

We propose a system to cover this gap and allow users to monitor and manage the budgeting risks caused by human factors.

In the following sections, we will elaborate the challenges of the problem and present how we model the situation so that our system will be able to conquer these challenges. Finally we will state our conclusion and describe the future work.

202

2   CHALLENGES

Since our goal is to provide a system to manage risks of budget allocated for projects within a whole organization, we view the organization from a project's perspective.

"A project in business and science is typically defined as a collaborative enterprise, frequently involving research or design, that is carefully planned to achieve a particular aim." (Project Definition on Oxford English Dictionary) The specific process for projects will differ depending on the organization, type of deliverable and the organization's industry. We will focus on the process of software development as it is well understood, but the model applies to others as well. In the software industry, a project typically is to create a piece of software based on an idea. Medium or large scale software projects usually comprises of many components and sub-functions. For each there is the option to choose an existing implementation from third party, create a new implementation from scratch, or outsource the component to another team. If a component that could be reused is chosen, the investment return can also be reflected in other projects. However, if a component with limited potential or even worse, design flaws, is chosen, not only is the investment wasted, the present project will be jeopardized. Even for a reusable component, there could be different solutions with different level of risks, costs and benefits.

These various ways of completing each project may lead to very different web of interactions between different departments. Various departments can collaborate on a project (i.e. project management, development and test may all come from different departments), or they can complete against one another (i.e. whether to reuse (and potentially modify) an existing component). In some cases, the departments will be well aware of one another, and other times one department may not know who is consuming their product directly or indirectly. The relationships could be short term, for instance a department conducts testing for another, or long term where two departments work together throughout the duration of a project, or even over multiple projects. Therefore it could be very challenging for a department to find out how the risk in another departments will impact itself, e.g. in what degree and for how long.

On an individual level, each person has a different degree of risk tolerance, according to their background, experiences goals, etc. Some people might be motivated by producing the best possible quality, while others just want to get a project done the quickest way possible. Finally, everybody has a different set of relationships with people in other departments. Since every person has so many facets to their personality, it also is very challenging to measure and predict how each affects their decision making, further more, budget risks in an organization.

3   THE MODEL

No matter what specific process a project has, it can always be viewed as a short term business with the owning department as the business owner. Since all the departments belong to the same organization, they share the same ultimate business goal, however they also have their own agendas. They thrive together as part of the organization, but some also compete against each other. The mixed relationships make them an ecosystem of businesses.

During the project lifecycle, the owning department interacts with other departments in different relationships, e.g. an upstream department creates a component that will be used by a downstream department, a testing department conducts testing on the product of a department, etc., however the relationship can always be simplified as a contracted service between a service provider and a service consumer. The provider delivers the service via a project of its own.

Therefore the whole organization can be modeled as a network of departments, each of whom has a role as a service provider, a service consumer, or both, connected with each other via the service (aka project). External customers can be viewed as service consumers and external vendors, service providers as well.

Figure 1. Organization Modeling

For example, as shown in Figure 1, a small organization XYZ has three departments, A, B and C. An external vendor provides a tool to A, A in turn provides a software component to B, who integrates it with its own software and delivers the final product to an external customer. Both A and B have a testing project outsourced to C, who therefore provides testing service to A and B.

In each department there may be a structure of leadership consisting of multiple key personnel, each of whom can make a decision that could impact a budget, either their department's or another department's, such as the manager who has a direct say in how an allocated budget is spent, architects and designers as they influence whether to buy, reuse or create project components. However, it is not rare for one person to wear multiple hats and take multiple responsibilities because he or she has expertise for all those functions. So we can simplify the whole structure as one virtual person. How a random person makes decisions is a process too complicated to be fully understood. But since we are looking at key personnel in a department, who typically have an important role in the department and can be assumed to have a proven capability to analyze and reason, therefore he or she can be considered rational. Therefore we model a department's key personnel as a rational actor.

"A rational actor is an individual with consistent preferences" (Herbert Gintis 2009), and his or her decision making process is "an appropriate set of Bayesian priors concerning the state of nature" (Herbert Gintis 2009, Eduardo Zambrano 2005).

In the organization environment, especially a software organization, an actor needs to make the following types of decisions regarding a project's budget; whether to choose a given vendor or not, whether to use a given sub-component or not, whether to hire a given team for testing or not, etc. Therefore the most important preferences that need to be considered are the actor's risk tolerance, previous experience on intra-team collaboration, technology preferences, current budget, current team's strength, and the future direction of the team. So the decision can be expressed in qualitative terms by:

$$P = f(R, C, T, B, S, F, t)$$

Equation 1. Relationship between Decision and Factors

where R represents level of risk, C collaboration experience, T technology, B budget, S team's strength, F team's future direction, t time.

Equation 1 illustrates the relationship between the decision regarding each of the questions and the preferences that influence the outcome. The result is the probability of an affirmative decision. Since the preferences are not something that can be derived mathematically, we will use a Bayesian method to approximate the relationship. This method will be elaborated in the next section.

4   THE SYSTEM

In this section, we will describe in detail how the model is created; when and how it gets updated; and finally how it can by used by an end user, aka, management, to gain business insights.

4.1   Creating the Model

Creating a model for a given organization can be split into two steps: creating models for the departments, and then creating a graph to represent the organization.

Step 1: Department Models

The most important part of a department is its rational actor. The actor makes decisions based on its preferences, which are a set of factors, as identified in Equation 1. When a person considers a number of "factors" before reaching a conclusion, each factor would have different level of importance to him or her. However, the relationship between the decision and the factors are not mathematically represented, thus we need to find a way to approximate the relationship.

If it is known in what order the factors will be weighed in, decision tree technology can be used to represent it. A decision tree consists of two types of nodes, decision nodes, which test a particular attribute against a given constant to decide which child node to route to, and leaves, which specify the ultimate decision of the tree. So the process of making a decision is to route down the tree from the root to a leave node. Once constructed, each route can be derived into a rule and the whole tree is easily converted into a rule set, which can be understood and validated by users. However it is not definitive how a rational actor prioritizes his preferences, aka, orders the factors. The decision making process in our case is complex, which suggests Bayesian networks will outperform decision trees and are better suited in this situation (Janssens, D 2004). The disadvantage is the rules derived from the network are more complicated and may result in less optimal decision-making.

"A Bayesian network, Bayes network, belief network or directed acyclic graphical model is a probabilistic graphical model that represents a set of random variables and their conditional dependencies via a directed acyclic graph (DAG)." (Bayesian Network Introduction on Wiki) It allows users to conceptualize the relationships between variables and making predictions, and as such is what is needed here. Learning Bayesian network from data falls into two categories: structural learning and parameter learning (Cheng, J. 1997). Structural learning is to determine the link structures, i.e. the dependency relationships between variables, while parameter learning is to determine the prior conditional dependencies of the variables, given the link structures and the data.

Since this is a matured technology, there are many existing implementation available. We are considering to use "Bayesian Network tools in Java (BNJ)" (BNJ Website), developed by Kansas State University KDD Lab.

The entire department will be modeled as a node in a directional graph, with each node containing one rational actor. The actor is modeled as a set of decision rules, which is produced with BNJ.

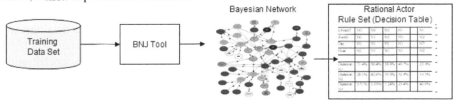

Figure 2. Rational Actor

The department node contains attributes to describe the department's team status, technology strength and history of projects.

Step 2: Organization Graph

In our modeling, an organization consists of the following: departments, external vendors and external customers. Vendors and customers play the same role as service provider and consumer, respectively; however their rational actor may not be visible, therefore need to be represented as an "external node". In most cases an external node is the same as a department node but missing the rational actor inside. Alternatively sometimes a vendor or customer is well enough understood (for example through surveys with those that interact with them) that a rational actor could be constructed, in which case they will be shown as a department node.

The whole organization can be modeled as a directional graph of nodes, where each node is either a department node or an external node and each arc is a project established between two nodes. The data associated with an arc is the detailed information regarding the project, including the budget.

So, for each department node, the incoming arcs describe how many projects it is sponsored to work on, and how much the funding has been, is or will be, received; the outgoing arcs how much the funding has been, is or will be, allocated to spend. Combining both, a user can have a clear view on the department's cash flow, presently or in the near future.

4.2   Preparing Data

Organizations typically do not have a data store describing how well people work together, what motivates a person or what their risk tolerance is, etc. However it is manageable to leverage a variety of existing data to profile a person, and then use this to train the Bayesian networks to model rational actors.

An organization usually archives each employee's resume, which provides detailed information on the person including technologies and skills the person acquired previously as well as the previous projects he or she was involved with. Resumes can be a source of information to initialize a starting profile. During employment, as the person grows and moves forward in his or her career path, the training received, meetings and conferences attended, publications, certificates, and yearly performance reviews can be used to update the person's profile over the time.

These sources of information are usually in free form format. Given the quite maturated Natural Language Processing (NLP) (NLP Definition on Wiki) and Machine Learning (ML) (ML Definition on Wiki) technologies, a parser leveraging existing NLP and ML software, such as Apache OpenNLP (OpenNLP Website) and Apache Mahout (Mahout Website), can be created to convert the raw data into well formatted data set.

Many companies allow employees to recognize each other's contributions through a sort of "thank you" award. This type of recognition also typically shows up in an employee's performance reviews. Such information can reveal the experience on cross team collaboration. An employee's email, instant messaging and phone call history can also be used (with appropriate permission) to tell what connections he or she has, and how established the cross team relationships are.

Using text analytics technology (Text Analytics Introduction on Wiki) this data can be gathered and interpreted to find out the person's communication patterns as well as the contact information on the other parties involved. Furthermore, data mining technologies (Data Mining Introduction on Wiki) can be employed to map the communication patterns into the person's profile to show the collaboration the person has with other specific departments.

Risk tolerance is harder to learn. One of the most straight forward ways to find out a person's risk tolerance is by simply asking them, and those they work with. There are a variety of ways to perform this task including using a survey, observing a project simulation etc. In addition, observing how a person makes decisions in their real-world projects can be used as well. The system could analyze how fast a person makes a project decision; whether they tend to select the option with more or less risk; whether they tend to fund experimental research, or only tried and true implementations.

4.3 Updating the Model

An organization is a dynamic ecosystem.

First of all, projects are just short term effort (not like a program or a routine operation), which starts and finishes within a specific frame of time. A project is typically created based on a requirement from customers. Customers change over time and so do their requirements. So from a department's perspective, projects come and go. Department themselves change as well. Other than reorganizations where departments could be merged, split, canceled or created, people in departments are moving in or out as well. Last but not least, the world outside the organization changes. For example, new technologies are invented, new software and hardware are released, new protocol or standard are published, new analysis on current users, prediction on next generations of the users are made, competitors change strategies, etc. All those changes have direct or indirect impact on the organization and therefore need to be reflected in our system. This means that we need a model controller to manage and maintain the model, and an agent to watch such changes and notify the model controller.

For external changes, the agent subscribes to the news source on the internet and uses text analytics technology to analyze the news. For example news on related technologies or competitors should concern the rational actors. Companies who provides or could provide support could become vendors, and news regarding those concern the model structure. The agent will constantly watch the news source and whenever it detects news of interest, it will generate events based on the characteristics of the news.

For internal changes, such as department structure changes (due to reorganizations), a team acquires a new tool, a key personnel moves leaves the company or transfers from one department to another, etc., the agent will learn from internal news sources and generate corresponding events as well.

There are also changes caused by projects. Over the time a project moves to a different stage in their life cycle. Some changes are planned, and are considered as internal changes. While some other changes are caused by external forces, e.g. when a financial crisis occurs, a company will terminate or halt all projects not critical to the business strategy or with a lower priority, and are considered as external changes. The agent learns about the internal change via the model controller who maintains the model as well as the status on each node and arc. It learns about the external project changes by watching internal sources, such as notifications, announcements, emails, etc.

All together the agent generates the following types of events and each will trigger one or multiple actions:

Table 1 Model Events and Actions

Event Type	Action(s)
Department Change Event	Graph Reconstruction Action
Actor Change Event	Rational Actor Re-profiling Action
Internal Project Change Event	Rational Actor Re-profiling Action, Nodes Re-connecting Action
External Project Change Event	Rational Actor Re-Profiling Action, Nodes Re-connecting Action

The model controller will receive the events and take the corresponding actions, such as re-construct the graph upon receiving the Department Change Event, Internal or External Project Change Events, or re-profile an actor upon Actor Change Event.

The system architecture is shown in Figure 3.

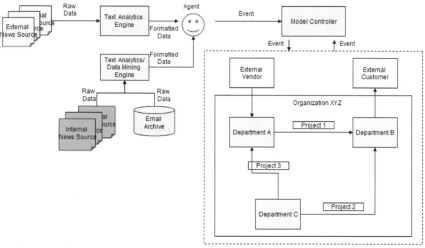

Figure 3. System Architecture

4.4   Applying the Model

Once established the system is capable automatically computing how a given change propagates through the whole ecosystem, identifying all departments involved, and measuring the impact on each of them. These impacts can be explored on a departmental basis, or can be further aggregated.

The system will also be able to expose non-obvious linkages between departments, and thus budget line items, as well as the risks and opportunities associated. This allows management to better prepare for or mitigate risks by for instance applying more oversight to projects with large aggregated impacts.

Finally, the system allows for analytics to gain insight into profitable and unprofitable patterns of decisions by predicting the risks to the budget for various types of (predictable) changes, such as a person changing roles, or moving to a different company. For internal personnel with large impacts, management can focus on retaining them before they decide to leave or transfer, and for external resources with large impacts management can expand extra effort to sustain the relationship. Vice versa, management can determine who it expands a lot of resources on (both time and or money), and evaluate that against the impacts that person has, and adjust accordingly.

5   CONCLUSIONS AND FUTURE WORK

In conclusion we demonstrated a model showing the relationships between departments and the rational actors that representing the key department personnel. Each decision to be made can be expressed as a function of the actor's level of risk, collaboration experience, technology preferences, budget, team's strength, and the team's future direction over time.

We also demonstrated a system built on top of this model including different options for both gathering the initial training data, as well as how the system can react to changes over time.

Finally we showed some of the advantages and insights the system can provide, such as the ability to better understand the complex interactions and inter-dependencies between departments, as well as the effects key decision makers have for directly and indirectly.

Having completed the detailed design, next we will create a prototype to realize the system and collect real data to test the system. We will then use the historical data to validate the results.

REFERENCES

History of Apple Inc. on Wiki http://en.wikipedia.org/wiki/History_of_Apple_Inc
Fortune 500 Ranking, http://money.cnn.com/magazines/fortune/fortune500/2011/snapshots/670.html
Project Definition, Oxford English Dictionary
Herbert Gintis, *The Bounds of Reason*, Princeton University Press March 16, 2009
Eduardo Zambrano,*Testable implications of subjective expected utility theory*, Games and Economic Behavior, 53(2), 2005.
Bayesian Networking Introduction on Wiki, http://en.wikipedia.org/wiki/Bayesian_network
Cheng, J., Bell, D., Liu, W., 1997. *Learning Bayesian networks from data: an efficient approach based on information theory*. In Proceedings of the sixth ACM International Conference on Information and Knowledge Management.
Janssens, D., Wets, G., Brijs, T., Vanhoof, K., Timmermans, H.J.P., Arentze, T.A., 2004. *Improving the Performance of a Multi-Agent Rule-Based Model for Activity Pattern Decisions Using Bayesian Networks.Journal of the Transportation Research board*, also in: Electronic conference proceedings of the 83rd Annual Meeting of the Transportation Research Board (CD-ROM)
BNJ Website, http://bnj.sourceforge.net/

NLP Definition on Wiki, http://en.wikipedia.org/wiki/Natural_language_processing
ML Definition on Wiki, http://en.wikipedia.org/wiki/Machine_learning
Open NLP Website, http://incubator.apache.org/opennlp/
Mahout Website, http://mahout.apache.org/
Text Analytics Introduction Wiki, http://en.wikipedia.org/wiki/Text_analytics
Data Mining Introduction on Wiki, http://en.wikipedia.org/wiki/Data_mining

CHAPTER 21

Computational Modeling of Real-world Services for a Co-creative Society

Takeshi Takenaka, Hitoshi Koshiba, Yoichi Motomura

National Institute for Advanced Industrial Science and Technology (AIST)
Tokyo, JAPAN
takenaka-t@aist.go.jp

ABSTRACT

This paper presents a discussion of strategies for modeling customer behaviors and business conditions using various data acquired through actual services. In services that have customer contact such as retail or restaurant services, the modeling of customer behaviors and customer needs is extremely important to improve actual service systems. However, it is not easy for managers to simulate customer behaviors using various business data because of technological and business constraints. This paper presents a discussion of service engineering technologies that are intended to support business decision making considering the dynamics of environmental factors, customer behaviors, and business decision making.

Keywords: service engineering, customer behavior, modeling, demand forecast

1   INTRODUCTION

While globalization and information networking are rapidly spreading, service industries for daily living such as retail and restaurant industries are confronting severe conditions along with demographic change and economic downturn. For instance, in Japan, the declining population has brought severe competition among domestic companies and has caused inconvenience and anxiety to consumers. In such industries, because many people are involved as customers or employees in the same local community, we must devote attention to the mutual benefit of those stakeholders when we consider sustainable growth of service industries in a co-creative society.

At a restaurant, for example, many and unspecified customers visit irregularly. Some customers will never come again if they do not like the restaurant. Nevertheless, it is difficult to ascertain customer needs or their satisfaction using simple purchase data. Moreover, because customer needs are diverse, merely simulating an average customer is insufficient for the modeling of customer behaviors. Customer behavior and satisfaction are also closely related to service provision processes. Therefore, those should be understood in a broad context including environmental factors, customers' motivations, and business actions such as in-store promotions, stock control, and employee scheduling.

Service engineering is a new and transdisciplinary research field that is intended to enhance service industry productivity and the productivity of service sectors of manufacturing industries (Sakao & Shimomura, 2007, Ueda, Takenaka, Váncza, & Monostori, 2009, Takenaka et al., 2010). Our research project on service engineering was started in 2008 with the support of the Japanese Ministry of Economy, Trade and Industry (METI). This project specifically examines human behaviors and values relevant to services, and develops technologies that can help customers, employees, and managers. Figure 1 presents our technological development strategy, which specifically examines human factors of services.

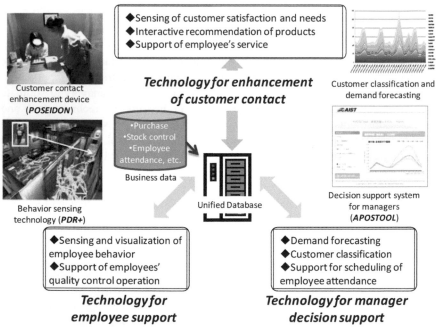

Figure 1   Service engineering technology development strategy: customer contact enhancement technology, employee support technology, manager decision-support technology.

Those technologies fall into three groups: technology for enhancement of customer contact, technology for employee support, and technology for manager decision support.

For those technological developments, we emphasize modeling of human activities. Regarding customer contact enhancement technology, we have developed an interactive device using a tablet computer, named the Point of Service Interaction Design Organizer (POSEIDON), which acquires customers' needs and satisfaction, and recommends suitable products through natural interaction. Regarding employee support technology, Kurata and his team developed sensing and visualization technologies for human behaviors, especially for use in labor-intensive service industries such as restaurants or nursing care services (Kurata et al., 2011). Those sensing and visualization technologies are also used for employees' voluntary quality control activities. For manager decision support technology, we have developed visualization methods for simulated customer behaviors and business conditions for managers at various levels (Koshiba, Takenaka & Motomura, 2011).

In this paper, data acquisition during customer contact and modeling methodologies of customer behaviors and business environment are discussed.

2   DATA ACQUISITION DURING CUSTOMER CONTACT

During customer contact in restaurants, menus and in-store promotion are crucial for a customer's choice of products. Many customers, especially new customers, choose products based on their expectations rather than their experiences. Therefore, even a best-selling product is not necessarily the most satisfying product. Moreover, for the modeling of customer needs and behaviors, we must consider the diversity of customer preferences and reasons underlying customer satisfaction.

POSEIDON, an interactive device using a tablet computer for customer contact, recommends products and elicits data related to customer needs and satisfaction through natural interaction. Figure 2 portrays screenshots of this device. The left one shows promotion of some products. The right one shows a customer satisfaction rating for a product after eating.

Figure 2 POSEIDON screenshots: promotion of products (left), customer satisfaction rating to each product after eating (right).

Figure 3 shows customer rating results for some products of a Japanese restaurant chain (Ganko Food Service Co. Ltd.). The products are best-selling products of the restaurant chain according to comprehensive purchase data. However customers' subjective ratings vary according to the product type. For example, product A is a dish that includes tofu, and over 50% people have awarded it a "Gold Prize". However, further investigation shows that satisfaction might differ between repeat customers and new customers. Figure 4 shows the difference of rating for the product. This product is more appreciated by repeat customers than by new customers. Moreover, other data acquired using POSEIDON show that repeat customers choose this product more often than new customers. Therefore, this product probably has the potential to enhance customer satisfaction.

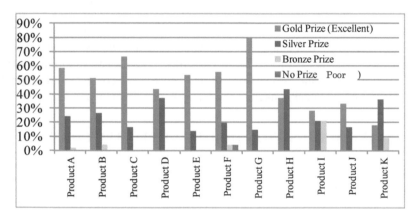

Figure 3 Sample of customer rating for products that are hot-selling products in the restaurant chain.

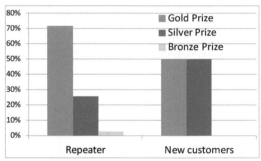

Figure 4 Results of customer ratings for a product, comparing repeaters and new customers.

Although POSEIDON data can provide rich information related to customer needs and satisfaction, applying POSEIDON to many customers in a restaurant is difficult because it requires additional work for employees. Actually, it was difficult to acquire hundreds of data in a few restaurants during a month because some customers did not like using this device.

Therefore we think it is important to combine such high-quality but limited data and quantitative data such as purchase data for the modeling of customer behaviors. In the next section, we briefly introduce a research example of simulation results about customer behaviors using aggregate purchase data and causal data.

3   MODELING OF CUSTOMER BEHAVIOR BASED ON AGGREGATE DATA AND CAUSAL DATA

The authors have developed technologies for modeling customer behavior using purchase data with or without identification in service industries. First, we developed a new category mining method that classifies items and customers automatically into some latent classes using probabilistic latent semantic indexing (PLSI) (Ishigaki, Takenaka & Motomura, 2010a, 2010b, 2010c). We also developed a demand forecasting method using aggregate purchase data and environmental factors as causal data (Takenaka, Ishigaki, Motomura, 2011; Takenaka et al. 2011). Using this method, we specifically investigated common factors underlying customer behaviors as causal data. We analyzed the effects of some factors on sales or number of customers (customer count) using actual restaurant and supermarket purchase data for 2–3 years. Our method estimates sales or quantities of customers using more than 50 parameters with a multiple regression model including stepwise selection of parameters.

This paper introduces a simulation result of customer behaviors in a restaurant considering not only the total number of customers but also the number of customers classified into several categories using our category mining method. In this case, we categorized all customers of this restaurant chain into 19 categories using the category mining method with two years of receipt data as a dataset. Details of this model will be provided in another paper by the authors presented at this same conference (Koshiba, Takenaka & Motomura, 2012). Figure 5 shows estimation results of the number of customers for a restaurant in Osaka owned by Ganko Food Service Co. Ltd.

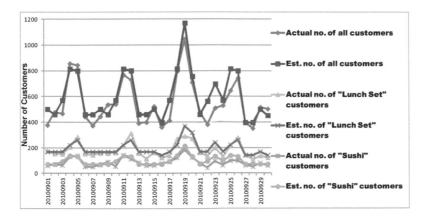

Figure 5 Sample of customer product ratings. The products are best-selling products in the restaurant chain.

We applied our demand forecasting model to all restaurant customers and also to the customers classified into those 19 customer categories. Figure 5 shows the actual and estimated numbers of customers in the two categories: "Sushi" category customers who come to the restaurant at night without reservations mainly order from the sushi menu with or without ordering beverages; "Lunch set" category customers prefer lunch sets in a lower price range.

For estimation of the number of customers, we used one year's purchase data (September 2009 – August 2010) and estimated customer counts of the succeeding month (September 2010). The coefficient of determination (r^2) is 0.84. We calculated this model's accuracy with actual numbers using the following formula (1).

$$p = \frac{a - |a - e|}{a} \tag{1}$$

The accuracy of the model p signifies the proportion of estimated customer count e to the actual customer count a. The average accuracy of this model for all customers of the same period (30 days) is 87.8%. However, the respective accuracies of this model for "Sushi" and "Lunch set" customers are 86.6% and 79.7%.

Table 1 presents selected parameters and their multiple regression coefficients for all customers. It shows the overall trend of this restaurant: many customers come to this restaurant especially on weekends and holidays and the number of customers is strongly influenced by seasonal and local events. This phenomenon might be related to the fact that the restaurant is located in downtown Osaka.

Table 1 Selected parameters and their multiple regression coefficients (all customers)

Parameter	Coefficient
Constant term	459
Saturday	356
Sunday	341
Sunday in consecutive holidays	711
Holiday in consecutive holidays	632
End-year party season	147
Holiday	299
Max. temp. below 11°C	125
Max. temp. 11–16°C	128
Jan. second	706
Friday	113
Jan. third	513
End of the year (last 3 days)	253
Rain over 10 mm	-62
Max. temp. 17–21°C	55
Bon (around mid-August)	195
Coming-of-Age Day	323
New year party Season	118
Wednesday	42
Day before holiday	104
Christmas Eve	-205

However, the model for "Lunch set" category customers seems to have a slightly different tendency, as shown in Table 2. Customers of this category are apparently less influenced by weekdays and seasons than all customers.

Table 2   Selected parameters and their multiple regression coefficients ("Lunch set" category customers).

Parameters	Coefficient
Constant term	169
Holiday in consecutive holidays	277
Sunday	94
Sunday in consecutive holidays	201
Last day of holidays	52
Saturday	52
Bon (around mid-August)	142
End of the year	106
Jan. second	183
Rain over 10 mm	-27
Holiday	98
New year party season	43
Welcome party season (weekday of April)	-25
Jan. third	106
Jan. first	-117
Max. temp. 11–16°C	16
Max. temp. below 11°C	12

Through those analyses presented above, we constructed computational customer models for each customer category. We confirmed that such models can also be used for purchasing or stock control through the interviews of managers. Because our research goal is to support actual stakeholders in service industries, we infer that customer models should be used for their daily business actions. The next section introduces a Bayesian network model using various business data and causal data.

4   MODELING OF BUSINESS CONDITIONS

A salient problem confronted by managers in service industries is making decisions about business actions based on various objective business data. However, it is not easy for them because environmental factors, customer behaviors, and management indicators are closely related. Moreover, those relations change dynamically. Consequently, actual managers' decision making often relies on their experience and intuition. Therefore we started with visualization of those relations and underlying business context.

Figure 6 presents a Bayesian network model that shows probabilistic relations among causal data, business decision making, customer behavior, and management indicators. For this figure, we used 6 months of causal data, purchase data (POS data), employee attendance data, and purchase data. All data were standardized and categorized as "-1, 0, 1", considering the distribution of values during the same period. In this figure, arrows show the conditional dependencies from parent variables to children variables. Those causal relations among parent and child variables were assigned in advance according to actual business operations.

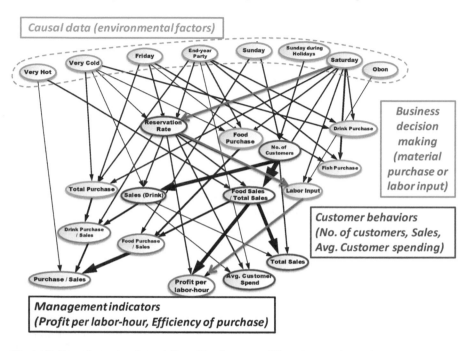

Figure 6 Bayesian network modeling of business conditions for a restaurant.

Taking red arrow connections in Figure 6 as an example, a manager can acquire knowledge that is useful for decision making related to work scheduling of employees as the following.

"In this restaurant, on a Saturday with fewer reservations, the manager tends to decrease labor inputs. However, on 57% of such Saturdays, restaurants can be crowded with customers who have no reservations. Consequently, although the gross profit per labor-hour of the day might well increase, both customer and employee satisfaction would be reduced because the restaurant is too busy."

The authors hope that knowledge of this kind will help support managers' decision making. For this purpose, we should customize our technologies for managers' needs of various types. We have developed a decision support system named POSEIDON. Another paper by the authors will introduce the system and the overall system structure.

5 CONCLUSIONS

This paper presents a discussion of computational modeling of customer behaviors and business conditions for service industries. It introduces a data acquisition method at the point of customer contact of restaurants. Subsequently, it introduces research examples of customer modeling and visualization of business dynamics through integration of various data acquired through actual services. Future work will explore technological developments for various stakeholders in services, particularly addressing their decision-making processes and co-creation of new services through mutual interaction.

ACKNOWLEDGMENTS

This study was supported as a project of service engineering from the Japanese Ministry of Economy, Trade and Industry (2009–2012). We appreciate Ganko Food Service Co. Ltd., which kindly provided fields for research and their business data.

REFERENCES

Ishigaki, T., Takenaka, T. and Motomura, Y. 2010. Customer-item category based knowledge discovery support system and its application to department store service. *Proc. IEEE Asia–Pacific Services Computing Conference*, 371–378.

Ishigaki, T., Takenaka, T. and Motomura, Y. 2010. Computational customer behavior modeling for knowledge management with an automatic categorization using retail service's datasets. *Proceedings of the ICEBE 2010*, 528–533.

Ishigaki, T., Takenaka, T. and Motomura, Y. 2010. Category mining by heterogeneous data fusion using pdlsi model in a retail service. *Proc. IEEE International Conference on Data Mining*, 857–862.

Koshiba, H., Takenaka, T and Motomura, Y. 2012. Service management system based on computational customer models using large-scale log data of chain stores. *Proc. of Fourth International Conference on Applied Human Factors and Ergonomics* 2012.

Kurata, T, et al. 2011, Indoor-Outdoor Navigation System for Visually-Impaired Pedestrians, Proc. of ISWC2011, 123-124.

Sakao T. and Shimomura Y. 2007. Service Engineering: a novel engineering discipline for producers to increase value combining service and product. *Journal of Cleaner Production*, 15: 590–604.

Takenaka, T., Ishigaki, T., and Motomura. T. 2011. Practical and interactive demand forecasting method for retail and restaurant services. *Proc. of International Conference on Advances in Production Management Systems (APMS)*.

Takenaka, T., Shimmura, T., Ishigaki, T., Motomura, Y., and Ohura, S. 2011. Process management in restaurant service – A case study of Japanese restaurant chain –, *Proc. of International Symposium on Scheduling*, 191–194.

Takenaka, T. et al. 2010. Transdisciplinary approach to service design based on consumer's value and decision making. *International Journal of Organizational and Collective Intelligence*, 1(1): 58–75.

Ueda, K., T. Takenaka, J. Váncza, and L. Monostori. 2009. Value creation and decision-making in sustainable society. *CIRP Annals – Manufacturing Technology*, 58(2): 681–700.

CHAPTER 22

Case Study Analysis of Higher Education and Industry Partnership Service Delivery

Doreen Nielsen, Anthony White*

The University of Greenwich
The Business School - QA306
Maritime Greenwich Campus, 30 Park Row
London SE10 9LS, United Kingdom
d.nielsen@gre.ac.uk

[ex] Middlesex University*
London, United Kingdom
anthony757white@btinternet.com

ABSTRACT

In the past, Higher Education Institutions (HEIs) provision specific mandates on delivering graduates with specific skills in general terms. However, the demands from the competitive environment in education and the graduates' future employers drive HEIs to redefine their strategies, institutional mandates and the redesign of curriculum delivery incorporating the use of technology simulations to deliver an enhanced level of service. This in turn redefines the HEI student-customers' experience and passage of rites.

The objective of this analysis explores the HEI and industry partnership service delivery using a longitudinal case study. For the students from the Business School (University of Greenwich) for example, HEIs and industry partnerships provide new and defining avenues of learning experience through strategic, innovative and creative use of simulations. Additionally, this also creates a tighter inter-organisational relationship that will facilitate and improve employability for the graduates.

Major changes such as these meant that the HEI lecturers (in this instance, as primary service deliverers) need to be system-aware and have the relevant IT skills to give the necessary support to student-customers. A number of key issues can and does arise from unclear agreements between the partners and may have detrimental impacts in the form of service gaps on student-customers' learning experiences.

There is evidence after several iterative runs of the curriculum delivery using this form of partnership that the overall benefits lead the way to more of such implementations. Lessons drawn from the implementation process, service gaps, collective feedback from staff observations and different student cohorts in the longitudinal case study provide the input for the iterative redesign, refinement cycle.

This new learning experience transformed many of the student-customers profiles from passive, rote learning or partially participating to active, collaborative learning students. Various other key dimensions such as the lecturer and student relationships, the learning engagement (resulting from the changes of control and dominance when using the externally implemented IT systems for learning) in the formative learning process, the extended social networking, and group collaborative learning provide further valuable lessons.

The future potentials of HEIs and industry partnerships curriculum delivery using technology simulations promise progressive sustenance in student-customers' learning engagement, thus improving learning. Positive findings and popularity of this elective curriculum prompted management mandate to raise the curriculum profile and extend its delivery from a single semester (half year run) to a full year, more intensive curriculum run.

Keywords: HEI Industry partnerships, service gaps, curriculum delivery, technology simulations, extended social networking, collaborative learning, student engagement, employability

1   INTRODUCTION

Past practices of Higher Education Institutions (HEIs) such as the provisions of mandates that govern and guide standard education delivery in general terms produced graduate cohorts with specific skills defined in general terms. A number of changes such as the continuous and progressive cuts in government funding and support (Bolton, 2012; Willets, 2010), the new demands from the graduates' perceived expectations and those of the future employers (Arthur and Little, 2010) influenced changes to these mandates. HEIs responded to these changes and demands by redefining education delivery as a service delivery, and therefore creating a competitive environment in education amongst the HEIs.

Dialogues by invitations between HEIs and the graduates' future employers provides the impetus for HEIs to redefine their institutional (internal university and school level) strategies, and institutional mandates. Noticeable changing trends and attitudes towards the use of technology in the social and work context, coupled with the reducing costs of technology, its ease of access and its availability allows the incorporation and use of technology in teaching and learning. This results in the modernisation and the redesign of graduate programmes and curriculum delivery to enhance the level of service delivery that fits students and future employers' needs better. It is expected that an enhanced level of service for the HEI student-customers means an enhanced higher education experience and passage of rites which could culminate in better employability (Arthur and Little, 2010).

2 THE PROJECT

A new third year undergraduate elective course, E-Business, developed for students from the Business School (University of Greenwich) started its run for academic year 2009-10. It continued to run in 2010-11 and is now running in the third academic year 2011-12. An agreement between the Business School and an industry partner, (Salesforce.com, (2012) also referred to as the software company,) allowed the students who signed up for this course to use the trial copy of the CRM software from Salesforce.com as part of their assessments. It was important to select and match online activities with tools that are fit for purpose and meet the learning outcomes for the particular course (Swan, Shen and Hiltz, 2006). Additionally, the students must also submit a reflective report critically assessing the software, their personal experience in using the software, its quality and limitations, functional purposes and innovative use in an "imagined" business environment.

2.1 Scope

The scope of this longitudinal project explored the HEI and industry partnership service delivery, the student-customers' learning experience and potential service gaps. Additional key areas of analysis include:
- What were the student-customers' perceptions of the enhanced level of service in the HEI?
- How did the enhanced higher education experience affect the student-customers' academic performance, their understanding of the course concepts, applications of their knowledge and deep learning?
- Did an enhanced level of education service mean an enhanced level of learning?

2.2 Methodology

The newly developed course incorporated the use of technology to enhance student-customers' learning experience so the criteria for evaluations looked at the students' learning processes, the results of their academic performance, their reflective reports and the potential service gaps in the HEI and industry partnership service delivery.

The students attended lectures for knowledge building and supervised laboratory sessions, working in groups of 3-5, for using the trial software to populate the database to simulate a fictitious, functional business environment. The student groups collaboratively created and defined their own fictitious E-Organisation and applied the relevant strategies, operations, controls and management processes. By going through this experience, the students explored the potentials of the trial software and followed this up with the mandatory submission of an individually written reflective report. The collated student-customers' assessments' results were included for a high level analysis of this case study.

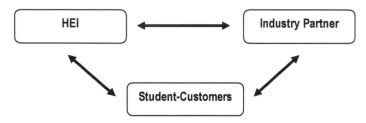

Figure 1   Illustration of direct, open communication channels between all participating parties. (Source: Author's, 2012.)

One limitation of this study was that the HEI and industry partner did not formalise the partnership with an agreement defining the service delivery expectations and fulfillment. It was loosely based on goodwill from the industry partner. During the first year run of the course, the industry partner offered two guest lectures and the running of two supervised laboratory sessions to students with three of their software trainers in attendance. Another limitation was the approach of using the student-customers' assessments' results at a high level analysis. It did not take into account the specific measurements of the students' ICT/technical abilities or aptitudes, their academic backgrounds and academic performances. In order to moderate the limitations, provisions of additional training materials and guidelines were provided by the industry partner, communication lines were open to all parties (Figure 1) and the students' performance results were measured by student cohorts by academic year.

2.3   Conceptual Framework for this Project

The cycle run for each academic year for the University of Greenwich comprises of three semesters. The first semester usually starts around end September and ends just before Christmas. The second and third semesters combined to form the second half of the academic year. The second semester usually starts in January after the New Year holiday break and ends just before the three weeks long Easter holidays. On returning from the holidays, the third semester runs until the first week of May when students have their exams.

The elective E-Business course only ran during the second semester each year. Since the licensing of the trial software carried a 30-day expiry date, during the 2009-10 first year project run, the industry partner organised a special extended expiry date via controlled password access managed by a specially dedicated staff member from their office. The academic staff teaching this course also attended a training session prior to the project run to ensure a relevant and reasonable level of student support at the HEI. The student-customers followed the HEI provisioned teaching plan for the course that was designed to guide them in their specific tasks. During the second semester, the student groups worked on the software and then captured evidence and documentation for a group submission, followed two weeks later with an individually written reflective report. Student-customers received feedback during the third semester. Any operational service gaps that arose during the course run were addressed jointly following consultations between the HEI and industry partner.

Student-customers' feedback, experience, lessons learnt and continuous review during the course run formed the collective motivation and service gap factors for revisions and adjustments to the course for the next run in the following academic year. This framework highlights the fact that in the HEI setting, the cycle of change and enhancements or improvements take longer than other service practitioners in industry. Additionally, the longitudinal nature of study permitted the need and opportunity for continuous cycle of improvement and refinement.

2.4   Findings and Discussions

This section examines contrasts and compares by discussions, the treatment and analysis of the evidence gathered over the academic years 2009-10 and 2010-11 for the elective E-Business course. The following are the topical investigations:

1)   HEI and industry partnership service delivery:-
The loosely organised informal agreement between HEI and industry partner depended largely on goodwill. This resulted in some disappointments for all participating parties at different stages of the project when a higher demand for technical support arose. There was a lack of clarity for the student-customers who needed to understand what the processes were, when something had to be done and by whom. Those student-customers, who lacked the relevant ICT/technical abilities or aptitudes and had a higher dependency on IT support, experienced some levels of frustration through the perceived lack of support. It must be noted that access to trial software requiring IT administration could not be managed by any staff from the HEI. This reaffirmed the need to establish more concrete agreements in partnerships that should include the 5-Stage process (Salmon, 2002) for a clear support for student-customers (Figure 2).

2)   The student-customers' learning experience:-
In the beginning, the student-customers did not comprehend the purpose and value of the use of technology simulations to deliver an enhanced level of education service even though the course specifications clearly explained it. So what were the student-customers' perceptions of the enhanced level of service in the HEI by using technology simulations? During the early stage of learning to use the trial software, the student-customers' main focus was on how to use the software, navigating it and how to populate the database. They perceived the group task as an unnecessary

chore and failed to develop a holistic view of the learning environment and process. A few of the student-customers thought it might be more useful for them to learn how to develop a website without realising the enormity of the task involved! These over-confident student-customers did not fully understand the purpose of using and evaluating a specific part of the online software. They also did not have the relevant programming skills background or aptitudes. Through consistent negotiations and dialogues, they subsequently developed a better understanding, with deeper thinking of the full spectrum of tasks involved with setting up a new online business.

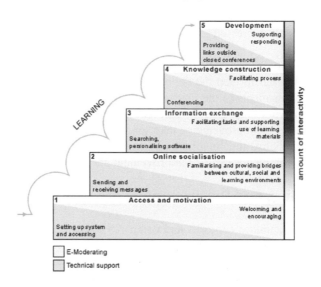

Figure 2   Illustration of the 5-Stage process; Stage 1 - Access & Motivation; Stage 2 – Socialisation; Stage 3 - Information Exchange; Stage 4 - Knowledge Construction; Stage 5 - Development (From Salmon, 2002; http://www.atimod.com/e-tivities/5stage.shtml)

To further understand how the enhanced higher education experience affect the student-customers' academic performance, their understanding of the course concepts, applications of their knowledge and deep learning, a high level comparison was made on the academic performance for the cohorts from academic years 2009-10 to 2010-11. In academic year 2009-10, 55 student-customers signed up for the elective E-Business course while in academic year 2010-11, 61 students did so (Figure 3). Only one student failed in each of the academic years due to partial submissions (the failed student only submitted one piece of work for assessment instead of the required two). It was generally expected that a small percentage of students might fail to deliver full assessments due to personal reasons.

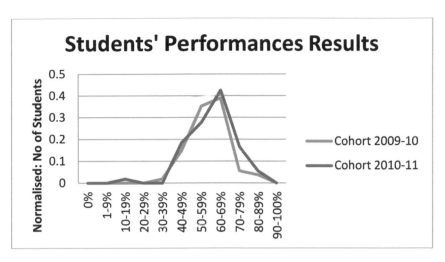

Figure 3 Comparison chart showing the distribution of grades achieved by both cohorts from 2009-10 and 2010-11. (Source: Author's, 2011)

The purpose of the assessments for this course was to test the third year student-customers' ability to reflect (as in deep thinking) individually, and to constructively and objectively work as a group in the technically simulated environment as well as one would expect in a real life environment. They should ideally be able to identify and define the details of a business goal or plan, possess the skills to plan their project, meet the deadlines, organise and negotiate terms and conditions in the business setting, able to intuitively learn and question the functionality and purpose of a given software tool, able to manage and decide like a manager on key administrative functions. Based on Figure 3, the 2010-11 cohorts perform better with a larger number of students scoring higher grades. The learning environment and preparedness of the teaching staff was better in that year after the learning experience in 2009-10. However, it must be noted that during 2010-11, the industry partner played no part in any direct or indirect student-customer contacts. While the first time experience was novelty for the industry partner, the organisation could not financially justify sending staff to the HEI to support student-customers. It was agreed that all the materials provided in 2009-10 could be reused by the HEI. The software available to the HEI for assessment purposes was the standard 30-day trial software available to all industry level customers. This did not require any special logins, registrations or contacts with any staff from the industry partner.

The HEA's (2012) "widely accepted definition of employability is a set of achievements, - skills, understandings and personal attributes – that make graduates more likely to gain employment and be successful in their chosen occupations, which benefits themselves, the workforce, the community and the economy". In this sense, the student-customers could claim relevant knowledge in the software and related skills and thus improve their employability (HEA, 2012). There was evidence from the first cohort of students (2009-10) where one of the students was offered a career defining and designing an online website for an organisation in industry. She was reportedly selected by the interviewer based on her having undertaken this elective E-Business course and was asked to start work in May 2010 ahead of the exams. Students want to be employable.

The external examiner for this course, who had assessed such similar courses in other HEIs in the UK, commented that the design and innovative use of technology simulations makes it useful for student-customers. There were only two elective course that student-customers' can choose from. In recognition of the value of this course building towards their employability, student-customer numbers have progressively increased every year. The current academic year 2011-12 (in its third run) accounted for 72 students registered for this course.

3)   ***Potential service gaps:-***

The use of the trial software from Salesforce.com (2002) which was external to the HEI IT systems setup and accessible via the internet, raised some difficulties that were not normally found in IT environments that were connected within the HEI's network. Some of the difficulties experienced by the student-customers and translated as potential service gaps, a few of which could be addressed were:

- Every time student-customers login to the HEI's IT network, the IT network systems generate a set of logical roaming IP address. When the student-customer login to the Salesforce.com's software site, it verified both the account number (generated by Salesforce.com) and the student-customers' logical roaming IP address together. Although this problem was unique in this instance, the inconveniences become potential service gaps. The HEI Technical Support team would not provide any IT supports because the choice to use this software was not made at the HEI level and they did not have ownership of it.
- The trial software comprised of limited tool functions and limited expiry dates so student-customers could not claim that they had fully used and experienced it. The 30-day trial created an intense formative learning environment.
- Continuous accessibility to the Salesforce.com website was fully dependent on the good external support. Student-customers would be rendered helpless if there were any system failures because this could result in a service gap.
- The student-customers' disparate ICT skills background meant that the support needs of the students had to be addressed in the beginning. Very often, ICT skills test was not a mandatory exercise because of the general wrong assumption that each subsequent cohort of student-customers was better skilled.
- The wide ranging cultures of student-customers meant that special support had to be organised to help students who need them.
- HEI lecturers facilitating the course acted as primary service support and needed to have the relevant ICT/IT skills to be able to support student-customers. This must be recognised as an important factor in influencing the success levels of the course run.

- Unclear agreements between the partners may result in detrimental impacts on the running of the course and learning process. These must be addressed with utmost urgency to minimise negative impacts.

Feedback from past cohorts of student-customers reaffirmed that the enhanced level of education service meant an enhanced level of learning. Learning engagement and interactions with the systems improved. When compared with student-customers attending other similar elective courses, those who attended this course discussed and negotiated more than they normally would do. The laboratory sessions were also more intense and focused with student-customers showing a higher level of urgency. The availability of a ready range of multimedia support like videos, online documents and web links in the trial site provided a very rich collection of online support materials for the student-customers. The extended online social networking and group collaborative learning provide more positive experience to student-customers.

3   CONCLUSIONS

This case study concluded that HEIs and industry partnerships could provide good service delivery in terms of learning experience for student-customers. Innovative and creative use of technology simulations built into course designs created a tighter inter-organisational relationship that would facilitate and improve employability for the graduates (HEA, 2012). Evidences gathered from the two previous years proved that the course in its present day is still popular, applicable and practical.

Iterative cycles of collective feedback from various participating parties and staff observations helped to redesign, refine and improve education service delivery for the following year. Various lessons drawn from the cycle run and any noticeable service gaps provide extended feedback.

Many student-customers' profiles changed from the shy, passive, rote-learner, or perhaps the partially participating to active, collaborative learning students. Other key dimensions like those of the lecturer and student relationships, and the learning engagement evolved in the form of changes of control and dominance. Salmon's (2002) 5-Stage model (Figure 2) and e-tivities which promoted social constructivism provided the framework for managing IT support to student-customers.

The promising findings of this case study encouraged more of such HEIs and industry partnerships that promise progressive sustenance in the increasingly competitive High Education industry.

ACKNOWLEDGMENTS

The authors would like to acknowledge the support of Mr. Chris Rauch (Director International Training and Certification, Salesforce.com(UK)) and his Trainer Teams during the trials in the academic year 2009-10 which completed successfully. Salesforce.com and the University of Greenwich: Business School (also thanks to Mr. Robert Mayor, Associate Dean/Director of Resources) jointly offered Certification scholarships to the best performing group of students reading this course. The University of Greenwich: Business School solely supported the offer of 2 scholarships and graduation day special mention for the best performing student in the academic year 2010-11.

REFERENCES

Arthur, L., Little, B., 2010, "The REFLEX study: exploring graduates' views on the relationship between higher education and employment - Higher Education and Society: a research report (March 2010)." Accessed February 29, 2012, http://www.open.ac.uk/cheri/documents/HigherEducationandSociety.pdf

Bolton, P., Changes to higher education funding and student support from 2012/13 - Commons Library Standard Note, 06 February 2012, Standard notes SN05753. Accessed February 29, 2012, http://www.parliament.uk/briefing-papers/SN05753

The Higher Education Academy (HEA), 2012, "Employability", Accessed February 29, 2012, http://www.heacademy.ac.uk/employability

Salesforce.com, © Copyright 2000-2012 salesforce.com, inc., Accessed February 29, 2012, http://www.salesforce.com/uk/?ir=1 .

Salmon, G., 2002, E-tivities: The Key to Active Online Learning, Routledge; Also http://www.atimod.com/e-tivities/5stage.shtml .

Swan, K., Shen, J., Hiltz, S.R., (2006) Assessment and Collaboration in Online Learning, Journal of Asynchronous Learning Networks, Vol. 10 Issue 1 Feb, http://www.sloan-c.org/publications/jaln/v10n1/v10n1_5swan_member.asp .

Willets, D., 2010, "Changes in funding for higher education", Accessed February 29, 2012, http://www.sconul.ac.uk/news/fundinghe

CHAPTER 23

Co-Producing Value through Public Transit Information Services

Aaron Steinfeld, Shree Lakshmi Rao, Allison Tran,
John Zimmerman, Anthony Tomasic

Carnegie Mellon University
Pittsburgh, PA USA
steinfeld@cmu.edu

ABSTRACT

Public transit riders are often unable to obtain timely information about transit service. At the same time, the complexity and scale of public transit systems makes it difficult for service providers to capture and manage rider concerns. To this end, we designed a system to foster a greater sense of community between riders and transit service providers. This system, named Tiramisu (www.tiramisutransit.com), focuses on crowdsourcing information about bus location and bus load, predicting the arrival time of buses, and providing a convenient platform for reporting problems and positive experiences within the transit system. Tiramisu is also designed to support the needs of riders with disabilities through the application of universal design principles. This system allows co-production of value from all users, including people who frequently encounter accessibility barriers with social computing and interaction techniques. This paper discusses co-production strategies used by the team and describes data contributed by the community.

Keywords: co-production, public transit, universal design, mobile systems

1   INTRODUCTION

We are interested in how to apply universal design, service design theory, and crowdsourcing technology to improve public services. Our intention is to help citizens who receive public services become active collaborators with transit agencies, thereby improving dialog and increasing overall service quality. Unfortunately, our early work suggests that citizens who receive public services, specifically transit riders, do not feel like co-owners (Yoo, et al. 2010). This work led us to focus on crowdsourcing bus location data in order to provide a platform from which a richer dialog could occur between riders and service providers (Steinfeld, et al. 2011; Zimmerman, et al. 2011). Tiramisu utilizes the location and communication services of devices commuters already carry, because commercial real-time bus arrival systems can cost tens of millions of dollars for medium sized cities. We hypothesized that engaging the users in the co-production of real-time information would also reduce the perceived cost of providing service quality feedback.

The general problem all riders face, especially people with disabilities, is uncertainty and anxiety around transit service delivery. For example, when riders arrive at a stop near a scheduled time, they have no way of knowing if they have just missed their desired bus, if the bus is seriously delayed, or if some unexpected event has temporary cancelled the service at this location. Riders feel "out of control." Under the status quo, drivers are the most natural point of communication for riders. However, drivers have many other concerns they might deem more critical to their job, such as safety and their own on-time performance. Furthermore, drivers lack safe and effective methods for conveying rider feedback into the system. Riders may also be unwilling to provide direct feedback to the driver on the driver's own actions.

There are limited options available to riders for mobile feedback about service quality issues. Riders can call their local transit agency, fill out web forms, and, in some instances, send a Twitter message. Unfortunately, these approaches often lack important scenario details needed for service actions. Similar systems in the public works realm (e.g., SeeClickFix; ParkScan.org; King and Brown, 2007; etc) suggest that crowdsourced smart phone methods can work. The use of onboard smart phone features, like GPS and cameras, can automate collection of key information.

Public services are an unusual case for service design, as many public services are monopolies that lack competition to drive service innovation. Researchers have speculated that co-design, where customers materially participate in the design of a service, could be the best path to public service innovation (Boyne, 2003).

2   THE TIRAMISU SYSTEM

In this section we describe the user interaction aspects of Tiramisu. Additional detail on the design process, system architecture, and rationale for various features can be found in earlier papers (Steinfeld, et al. 2010; Yoo, et al. 2010; Steinfeld, et al. 2011; and Zimmerman, et al. 2011).

Tiramisu allows users to find nearby bus stops through a map or list, as shown in Figure 1 (A, Main Map). The system provides predictions of arrival times based on scheduled, historical and real-time location data and ratings for bus load ("No

(A) Main map

(B) Select route

(C) Before recording

(D) During recording

(E) Agency report categories

(F) Filing a report

Figure 1. Tiramisu interface screens.

Seats" in Figure 1 (B)). Users identify the bus they are boarding, their destination, and rate the bus load (C) and share a location recording while riding (D). When users wish to file a report, they first select whether the report is specific to the Tiramisu system, the schedule and predictions, or the local transit agency. For the latter, they are offered high-level categories (E) prior to the actual report screen (F). On the report screen, tags are listed and the text field is pre-populated with location and route information so users will realize they do not need to provide such details.

We considered several interface designs that emphasized either completely automatic or completely manual use. We elected to focus on the user contributing data about their current trip (a bus on a particular route at a particular time and day).

234

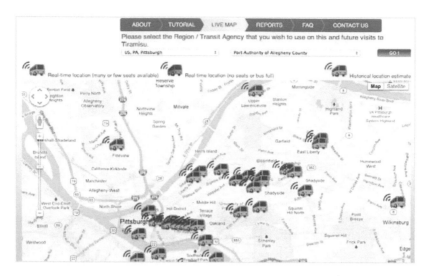

Figure 2. The Tiramisu website includes a live map and access to submitted reports

When providing a trip, the user explicitly chooses to share a location trace during her ride. From this information, the system can construct a reliable model to predict bus arrival times. This user interaction model also mitigates privacy concerns since the recording is opt-in, rather than automatic.

Earlier work on reporting modality preferences (Steinfeld, et al. 2010) revealed that users prefer still images to video and text entry over spoken recordings. Analysis of report modality quality (Steinfeld, et al. 2011) concurs with the combination of photo images and text details. However, we retained the option for spoken feedback through a shortcut button that starts a regular phone call to the local agency's customer service line. We did not record these calls. This alternative supports users who have difficulty entering text on smart phones and those who wish to speak directly with a transit agency representative.

In addition to a mobile client, Tiramisu offers limited information on web by visualizing the model predictions for the current time on a map (Figure 2). This map also provides route information and load information when the cursor is hovered over a specific bus. Reports are listed sequentially in the Reports tab.

3   USER CONTRIBUTIONS

Our initial work revealed that the integration of crowdsourcing real-time information and service quality reporting did, in fact, increase the rate of feedback about the service (Zimmerman, et al. 2011). The system has now been in public operation in Pittsburgh, Pennsylvania since late July 2011 and users have recorded over 20,000 trips. There is also high utilization of the service quality reporting function, with over 1,000 reports to date. In this section we provide some sample reports and the outcome of our analysis of behavioral trends, emergent themes, and the implications of these findings.

3.1   Report examples

As seen before (Zimmerman, et al. 2011), riders provided transit agency feedback through the system at rates much higher than would normally be expected. There was strong representation for reports related to "point-of-pain" experiences, often related to missing or late buses. Other comments included driver feedback, maintenance issues, and thoughts about fellow riders (e.g., Table 1). These reports suggest a willingness of users to report from bus stops and other transit environments in real-time.

As expected, the system captured instances of where drivers would not let wheelchair riders onto the bus (e.g., Table 1, reports 735 and 936). This confirms laboratory results that implied riders would use mobile reporting to document such events (Steinfeld, et al. 2010).

Some users demonstrated good use of the photograph option, often for documentation of damage and maintenance concerns (Figure 3, left images). Others took advantage of the screenshot feature to illustrate problems with the software, both interaction design and schedule information (Figure 3, top right). Finally, there were also images where users included screenshots of the report screen itself, suggesting an interaction breakdown (Figure 3, bottom right).

3.2   Behavioral Patterns

User reporting behaviors reveal concerns about user's mental models for the reporting function and how they utilize the function with the application. Language and report content can help identify user feelings and concern regarding reporting. For example, some reports take the form of questions, which demonstrates that the user is looking for a reply, whether immediate or over time. The prevalence of questions in reports suggests that users do not understand the destination and handling of their reports after submission. In some cases there were clear demonstrations that users were unaware they could learn the outcome of their report by visiting the Tiramisu webpage. For example, some reports included contact information and requests for replies concerning an issue or an application suggestion. We admit that we did not effectively communicate that feedback could be obtained by visiting the Tiramisu website.

We also observed varying levels of proactive language. Reporting provides the user agency to comment and make suggestions, so this was expected. Feedback on Tiramisu ranged from user-identified problems to suggestions on improving information flow. Suggestions about routes and schedules implied feedback based upon their own transit experiences. These ranged from recommendations for improving existing routes to insights about bus schedule patterns.

Mood was also apparent in reports. Users that expressed frustration in reports often complained about chronic service issues (e.g., lateness and no-show buses) or

Table 1. Example reports describing service delivery

Number	Report description
9	on 71A, inbound, towards (null) - Bus 5338 smells like there's been someone smoking cigarettes recently.
194	on BLLB, inbound, towards Subway at Steel Plaza Station - T was late, previous one passed up a long line of people waiting at St. Anne even though we could see passengers on it and seats available. Driver of Library apologized for being late and said there was a disabled car at Washington Junction.
678	on 61B, inbound, towards 6th Ave at Smithfield St - super nice driver, he saw a girl running to catch another bus and he called out "get on, I will see if I can catch him!" tried to help her out!
735	on 61D, inbound, towards Forbes Ave opp Hamburg Bldg - a disabled rider in a wheelchair could not get on this bus as the bus was not wheelchair accessible. Everyone else could get on the bus but this passenger was made to wait.
881	Can you please get the route 58 bus onto a regular evening schedule? I routinely end up standing at the stop for 45 minutes -- which means your printed schedule is complete fiction. It is frustrating and, as the weather gets cold, unhealthy. Thank you
936	on P16, outbound, towards Main St at Thompson Run Rd - ten minutes late. Wont lower step for disabled passenger.
952	on P67, inbound, towards East Busway at Wilkinsburg Station D - driver pulled up early on first stop of a bittter cold morning to let passengers wait in the warm!

expressed impatience. The majority of reports displayed a neutral tone and uncertainty on whether they were reporting to PAT or to Tiramisu. In addition to language and content, we identified several emergent user behaviors. There were 51 instances of multiple consecutive reports by a user. These sequential, and sometimes identical, reports sent to Tiramisu suggest two possibilities. First, users may have been testing the reporting function. Second, they may have misunderstood the submission status. There were also double reports from users attempting to ensure their report would not be missed. Other multiple try attempts were seen. For these either (i) an empty report was sent first, followed by a report with the desired comment or (ii) the user sent a report with a comment, and followed with a second report to correct or update the comment. As with before, these behaviors suggest breakdowns in the report feedback loop.

We also looked for content breakdowns to reveal experience problems before report submission. As mentioned, route and destination are pre-filled in reports.

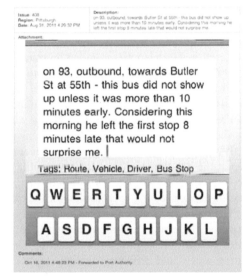

Figure 3. Example report images

There are a large number of "towards (null)" reports, which have route information but have (null) in place of a destination. We suspect this is due to users selecting a desired bus, but not a destination. This could be the result of riders reporting from bus stops before buses arrive or users boarding but not completing the bus trip recording sequence. There are also reports where the user entered the route information. This type of report could occur if the user either (i) did not select a bus prior to reporting or (ii) submitted a report after completing a trip. Similarly, there were reports with content that was completely user generated. These reports contained general comments and suggestions about transit service as well as specific descriptions of things affecting transit service.

3.3   Emergent Themes

A majority of agency-related reports were focused on the difference between agency **schedule and the observed bus arrival time**. Most of the reports in this category emphasized late buses, sometimes revealing a pattern of lateness for specific routes. The content of these reports suggest that most of these were sent by users waiting at bus stops. However, there were cases where users tried to report real-time bus lateness through a report, rather than the recording function. As of this writing, there were approximately 66 reports about late buses, 22 for early buses, and only 6 for on-time buses. Riders also use the reporting system to complain about buses that do not stop at the designated stop, leaving commuters stranded, and buses that are too crowded to carry more passengers. These complaints were universal problems and not specific to wheelchair users.

Users also showed willingness to comment on **route design and data quality**. There were 64 reports suggesting detours and routes that would benefit the agency. Some daily commuters also provided feedback about errors in the schedule data. Examples included missing stops and out of date information. The raw schedule files used in Tiramisu are provided by transit agencies in a standard form. Reports about these errors could be used to improve agency schedule data.

There were many examples focused on the **cleanliness and maintenance** of the transit system. Users complained of cleanliness of a bus and identified issues that required attention. Maintenance problems were reported for buses, bus stops, and bus shelters.

While users reported issues with **drivers** (e.g., "rude driver", "driver did not stop at the bus the stop") they also complimented drivers who demonstrated good service. For example, one user submitted three reports over different days for a specific trip expressing appreciation for the clean bus and friendly driver.

The reporting function is used extensively to provide **feedback to the Tiramisu team** for suggestions and problems with the application. Feedback to the team was sorted into feature requests and concerns. The latter included application crashing, battery usage, and difficulty loading stop and bus information. These issues were addressed in the subsequent versions of the application. There were approximately 33 feature suggestions made by users. A majority of these have been incorporated into the design over time.

The large quantity of reports indicates that **users want a method for communicating with service providers**. There was also evidence that riders viewed Tiramisu as a valuable tool for public transit users.

3.4   Implications

The data suggests our current methods for handling reports need adaptation. Users desire to participate in a feedback loop with the agency and the Tiramisu team. The current system is weak in this regard; it is hard for users to understand how reports were handled and obtain feedback on their contribution. The data suggests a need to enable, cultivate, and support discussion and dialogue between users and service providers.

Prior to analysis we developed a simple report handling methodology to manage and route reports. Currently, all reports are instantly shown on the Tiramisu website. The team manually reviews and replies through the website – not the mobile application. All reports receive one of the following answers: *Thank you, Forwarded to Port Authority, Forwarded to App Team,* or *Added to Feature Request List.* All reports regarding transit service issues are forwarded to agency staff via email. We identify these by examining user-selected categories and reviewing incoming reports for comments specific to transit service delivery. Reports specific to the Tiramisu application are examined for problems that need to be fixed and user interaction breakdowns. For reports where the user has difficulty with the application, we document the problem within our internal development system and reply to the report with an explanation and advice. Reports about bugs or functionality issues are forwarded to the software team, also through our internal development system. Any suggestions made about future features or improvements to the application were added to a feature request list and discussed internally. While many Tiramisu concerns have been addressed with changes to the software, there is no feedback to users that their input made a difference. Likewise, we lack good methods for feedback from the local transit agency.

It is evident that users desire timely feedback from the service provider (either Tiramisu or the transit agency, depending on the problem). Some reports also illustrate a desire to broadcast bus location and schedule discrepancies to other users on the route. Facilitating real-time interactions will improve the quality of the user's experience and possibly the quality of the transit system as a whole. We have plans to move in this direction and are working towards this goal (Steinfeld, et al. 2012).

4   DISCUSSION

The results show that strategies used by the team to ensure co-production can be effective for all citizens. This finding is counter to the status quo where transit riders without disabilities rarely file reports through conventional options (Steinfeld, et al. 2010; Yoo, et al. 2010; Zimmerman, et al. 2011). This finding implies that smart phones and other social computing models may be successful in public services.

However, it is clear that a feedback loop is desired and methods for user contribution need to be carefully designed. Weakness in either of these interactions can lead to breakdowns. Work in other public service co-production systems shows it is important to provide methods for service providers to easily, quickly, and asynchronously provide feedback on reports (e.g., Neighborhood Parks Council, 2007).

One unforeseen aspect is the interaction between user contributed data and existing service quality policies. For example, third party reporting systems, like Tiramisu, may not be considered acceptable documentation for legal or disciplinary action under existing contracts. However, such data may allow agencies to triangulate actionable data within their own information systems. It is also likely that positive feedback requires less stringent documentation. Route planning, non-contractual service modifications, and other co-production efforts are also less likely to encounter resistance to third party data contribution.

ACKNOWLEDGEMENTS

The authors would like to thank colleagues in the local Pittsburgh area who facilitated participant recruitment and the Port Authority of Allegheny County for their continued cooperation and insight.

The Rehabilitation Engineering Research Center on Accessible Public Transportation (RERC-APT) is funded by grant number H133E080019 from the United States Department of Education through the National Institute on Disability and Rehabilitation Research. Additional funding was provided by Traffic21 at Carnegie Mellon University, a program developed with the support of the Hillman Foundation.

REFERENCES

Boyne, G. A. 2003. Sources of public service improvement: a critical review and research agenda. *Journal of Public Administration Research and Theory,* 13(3), 367-394.

King, S. F. and Brown, P. 2007. "Fix my street or else: using the internet to voice local public service concerns." In Proceedings of Theory and Practice of Electronic Governance, ACM Press. 72-80.

Neighborhood Parks Council. 2007. "2007 ParkScan.org annual report." Accessed March 20, 2008 from http://www.parkscan.org/pdf/2007/ParkScan_Report_2007_web.pdf

Steinfeld, A., Zimmerman, J., Tomasic, A., Yoo, A., and R. Aziz. 2012. Mobile transit rider information via universal design and crowdsourcing. *Transportation Research Record - Journal of the Transportation Research Board* 2217, 95-102.

Steinfeld, A., Aziz, R., Von Dehsen, L., Park, S. Y., Maisel, J., and E. Steinfeld. 2010. The value and acceptance of citizen science to promote transit accessibility. *Journal of Technology and Disability,* 22(1-2), 73-81. IOS Press.

Zimmerman, J., Tomasic, A., Garrod, C., Yoo, D., Hiruncharoenvate, C., Aziz, R., Thirunevgadam, N., Huang, Y., and A. Steinfeld. 2011. "Field trial of Tiramisu: Crowdsourcing bus arrival times to spur co-design." In Proceedings of the Conference on Human Factors in Computing Systems (CHI).

Yoo, D., Zimmerman, J., Steinfeld, A., and A. Tomasic. 2010. "Understanding the space for co-design in riders' interactions with a transit service." Proceedings of the Conference on Human Factors in Computing Systems (CHI).

CHAPTER 24

Integrating a User Interface Design Environment into SOA and Service Engineering

David C. Dunkle

Lockheed Martin Enterprise Business Services
Orlando, FL 32825-5002, USA

ABSTRACT

One of the core responsibilities of Service Engineering is to provide a set of traditionally "systems-oriented" methodologies (e.g., specification, modeling, verification, etc.) throughout the service-oriented architecture (SOA) project lifecycle. However, in many cases, UI design, an important element of systems-oriented methodologies, is eliminated from a development role or relegated to a minor, unempowered review status during SOA project development lifecycles.

There are multiple drivers within a SOA environment that contribute to this reduction of focus on UI design. They include the reusable nature of SOA components, the dilution of UI design responsibility in SOA projects, and attempts to minimize development costs. These factors often drive SOA projects toward common risk areas, including misalignment with business processes and application inconsistency.

A method for integrating UI into the SOA lifecycle, and linking UI to SOA modeling, assembly, deployment and management phases is described. The benefits of this integration are then explained.

.

Keywords: SOA, service engineering, user interface design

1   SERVICE ENGINEERING AND THE DEARTH OF USER INTERFACE DESIGN

One of the core duties of Service Engineering is to provide a set of traditionally "systems-oriented" methodologies (e.g., specification, modeling, verification, etc.) throughout the service-oriented architecture (SOA) project lifecycle. Modeling, by its nature, is often considered to be at the very core of every service engineering method (Karakostas and Zorgios, 2008). However, in many cases, UI design and associated user modeling, both important elements of systems-oriented methodologies, are eliminated from a development role or relegated to a minor, unempowered review status during SOA project development lifecycles.

It is curious as to why UI design suddenly disappears from the lifecycle. Service engineering and service-oriented architecture are both efforts in model-driven development. Often they are designed in an object–oriented fashion (perhaps using the Unified Modeling Language (UML)).

Likewise, user interface design is also an exercise in modeling. Mental models, user and task modeling, etc. are all tools by which a user interface is effectively created. At first glance, one would expect there to be a natural symbiosis of SOA and UI models, with both modeling efforts combining to produce a well-modeled product, from a business and user perspective. UML, for example, provides a number of opportunities to incorporate user interface modeling within its lifecycle (e.g., use cases, abstract prototyping, activity diagrams, interaction models, etc.) (Van Harmelen, 2001).

Yet, all of these tools can be specified and pushed down into the lower levels of a product, where technology is described instead of human interaction. The tools can focus on data architecture and communication protocols, programming languages and database queries, to the point where the user is no longer a real part of the use case, the activity diagram only illustrates network activity, and the product is now complete without a concern for usage.

1.1   Elements That Reduce Focus

Service engineering never intended to forget about the user interface. UML has always provided avenues for user interface incorporation. But something happens along the product lifecycle. What are the drivers that make use lose focus on the user interface?

One of the basic assumptions within SOA sometimes leads to reduced focus on the user interface, through a mistaken extrapolation of its concept of "reuse". One of the strengths of SOA is the capability to reuse (i.e., transform and refine) components across different areas, for similar purposes. This capability, while very useful for repeated patterns of similar services, can be abused when it comes to the user interface.

How can reuse be so beneficial in one perspective, and detrimental in another? It is because of the differing nature of the operations and the operator; the difference between the product and the user.

Software is a highly customizable entity. If you need a software service to calculate the derivative of a function, you can simply add code that includes derivative calculation capabilities. You can probably reuse code from a different but similar service to do this. What's more, if you've done your job right, your service will forever calculate derivatives quickly and correctly. You've crafted your service to perform exactly how you want it to perform.

It's different with people. Suppose you need a person to calculate the derivative of a function. Do you simply add code to the person? Do you instruct them with a reusable resource that you have used elsewhere? [Actually, you probably do start this way, and when it fails, you switch to a different training resource]. If you happen to surmount the obstacle of embedding calculus knowledge within your person, will they consistently perform optimally, correctly calculating derivatives every time?

This illustrates an important, but often overlooked difference between people and computer systems. People cannot be crafted to a desired state, at least not within the span of our ever-shorter development lifecycles. In the same way, the user interface- that portion of the system that actually "touches" the user, cannot be optimally constructed by copying components from other areas. Unlike services, people do not respond to the same presentation in the same way. People do not seamlessly transform the functionality of one operation into a refined functionality of a similar operation. It's all too easy to copy a user task pattern from one system and inappropriately apply it to your own. It's also easy to show how inappropriate this can be. For example, you wouldn't do your taxes the same way a large corporation does, would you? You wouldn't drive your car the same way an ambulance driver, or a race car driver does. How boring it would be if we did, in fact, all perform the same task the same way. What we might gain in efficiency, we would lose in creativity and uniqueness.

The attempt to reuse user elements is not the only element that reduces focus on the user interface. Another element resides in the lifecycle itself.

Military and government development programs, for all their bureaucratic challenges, often take a firm grasp of the importance of user interface needs. Specific tasks are performed to create the user interface, and a qualified user interface designer is expected to perform those tasks. This aspect of concrete design tasks and assignees is lost in the service engineering development lifecycle.

Table 1: Elements That Reduce Focus

Element	Focus Issue
Reuse	Reuse cannot be applied as effectively to user interfaces or users, as it can within other areas of system development.
Dilution	Dilution of user interface responsibilities distribute design tasks informally, often not bothering to monitor progress.
Cost	When user interface design is not valued, the cost can always be reduced.

Amazingly, even when a project or product does not have specific design development steps, or no assignee to perform those steps, this will not always create a roadblock to development progress. User interface design has always been saddled with the unearned reputation of being easy to perform. Often, team members will jump at the chance to create a user interface, using their own techniques and mental models. As multiple "designers" each create new perspectives on a product's user interface, soon you have a conglomeration of different designs, and in the end, an exasperated user.

This is exactly what has happened over time in service-oriented environments. Multiple developers have created components, and with them, component-level interfaces. Sometimes, this includes the user interface. When another project needs to reuse a component, the user interface comes along for the ride, even if the application of the component now involves a completely different user.

1.2 Cost Elements That Reduce Focus

The budgetary aspects of many projects can reduce focus on the user interface. When the discipline itself is not valued (as exhibited by a diluted design responsibility, and a heavy reliance on UI reuse) a project manager can easily justify removing the costs of UI design activities from the project budget. Because the end result of user interface design is not apparent until late in the product lifecycle, a manager may not be able to visualize what he/she is losing by removing UI tasks from his product. The lifecycle and standards that are being followed for the Service engineering project most likely have little or no instruction on user interface activities, so they are often considered "outside of plan" by managers.

2 COMMON RISKS DUE TO UI DESIGN DILUTION

As pointed out, it is fairly easy to dilute the focus on user interface design for service engineering projects. What is the trade off from this? Does a project take on any risks from doing this? Unfortunately, diluting the user interface design initiatives within a service engineering project can add risk across the project lifecycle, affecting business, application, information and technical architectures (Mazumder, 2006).

2.1 Business Architecture Risks

Establishing a correct business architecture is a very visible core task for service engineering projects. The designed service must be able to work well with the existing business that surrounds it. Often, the users themselves represent the majority of the surrounding business, and the key architecture that the service must accommodate. Users bring with them existing policies, legacy methods, and nonspecific resolution paths. Any of these alone can create a mismatch between the project and the business.

2.2 Application Architecture Risks

The application architecture defines and provides the product service. This depends of proper identification and implementation of business services. Without the interaction with business users that UI design affords, there is increased risk that the service is incorrectly defined and/or implemented. As mentioned for business architecture risks, the users are typically the source for real-life resolution paths, (or error scenarios) critical to a robust service.

2.3 Information Architecture Risks

The value of an information architecture that accurately represents the users' information models is profound. In the development stages, as long as there is an easy translation between the service's information architecture and the user's, the project will develop in concert with the user's needs. However, this is not typically the case. Information architecture is often generated behind closed doors, with an emphasis on high service efficiency. The risk that rises here is one of a overly complex service information architecture that does not represent the correct interaction of information as the user's portray it.

2.4 Technical Architecture Risks

The technical solution for a service can be affected by many unhelpful forces, notably, a lack of review and selection rigor. Part of this selection rigor must include user interface considerations- for it is the only part of the service of which the users are aware. Improper product selection risks are always increased without user interface evaluation.

3 INTEGRATING UI INTO THE SOA LIFECYCLE

As demonstrated, many factors seem to work together to dilute the incorporation of user interface design into the SOA lifecycle. Knowing the value that user interface design brings, it becomes our effort to place it back in the lifecycle. There are a number of different points where this can be accomplished, but first, we must examine the argument that UI design even belongs within SOA. Is it possible to accomplish our goals by following processes wholly external to SOA; in effect, effecting change and modification from outside the confines of an unfriendly environment?

3.1 Why Separation From SOA Hurts UI Design

Separation from SOA, for UI design, seems to make sense at first. Outside of the SOA environment, UI design can be performed correctly, unaffected by the dilution and reuse quandaries that SOA breeds. The trouble begins when you attempt to bring that design back into the SOA environment. Where is it supposed to reside? How do you apply your design? Sadly, by removing the user interface focus from SOA, you will have left yourself nowhere to return to, once your design is ready. In most likelihood, the SOA project will have moved on without you.

The effort has to be in embedding the user interface within the SOA lifecycle. It has to touch every phase, and link to itself across phases, almost as if the user interface was a lifecycle within a lifecycle. Only as an insider will the user interface be recognized and accepted.

Figure 1: SOA Lifecycle with UI Embedded

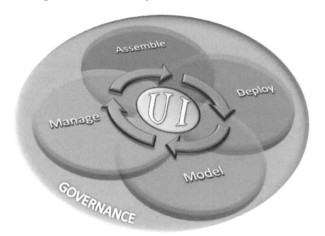

3.2   Linking UI Design to Modeling

There is not a more important phase than modeling, within which to incorporate user interface design. Often, the SOA project kicks off with the modeling phase, so it is of utmost importance to for UI designers to become involved in the project early. This is the phase where requirements are gathered, modeling is performed and design takes place. Each of these tasks involves user perspectives. User requirements, modeling, and UI design should take place alongside of service modeling tasks, complementing each other, and setting the foundation for further system and UI refinement.

3.3   Linking UI Design to Assembly

The Assembly phase often focuses on service refinement, testing different options for fulfilling service requirements. As such, the same should be done with the user interface, prototyping different options for designer review.

It is always important to note that rapid prototyping of any sort must perform as described; that is, it must be rapid. Often such effort is given to prototyping, that an actual interface or component could have been assembled within the same time and effort. In the UI realm, specifics and concrete designs are not required in this phase. Focus on getting the user interface abstractly organized, and refined it in the Deployment phase.

3.4   Linking UI Design to Deployment

The focus of the deployment phase is integration. All people, processes and information must be integrated into a single architecture. At this point, you need to have solid UI concepts to integrate as well. Plan to revise and refine your designs to coordinate with the other integration components. As long as there has been UI efforts along the way, bringing the design to the product will not be a surprise, and more likely will be welcomed by other developers (who probably were wondering what the front-end might look like).

3.5   Linking UI Design to Management

Do not forget about user interface design during the management phase. This is when metrics are monitored, so it is exactly the time to perform evaluative efforts to rate the performance of users within the designed user interface. Collect real-time data, so as opportunities for UI revisions appear, you have already performed the legwork necessary to justify new UI improvements.

4   BENEFITTING FROM USER INTERFACE DESIGN IN SOA

Control, agility and cost have been cited as the core benefits of a SOA deployment (Freivald, 2008). Certainly, SOA brings the capability for reduced cost of development, and quicker speeds of deployment. But a service that is delivered without user interface design incorporated within it offers no consolation in savings or speed. Services are to be used, and services without a well-designed user interface are not usable. To offer true and valuable service engineering, we must make sure that user interface design has not been removed from the development cycle, and we must add our efforts in performing the user interface design tasks that are critical to system and service usability.

REFERENCES

Freivald, J., 2008 "The Business Benefits of SOA Implementation" Accessed 1/20/12
http://www.information-management.com/infodirect/2008_57/10000665-1.html

Karakostas, B. and Y. Zorgios. 2008. *Engineering Service Oriented Systems: A Model Driven Approach*. IGI Global.

Mazumder, S., 2006 "SOA: A perspective on Implementation Risks." Accessed 1/20/12
http://www.infosys.com/consulting/systems-integration/white-papers/Documents/soa-perspective-implementation-risks.pdf

Van Harmemen, M. (editor) 2001. *Object Modeling and User Interface Design*. Addison Wesley.

Woods, D. and T. Mattern. 2006. *Enterprise SOA – Designing IT for Business Innovation*. Sebastopol, CA: O'Reilly Media.

CHAPTER 25

BtoB (Business to Business) Product Planning and Platform Making by the Business User Model

Toru Mizumoto, Kazuhiko Yamazaki

Wakayama University, Chiba Institute of Technology
Wakayama, JAPAN, Chiba, JAPAN
Mizumoto.Toru@sysmex.co.jp, designkaz@gmail.com

ABSTRACT

A lot of product planners know the methodology and effectiveness of the persona method as very effective to improve customer's satisfaction on their products. The product planners want to surely use the persona method. However, the examples shown in technical books are almost only about BtoC (Business to Consumer) products. Therefore we made the personas for BtoB (Business to Business) product planning. We call it the Business User Model. We report the profit of using the Business User Model.

Keywords: personas, product planning, user centered design

1   BACKGROUND

BtoB product means the instruments for office work, the machine tools, the medical instruments and the other instruments for professional use that are delivered to offices, factories, hospitals and other facilities. These are used by users with special knowledge. The users with special knowledge have pride such as "we are professionals." and "we can use the product even though it is difficult to use." So, those users tend to prefer function and performance than usability. In addition, because BtoB products are purchased as instrument of facility, operators and a person who decides which products to purchase are different. Therefore, the person without experience for using products selects which to buy by only looking catalog. In that case, usability of the products is not considered.

In such a situation, function oriented design is emphasized on development of BtoB products. If the product just has a lot of functions or good performance than the competitor's product, the business is succeeded. However, if the highly efficient succeeding model is put on the market one after another, the market will be saturated. Then function and performance of the products will reach the ceiling and it will become impossible to keep the competition predominance by function and performance. Moreover, user's special knowledge will be no longer necessary because a product will become highly efficient. And users with poor knowledge will be increased in the same way as the market of BtoC products. In such a situation, if we want to have the advantage over the competitors, our BtoB products have to have the high quality usability which satisfies user needs than function and performance. User centered design becomes important to BtoB product planning.

2   PROBLEM OF THE PERSONA METHOD

Therefore, we considered to introduce the persona method in order to raise customer's satisfaction. The "Persona" is the concrete person image which set up name, mug shot, role, goal, etc. supposing a typical user of the products. The technique for planning and development of the products which satisfies the persona's goal by always using the persona is called "Persona method". The persona method is often published in technical books and magazines. Then, a lot of product planners know the methodology and effectiveness of the persona method as very effective user centered design method to improve customer's satisfaction on their products.

However, the examples shown in the technical books and magazines are almost only about BtoC products like home electronics, food, etc. Of course it is important that we understand the end user for BtoB products too. But, in the case of BtoB products that is purchased by facilities budget, there are additional important factors such as the relation between end users and purchasing decider, the stakeholders of the facility, etc. Thus, the persona for BtoC products is inapplicable to BtoB products.

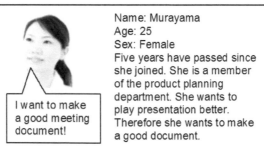

Name: Murayama
Age: 25
Sex: Female
Five years have passed since she joined. She is a member of the product planning department. She wants to play presentation better. Therefore she wants to make a good document.

I want to make a good meeting document!

Figure 1 The persona for BtoC products

3   MAKING OF THE BUSINESS USER MODEL

As stated above, the user's attributes that we have to consider for BtoB product planning are different from the BtoC product planning. Therefore we gathered the well-informed people of our company and carried out brainstorming to extract the user's peculiar to BtoB products. If there are similar BtoC products, it is easier to extract attributes in comparison with it.

We show the example of the combination printer-copier-scanner-fax for business use. We were able to extract the following attributes. "place to use, way of the setting, end user, purchasing decider, number of copied paper, paper size, how to learn, necessary speed, frequency to use, purpose to use, number of the machine, troubleshooting, favorite design, etc." Then, we arranged the attributes and made the flow of the user interview. And, we investigated elements in every attribute by user interview.

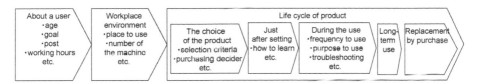

Figure 2 Flow of the user interview

The model of the target user for BtoB product planning is completed by performing a classification of the result of the user interview by KJ method. This model is the new type of the persona for BtoB product planning. We call it the "Business User Model" in distinction from the conventional persona.

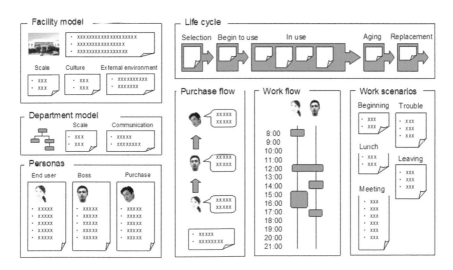

Figure 3 The Business User Model

Table 1 Items of the Business User Model

Item	Contents
Facility model	location, scale, basic principle, corporate culture, external environment, etc.
Department model	area, organization, shift, frequency to use, role, the number of staffs, communication, etc.
Personas	End-user, Boss, Purchasing decider
Life cycle	events between selection and replacement
Purchase flow	flow and criterion of the approval
Work flow	flow of report, communication, consultation between employees, etc.
Work scenarios	events of each task

The personas and the department model act based on the corporate culture of the facility model. And they are affected by the external environment of the facility model. The communication of the department model affects the action of the personas.

In the product design, we adapt basic specifications to the scale of the facility model. And we examine the specifications that can appeal to the "purchasing decider" persona. Furthermore, we examine usability to satisfy the "end user" persona from work flow and scenarios. We show the example which extracted the idea for BtoB products by the Business User Model.

Table 2 Product design by the Business User Model

Problem	without the Business User Model	With the Business User Model
User said: "Many meetings begin at 13:00. Many staff print and copy their documents before 13:00. So I have to wait. And I am upset very much."	Product design: The combination printer–copier–scanner-fax which works faster than the current model	Facility model tell: "There are two combination printer-copier-scanner-fax in the office. There is no place for put 2 machines together. Reluctantly, they are placed 10 meters away" Department model tell: "The machines are used by 30 people of the product planning department."

		Personas tell: "We want to make a good document. So, we print documents just before the time limit." Product design with new ideas: "Because 30 people print documents at the same time, a drop in the bucket even if the machine is faster. The user can confirm the congestion situation of each machine with users PC. The user can choose the machine which is not so busy. And the user can print the documents.

We can add the high usability that satisfies user needs to our BtoB products by the Business User Model. And, we can add the new idea that the competitors cannot have to our BtoB products by the Business User Model.

4   PLATFORM MAKING BY THE BUSINESS USER MODEL

When we want to expand overseas market share of our BtoB products, the understanding of overseas user needs is a problem to be solved. Even if country and culture are different, the making procedure of the Business User Model is the same. Therefore we made the Business User Model of each target countries. When we make the Business User Model of target countries, we will find the same items in every target country and we will find the different items in each of target country. We will notice that we can make a common design about the same items to cost-cut. We will notice that we can make a modifiable design about the different items to support individual needs. In this way, we can use the Business User Model for the platform making of the BtoB products.

Country variations

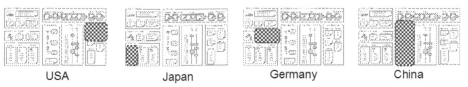

USA   Japan   Germany   China

Figure 4 The Business User Model of each country

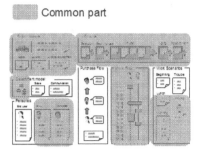

Common part

Figure 5 Common part of the Business User Model

5 CONCLUSION AND VISION

We showed that the Business User Model was very important for BtoB product planning with user centered design. However, the Business User Model is made based on the user interview. Therefore, it expresses the present target user, but does not express the future user image. Because BtoB product is purchased by the budget for facilities, the user cannot buy a new product immediately even if the user is interested in it. The user may get a budget several years after release of a new product. If we develop BtoB products using the Business User Model that we made when we began to examine the product planning, our product may become outdated when the user chooses a new product. Therefore, we have to think about how to make of the Future Business User Model.

REFERENCES

Donald A. Norman. 1998. The Psychology of Everyday Things. *Basic Books*

Masaaki Kurosu, Masako Ito and Tomoko Tokitsu. 1999. User kougaku nyumon -tsukaigatte wo kangaeru ISO13407 eno gutaiteki approach. *Kyoritsu Shuppan Co., Ltd.*

Toshiki Yamaoka. 2000. User yusen no Design sekkei –atarashii shouhin kaihatsu no kangaekata. *Kyoritsu Shuppan Co., Ltd.*

Kazuhiko Yamazaki, Minako Matsuda and Ryoji Yoshitake. 2004. Tsukaiyasusa no tame no Design -User centered design. *Maruzen Company, Limited*

John S.Pruitt and Tamara Adlin. 2007. THE PERSONA LIFESYCLE Keeping People in Mind Throughout Product Design. *DIAMOND,Inc.*

Calol Righi and Janice James. 2007. USER-CENTERED DESIGN STORIES Real-World UCD Case Studies. *Morgan Kaufmann Publishers*

Hiroki Tanahashi. 2008. The Process of Persona Design. *SOFTBANK Creative Corp.*

Toshiki Yamaoka. 2010. Design no zoukei kogaku. *Kougyou chousa kai*

Toru Mizumoto, Shunsuke Ariyoshi and Kazuhiko Yamazaki. 2011. Introduction and use of "Persona method" in product development. *PROCEEDINGS OF THE 58TH ANNUAL CONFERENCE OF JSSD*:344-345

Section V

Service Design

User eXperience in Service Design: Defining a Common Ground from Different Fields

Karim Touloum[1], Djilali Idoughi[1], and Ahmed Seffah[2]

[1]Department of Computer Science, University A-Mira, Béjaïa, Algeria
{touloum.k, djilali.idoughi}@gmail.com
[2]TechCICO Laboratory, University of Technology, Troyes, France
ahmed.seffah@utt.fr

ABSTRACT

The emerging field of service design combines several methods mainly from the fields of product, software and interaction design, for designing the experience and interface to services. However, User eXperience design is not easy to understand because it is a new approach and covers many different fields (usability, psychology, marketing…). More often the design team members come from various domains with different cultural backgrounds such as engineers and marketing. Therefore, it is necessary to have a common definition to share the same understanding of this concept. In this paper, we investigate the main definitions proposed in the literature and the use of User eXperience in many disciplines other than computer science. The main goal of this paper is to find important ingredients from different fields allowing to defining and characterizing the User eXperience in service design. Thus we propose to extend an existing design tool (*persona*) in order to cover the whole of User eXperience components (hedonistic, aesthetic interaction, social factors...).

Keywords: User eXperience, Service design, Persona, Interaction design, Human Computer Interaction

257

1 INTRODUCTION

User eXperience has become a lighthouse subject for many practitioners and researchers from different disciplines. This trend is justified by the impact of this new concept on the success of the product or the interactive service. A great effort is made during the last decade to explain the concept of the User eXperience, its components with their influential factors and also its scope to the current disciplines of computer science (HCI, software engineering, etc.). However, the wide meaning given created some ambiguity and confusion about the semantic of that concept and its characterization for effective use in interactive service design. Therefore, to make effectively the User eXperience as a design tool in service development project remains a difficult task in mind of the design team to carry out. Thus we believe that a deep study of the concept of User eXperience as a design tool is required today. This is expected to understand the attributes of this experience and how to retrieve and use it in interactive service design. The understanding of the internal behavior and reactions of individuals through the experience was widely adopted in several areas before the field of computer science. The fields of sociology, psychology, Criminology, and marketing are good examples on the use of the concept of the human experience. A literature review will be presented in this article in order to show how practitioners from different disciplines address the subject of the experiment. There is an interest in particular to the main elements of the experience which are exploited.

Several frameworks have been proposed to explain the components of the experience that can positively influence the design of products, which leads to a totally positive experience. (Hassenzahl, 2003) provides a model to describe people's goals and actions when interacting with a product. This model is mainly based on pragmatic attributes (e.g. Manipulation and usability) and hedonic attributes (e.g. Stimulation and evocation). On other hand, (Forlizzy and Battarbee, 2004) present a framework based on three types of user-product interactions (fluent, cognitive and expressive) for the design of the experience and three types of experience (experience, an experience and co-experience). In his dissertation, (Arhippainen, 2009) present the U^2E-frame, a framework to planning and conducting tests by using User eXperience heuristics. Other frameworks were proposed to measuring and assessment the human experience in interactive system design; for example (Nacke, 2009) for media enjoyment (gameplay experience) and (Mahlke, 2008) to experiencing the portable audio players. Despite all these efforts, it remains that little work has put emphasis on the applicability of a model of the User eXperience in the design process phases. Among these works, we cited (Schaik and Ling, 2008) who presented an experiment tested and extended Hassenzahl's model of aesthetic experience, and (Jääskeläinen, 2009) who presented a set of User eXperience tools for developers to help them solve some User eXperience problems, such as the gap between test and marketing and design phases, and lack of development tools for supporting User eXperience designed features. However, most of the related work was based only on the assessment of the experience when we have the finite product (or service). Therefore, the gap still exists between theoretical frameworks with practical design process like UCD

(User-Centred Design). This can justify not only the subjectivity of experience components (Karapanos, 2010) but also the lack of precise definition (Hassenzahl, Diefenbach, and Göritz, 2010), and characterization of this concept.

In this paper, we investigate the main proposed definitions in literature and the use of User eXperience in many disciplines other than computer science. We also discuss about some design tools in order to characterize the User eXperience. In section 2 of this paper, we explore some UX definitions and propose our working definition. The role and position of UX in HCI and related fields will be discussed in section 3. Viewpoints from fields elsewhere and their use of User eXperience are presented in Section 4. The section 5 addresses some challenges and perspectives of the UX in services design.

2. DEFINING THE USER EXPERIENCE

In recent years, the term User eXperience (UX) was like ambiguous buzzwords in product or service design (Hassenzahl and Tranctinsky, 2006; Forlizzy and Battarbbee, 2004). Defining the User eXperience term remains difficult because it has both dynamic and paradoxal nature (Arhippainen, 2009). Dynamic as people experience the product all times and also the changing of the experience over time. Paradoxal because the meaning of the experience can be at the same time subjective, i.e. personal perception, personal experiencing (Senders, 2000) and collective, i.e. sharing the experience, social experience as claimed by (Battarbee, 2004) work about *co-experience*. Moreover, some researchers tend to deduce a unique definition of User eXperience from literature review (Marcus, Ashley, and Knapheide, 2009; Law, Roto, and Hassenzahl *et al.*, 2009). All this work aims to sharing a common definition between a large community of designers and practitioners of User eXperience.

2.1 Definitions review about User eXperience

(ISO 9241-210, 2009) describes User eXperience as a *person's perceptions and responses that result from the use and/or anticipated use of a product, system or service.* According to this definition, the User eXperience concept includes some attributes related to user's emotions, beliefs, perception, behavior, physical and psychological responses that occur before and after of use. This standard definition claims that User eXperience is related to usage. (Nielsen and Norman, 2007) gave another definition where *User eXperience encompasses all aspects of the end-user's interaction with the company, its services, and its products.* The authors claim that an exemplary User eXperience is to meet the exact needs of the customer, without fuss, follow-up with simplicity and elegance that produce products that are a joy to own, a joy to use. For (Pabini Gabriel-Petit, 2008), *User eXperience design takes a holistic, multidisciplinary approach to the design of user interfaces for digital products. It integrates interaction design, industrial design, information architecture, visual interface design, instructional design, and user-centered design, ensuring coherence and consistency across all of these design dimensions.* They add that the *User eXperience design defines a product's form, behavior, and content.* Whereas (Javahery, 2007) describe User eXperience as a generic term referring to a collection of information on user behavior, expectations, and perceptions influenced by user characteristics (knowledge, expertise, personality and demographics information) and service characteristics (domain, content, visual design and interaction type).

According to (Hassenzahl and Tractinsky, 2006) UX can be a *consequence of a user's internal state (predispositions, expectations, motivation, mood, etc.), the characteristics of the designed system (e.g. usability, functionality, etc.) and the context (or the environment) within which the interaction occurs (e.g. organizational/social setting, meaningfulness of the activity, etc.).* The three factors described by the authors (user's internal state, system characteristics and context of use) affect directly the key elements of User eXperience such as usability, flow interaction, pleasure and hedonistic aspects of use. (Hassenzahl, 2008) propose another two parts UX definition which defines UX in itself and states how UX is made respectively. In the former, UX is defined as a *momentary, primarily evaluative feeling (good-bad) while interacting with a product or service.* The later describes that a *good UX is the consequence of fulfilling the human needs for autonomy, competency, stimulation (self-oriented), relatedness, and popularity (others-oriented) through interacting with the product or service (i.e., hedonic quality). Pragmatic quality facilitates the potential fulfillment of be-goal".*

In (Schulze and Kromer, 2010), a User eXperience is defined as *the degree of positive or negative emotions that can be experienced by a specific user in a specific context during and after product use and that motivates for further usage.* Moreover, the User eXperience can be also a result of a motivated action in a certain context. For example, (Väänänen, Väätäjä, and Vainio, 2009) suggest that User eXperience promotes broader views of user's emotional, contextual, and dynamically evolving needs, and the impact of user' previous experiences on the new experiences. Finally; In his recent book, (Kuniavsky, 2010) stated that the User eXperience was the totality of end-users' perceptions as they interact with a product or service. These perceptions include effectiveness (how good is the result?), efficiency (how fast or cheap is it?), emotional satisfaction (how good does it feel?), and the quality of the relationship with the entity that created the product or service (what expectations does it create for subsequent interactions?).

2.2   Our work definition

The main highlighted definitions, yielded some differences in purpose: evaluation (Hassenzahl and Tractinsky, 2006), company (Nielson and Norman, 2007)), and also in how and when the experience occurs (during the interaction (Kuniavsky, 2010) or after (Hassenzahl, 2008)). Therefore, from our point of view, we can define the *User eXperience as something felt by the user, or by a group of users, following the use of a product (or service), or during its interaction with the product (usability and aesthetics), or even a possible use (or purchase) of a product. The result of this experience expresses the judgment made by the user (positive or negative) on all the moments experienced with the product (or service).* We use the word "something" to refer to the broad meaning that covers the term experience (emotions, perceptions, reactions...). Moreover, we mention that the experience occurs within each individual or shared by a group of individuals and this not only during or after the interaction, but also can occur before the acquisition of the product through brand, his previous experiences and opinions of others. This definition takes into account four characteristics of a User eXperience (Figure 1): the information constituting an experiment: degrees of acceptance (Schulze and Kromer, 2010), pleasure and sensation (Hassenzahl, 2008); the influential factors of experience: characteristics of the product, context of use, social interaction and previous experiences; the physical existence of the product: exists (ownership) physically, virtually exists (online service) and psychological existence through the image of the product; the moments of experience (or lifecycle): before, during and after the interaction with the product or service.

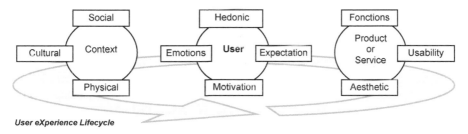

Figure 1 User eXperience Characterization

2.3   Abuse of UX

During this last decade, a lot of works on UX has tried to adapt this new concept of design compared to the already existing techniques and disciplines, in particular in usability and interaction design. Thus, some authors define and exploit the UX term depending on the flavor of the day, which leads to abuses on the use of User eXperience term, especially in HCI field. The concept of the user's experience is still being developed by many researchers from different areas of knowledge. Many times, the term experience is used as a synonym of pleasure or emotion. Other times, like in Garrett studies, it may be used to describe the result of the usability of a product, not considering emotional factors (Garrett, 2003). In fact, the User eXperience is beyond the satisfaction of basic needs. Thus, it is something related to positive experiences and to the enjoyment of life. Garrett offers a model for information design that connects structuring the User eXperience of hypertext and interface with creating the interface for a website. The design of experience happens on five levels: strategy, scope, structure, skeleton and surface. However, this model of User eXperience for information design fails to include sensory qualities of experience, which can be included as an elaboration of Garrett's model towards all kinds of online experiences. Borrowing Norman's view that UX pertains to *all aspects of the user's interaction with the product, how it is perceived, learned, and used* (Nielson and Norman, 07). UX, therefore, includes usability and perceptions of utility, but it goes further to consider emotional responses.

3. USER EXPERIENCE IN HCI AND SERVICE DESIGN

Over the last decade, many methods, practices and designs have been developed in human-computer interaction (HCI) to cover the full range of User eXperience. The relationship between HCI and UX is even complex to define because two fields share some aspects. On one hand, HCI addresses the topics such as the aesthetics of interaction, affective computing, and ludic engagement (Wright and McCarthy, 2008). On the other hand, customer relationship management and marketing play a large role in actual day-to-day experiences with products and services. Therfore, designing the User eXperience for interactive systems is even more complex, particularly when conducted by a team of multidisciplinary experts (Forlizzi and Battarbee, 2004). Thus, several frameworks have been proposed to take into account the whole User eXperience in interactive system design and their user interfaces. In the remainder of this section, we explore a set of frameworks aiming to improving the User eXperience in interactive system design.

(Hassenzahl, 2003) propose a model which views User eXperience from a designer and a user perspective making a distinction between the intended and apparent character of a product (figure 2). But he claims that there is no guarantee for designers to ensure their products are used or perceived as intended. The emotional personal response to a product is based on the situational context.

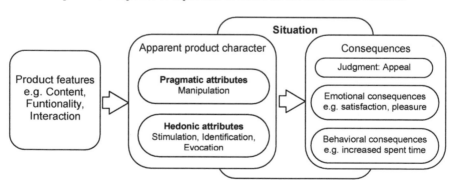

Figure 2 Key elements of Hassenzahl's model of User eXperience

In this model, product can have pragmatic (e.g., utilitarian value) and hedonic (e.g., knowledge/skill stimulation, communication of identity, memory evocation) attributes. On other hand, the experience is formed from the iconic value and prior memories the product triggers. (Wright, Wallace, and McCarthy, 2008) elaborate conceptualized framework for aesthetics experience of human-computer interaction and interaction design This framework used to critically reflect on research into the aesthetics of interaction and to suggest sensibilities for designing aesthetic interaction. (Forlizzi and Battarbee, 2004) present a framework for designing experience for interactive system with three types of user-product interactions (fluent, cognitive and expressive), and three types of experience (experience, an experience and co-experience).

The usability term is often confused with User eXperience. In fact, there is similarity between these two concepts in evaluation measures, but the difference in concerns of such discipline during development (Bevan, 2009). In one hand, the usability aims to designing for evaluating overall effectiveness, efficiency and user comfort and satisfaction. In other hand, the concerns of User eXperience include understanding and designing the user's experience with a product and also maximizing the achievement of the hedonic goals of stimulation, identification and evocation and associated emotional responses. Nowadays, it has been understood that even a product with good usability can cause negative experience or dissatisfaction (Jokola, 2004). It is thus imperative to study all the factors that can positively influence the experience outcome by focusing on the perception and the user's emotional responses in addition to the usability aspects. Therefore, the interactive product design has crossed the step of the usability toward aspects to assess the comprehensiveness of UX, such as pleasure, intellectual stimulation, pride and affection (figure 3).

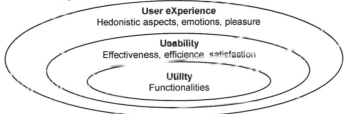

Figure 3 The evolution of system interactive design

In his thesis, (Arhippainen, 2009) presents a novel framework called U2E-Frame for planning and conducting User eXperience test. It depicts that when a user is using some product, there are several influencing factors in all parties. It elicits what factors have an impact on User eXperience in a user-product interaction.

3.1 User eXperience in Service Design

A service design is a multidisciplinary approach to creating useful, effective, and efficient services (Seffah, Kolski, and Idoughi, 2009). It tends to develop better ways for people to access the services they need. Today, most products are combined with services, thus it is the overall experience that counts and that is finally judged by customers. The emerging field of service design combines some design methods mainly from product, software and interaction design, for designing the experience and interface to services. In addition, emergence of the internet arise a several new possibilities for services and experiences.

In this section, we discuss the consideration of User eXperience in service design. In Internet services experience, (Chen and Chang, 2003) have identified three components in the online shopping experience: interactivity, transaction and fulfillment. Recently, (Väänänen, Väätäjä, and Vainio, 2009) presented the affecting factors specific to service User eXperience (SUX) and their corresponding key elements which are: the *composite nature of services* that affect the *trust and coherence of service interaction*; the *presence of social environment* that affect a *social navigation and interaction*; the *dynamically changing of user interface* affecting the *temporal nature* of SUX, the *intangibility* of interaction effects lead to *nonphysical interaction* and the *multidevice access* devices affect the *styles of multiple interaction*.

4. USER EXPERIENCE ELSEWHERE

The concept of experience was widely used by different practitioners from other areas, and this well before the advent of User eXperience concept in the design of new technologies (services, digital product, etc.). We cited below some works that focused in human experience to achieve their goals (costumer attraction, profiling of population, criminal profiling, etc.). We address UX in marketing and branding, UX in sociology and psychology and UX in criminology.

Branding is a broad ideal which encompasses a multitude of elements (identity, usability, customer service, delivery, desirability, follow-through, etc.). (Rondeau, 2005) explores in his work the relationship between branding and mobile development applications. He claims that brand is experienced indirectly when others tell us about the product or service, and the best way to establish a brand is to create a positive direct experience that can only be achieved through the design of the application. In marketing field, marketers spend often more and more money trying to influence consumers, especially on Internet services. (Dahlen, 2002) discuss how Internet users change their behaviors and responses to marketing with increasing experience. Several factors were cited when experienced users change their behaviors such as automaticity and changing advertising context effects (i.e., involvement and pleasure). All this effects should influence user behavior that come from increasing Internet experience. A user profiling is another subject which interested marketers to improving a consumer knowledge management. (Vertommen, Janssens, and Moor, 2008) describe an algorithm to automatically construct expertise profiles for company employees, based on documents authored and read by them They suggest that a good user profile should represent user expertise at different levels.

(Forest and Arhippainen, 2005) describe a methodology called *sociology of User eXperience* to observing and anticipating cultural aspects in society. (Battarbee, 2004) has also elaborated the concept of User eXperience by situating it in social interaction called *co-experience*. Whereas, in psychology, (Mandryk, Inkpen and Calvert, 2006) describe two experiments designed to test the efficacy of physiological and psychological measures as evaluators of User eXperience with entertainment technologies. They suggest that by using physiological and psychological responses lead to objectively evaluate a user's experience with entertainment technology.

(Turvey, 2000) describes criminal profiling as any process elucidating notions about a sought-after criminal. He differentiates between two kinds of criminal profiling: inductive and deductive. The former involves broad generalizations and/or statistical reasoning (subjective approach). However, a deductive method of criminal profiling (objective approach), in which the criminal profiler possesses an open mind; questions all assumptions, premises, and opinions put forth. According to (McCord, 2002) criminal profiling offers a paradigm of understanding crime through offenders' behavior and personality characteristics. Furthermore, (Rogers, 2003) discuss the role that profiling can and should play in the computer forensics process. His says that although the nature of evidence may be evolving we need not totally abandon traditional investigative approaches, but merely allow them to evolve.

5. DISCUSSION

By studying closely the presented works previously, there is the importance of understanding experienced users behavior and causes leading to such behavior.

Personality, social and cultural environment, different physical and psychological individual situations are crucial for, not only the assessment of the quality of the resulting experience (purchase of a new product, adaptation of the people in a new society, operating modes used by criminals, etc.), but also in the prediction of this experience and its consequences before even its trigger. We believe that the techniques used in these areas (measure of the influence of branding on the consumer, construction of hierarchy for a population, profiling of criminals) can enrich positively the exact understanding of the components of the User eXperience, their interactions and also their involvement in the construction of the designs exciting a good User eXperience.

In table 1 we summarized the roles and benefits of some elements coming from different fields in service User eXperience design. This can bring new related service design opportunities and challenges.

Contributing fields	Relevant models and techniques	Related service User eXperience design opportunities
Sociology, anthropology and psychology	Co-experience framework: collaborative User eXperience (Batterbee, 2004) Pragmatic/Hedonic model (Hassenzahl, 2003)	Social navigation and interaction : improve participant design (Mahlke, 2008); Co-creation of services by users and service designers (Väänänen, Väätäjä, and Vainio, 2009).

Marketing and branding	Consumer clustering (Vertommen, Janssens, and Moor, 2008) Persona (Idoughi, Seffah, and Kolski, 2011)	Trust and coherence of services interaction: trust created using an open communication with users communities; cooperation between users and trust service providers.
Criminology and profiling	Hierarchical criminal profiling (McCord, 2002) Operating modes (Turvey, 2000).	Social navigation and interaction; Create coherent design compared to dynamic (or temporality) of the experience (Karapanos, 2010).

Table 1 Related service User eXperience design opportunities with different research fields elements

6. NEW CHALLENGES AND PERSPECTIVES FOR UX DESIGN

Researchers in HCI fields have understood that the user has to be taken into account in design as a comprehensive person, not just a user of product or service. However, there have been critical discussions on whether designing experience is even possible. According to (Sherman, 2007), the designer can only try to influence the User eXperience. (Sanders, 2001; Roto, 2006) were also arguing that designing User eXperience is impossible because experiencing is in people and we cannot know how user will actually perceive and appreciate the product. However in recent studies in UX, some researchers focalize in UX modeling that take account some existing models, as pragmatic/hedonic model (Hassenzah, 2003), or (Schain and Ling, 2008) UX model with web sites. Recently (Zhou, Jiao, and Xu, et al., 2011) propose a systematic approach to UX modeling for product ecosystem design, by using fuzzy reasoning petri net (FRPN). Others works tend providing a set of design tools for capturing and modeling User eXperience. For example, *TUX Modeler* (Wolff and Seffah, 2011) is proposed to model the User eXperience using persona and design patterns methods. We cite also a *Design tool for UXD* proposed in (Yamazaki and Furuta, 2010) that focused on lifecycle, environment and user viewpoints. For all those works, it demonstrates that UX design is possible when we give an appropriate formalism to overall variables that form UX.

6.1 Capturing User eXperience by persona for service design

A persona is a fictional and typical characterization of a user created to represent a user group (Seffah, Kolski, and Idoughi, 2009). Persona often include a name, photo, likes and dislikes, habits, background and other information collecting from real users, and using many techniques (by observing interviewing, focus group, etc.) (Idoughi, Seffah, and Kolski, 2011).

In User eXperience design, the persona can to be a helpful technique, with many resulting benefits, in closing the gap between service design tool's functionalities and the intend users' tasks and also their experiences.

In our context, the persona as HCI design tool, can not only communicate and share user's needs, but also depicts the elements that creating a good User eXperience. For example, the degrees of pleasure, mood and joy when using some functionalities of interactive service. The psychophysiological responses of users about the brand of services' owners can also be integrated to the persona description. Thus, using persona to support a service design has great benefit to evaluate new features and to help make pertinent design decision.

7. CONCLUSION

The academic research in User eXperience design field is very recent and the definitions and frameworks are still being designed. Thus, value should be given not only to theoretical researches, but also to create a common ground between all related User eXperience fields, which aim to build solid knowledge about service experience design. This study intends to give some purpose to improve this knowledge. We have tried to give in first our work definition about User eXperience after reviewing the main existing definitions of this concept. Moreover, we have presented some work from other areas (psychology, sociology, Criminology, etc.) to take some relevant practices to apply in the service design. Finally, we have presented a persona technique that can be extended for capturing the most important information about User eXperience and to sharing them between members of multi-disciplinary design team. Through this work, we hope to provide some answers about the scope of the User eXperience in different scientific fields and that can greatly enrich the service design.

REFERENCE

Arhippainen, L. 2009. "Studying User eXperience Issues and Problems of Mobile Services-Case ADAMOS: User eXperience (im)possible to catch?". Thesis dissertation- May 2009, University of Oulu, Finland. Accessed October 25 2011, www.digibusiness.fi/uploads/reports/1265363661_arhippainen.pdf.

Battarbee, K. "Co-Experience: Understanding User eXperiences in Social Interaction". Thesis dissertation- *Publication series of the University of Art and Design Helsinki* (2004).

Bevan, N. "What is the difference between the purpose of usability and User eXperience evaluation methods?". In *UXEM'09 Workshop*, INTERACT 2009, Sweden (2009).

Chen, S. J., and Chang, T. Z. "A description model of online shopping process: Some empirical results". In International Journal of Service Industries Management, 14, pp. 556-569, (2003).

Dahlen, M. 'Learning the Web: Internet User eXperience and response to Web Marketing in Sweden'. In *Journal of Interactive Advertising*, Vol 3 No 1 (Fall 2002), pp. 25-33.

Forest, F., and Arhippainen, L., "Social acceptance of proactive mobile services: observing and anticipating cultural aspects by a Sociology of User eXperience method". In *Joint sOc-EUSAI conference Grenoble, Oct. 2005*. ACM New York, NY, USA (2005).

Forlizzi, J., and Battarbee, K. 2004. Understanding Experience in Interactive Systems. In. *Human-Computer Interaction Institute*. DIS2004, August 1–4, 2004, Cambridge, Massachusetts, USA. ACM (2004).

Garrett, J. J., "Elements of User eXperience", book, New Riders Press, 2002.

Hassenzahl, M., The thing and I: Understanding the relationship between user and product. In *Funology: From Usability to Enjoyment*, pp.31-42, Kluwer Academic Publishers, 2003.

Hassenzahl, M., and Tractinsky, N. (2006). "User eXperience – a research agenda". In Behaviour & Information Technology, Vol.25, pp. 91-97.

Hassenzahl, M., "User eXperience (UX): Towards an experiential perspective on product quality". In *IHM'08*, Metz, France, pp. 11-15, ACM 2008.

Hassenzahl, M., Diefenbach, S., and Göritz, A. "Needs, affect, and interactive products – Facets of User eXperience". In *Interacting with Computers*, Vol.22 (2010), pp.353-362.

Idoughi, D., Seffah, A., and Kolski, C. 2011. Adding User eXperience into the interactive service design loop: a persona-based approach, In. *Behavior & Information Technology*, 2011, 1–17.

ISO FDIS 9241-210 (2009). *Ergonomics of human system interaction* - Part 210: Human-centred design for interactive systems (formerly known as 13407). International Organization for Standardization (ISO). Switzerland.

Jääskeläinen, A. User eXperience: Tools for Developers. In T. Gross et al. (Eds.): Interact 2009, Part II, LNCS 5727, pp. 888–891, Springer 2009.

Javahery, H. Pattern-Oriented UI Design Based on User eXperiences: A Method Supported by Empirical Evidence. PhD Thesis, Concordia University, Montreal, 2007.

Jokola, T., "When Good Things Happen to Bad Products: Where are the Benefits of Usability in the Consumer Appliance Market?". In *Interaction* 2004, Oulu University, Finland.

Karapanos, E. 2010. "Quantifying Diversity in User eXperience". Thesis dissertation at *Eindhoven University of Technology*, Netherlands, ISBN: 978-90-386-2183-8 (2010).

Kuniavsky, M. "Smart Things: Ubiquitous Computing User eXperience Design", book. Morgan-Kaufman, 2010.

Law, E., Roto, V., Hassenzahl, M., Vermeeren, A., Kort, J., Understanding, Scoping and Defining User eXperience: A Survey Approach. In *Proceedings of Human Factors in Computing Systems conference*, CHI'09, Boston, MA, USA. ACM (2009).

Mahlke, S. "User eXperience of Interaction with Technical Systems". Thesis dissertation at University of Technology of Berlin (2008).

Mandryk, K., Inkpen, M., and Calvert, T. W. "Using psychophysiological techniques to measure User eXperience with entertainment technologies". In *Journal of Behaviour and Information Technology*, Vol. 25, No. 2, 2006, pp. 141 – 158.

Marcus, A., Ashley, J., Knapheide, C., Lund, A., Rosenberg, D. and Vredenburg, K., A Survey of User-Experience Development at Enterprise Software Companies. In *Human Centered Design*, HCII 2009, LNCS 5619, pp.601-610, Springer 2009.

McCord, E. S., "Book Review of Criminal Profiling: An Introduction to Behavior Evidence". In *Western Criminology Review* 4 (1), pp.83-85 (2002) San Diego: Academic Press.

Nacke, L. E. "Affective Ludology: Scientific Measurement of User eXperience in Interactive Entertainment". Thesis dissertation at Blekinge Institute of Technology, Sweden (2006).

Nielsen and Norman Group (2007). "User eXperience – Our Definition". Available at : http://www.nngroup.com/about/userexperience.html (Accessed October 25, 2011).

Pabini, G. P. "What is User eXperience?". Available at : http://www.uxmatters.com/glossary, UXmatters (2008).

Rogers, M., "The role of criminal profiling in the computer forensics process". In *Computers and Security*, Volume 22, Number 4, May 2003, pp. 272-298, Elsevier 2003.

Rondeau, D. B., For Mobile Application, Brandin is experience. Communications of the ACM July 2005/Vol. 48, No. 7. ACM (2005).

Roto, V. (2006). "Web Browsing on Mobile Phones-Characteristics of User eXperience". Thesis dissertation at Helsinki University of Technology, Espoo, Finland (2006).

Schaik P., and Ling, J., Modelling User eXperience with web sites: Usability, hedonic value, beauty and goodness. In Interacting with Computers 20, pp. 419–432. Elsevier 2008.

Schulze, K., Krömker, H., "A Framework to Measure User eXperience of Interactive Online Products". In *MB'10 August 24-27, 2010*, Eindhoven, Netherlands. ACM 2010.

Sanders, E.B.-N. "Virtuosos of the experience domain". In *Proceedings of the 2001 IDSA Education Conference* (2001).

Sherman, P.J. (2007) "Envisioning the Future of User eXperience". UXmatters. Available at : http://www.uxmatters.com/MT/archives/000184.php. Cited 2007/08/30.

Turvey.B., "Criminal Profiling: An Introduction to Behavioral Evidence Analysis", In *Am J Psychiatry* 157: pp.1532-1534, Sep. 2000.

Väänänen, K., Väätäjä, H., and Vainio, T., Opportunities and Challenges of Designing the Service User eXperience (SUX) in Web 2.0. In Future Interaction Design II, pp. 117-126. Springer-Verlag 2009.

Vertommen, J., Janssens, F., Moor,B. D., and Duflou, J. R. 'Multiple-vector user profiles in support of knowledge sharing'. In *Information Sciences* 178 (2008), pp.3333-3346.

Wolff, D., and Seffah, A. "UX Modeler: A Persona-based Tool for Capturing and Modeling User eXperience in Service Design". In IFIP WG 13.2 Workshop at INTERACT 2011 : Patterns, Usability and User eXperience (PUX 2011), pp.7-16, Lisboa, Portugal.

Wright, P., and McCarthy, J., "Empathy and Experience in HCI", In *CHI 2008 Proceedings - Dignity in Design*, April 5–10, 2008, Florence, Italy. ACM (2008).

Wright, P., Wallace, J., and McCarthy, J., "Aesthetics and experience-centered design". In *ACM Trans. Comput.-Hum. Interact.* 15, 4, Article 18 (Nov. 2008), ACM (2008).

Yamazaki , K., et Furuta, K.,"Design Tools for User eXperience Design". In J. Jacko (Ed.): Human-Computer Interaction, Part I, HCII 2007, LNCS 4550, pp. 298–307, 2007. Springer-Verlag Berlin Heidelberg 2007.

Zhou, F., Jiao, R. J., Xu, Q., and Takahashi, K. "User eXperience Modeling and Simulation for Product Ecosystem Design Based on Fuzzy Reasoning Petri Nets". In *IEEE Transactions on Systems*, Man, And Cybernetics—Part A: *Systems And Humans*. IEEE (2011).

CHAPTER 27

A Proposal of the Service Design Method and the Service Example Based on Human Design Technology

Keita Yasui, Toshiki Yamaoka

Wakayama University
Wakayama, Japan
s145051@center.wakayama-u.ac.jp

ABSTRACT

The purpose of this study is to propose the service design method and the service example using Human Design Technology (HDT) as a logical design method. And a GUI design example is also shown. Proposed method is composed of HDT, the service design Framework, and Service design items in B2C.

According to the result of the evaluation, the example using this method was evaluated highly. The logical design method causes the design proposal to be an original design efficiently.

Keywords: Service design, Human Design Technology (HDT), GUI design

1   INTRODUCTION

The purpose of this study is to propose the service design method and the service example using Human Design Technology (HDT) as a logical design method. And a GUI design example is also shown using this method.

2   PROPOSED METHOD

The service design framework and 3 service design items as service design method based on HDT were proposed.

2.1   Human Design Technology (HDT) (Yamaoka,2000)

HDT (Human Design Technology) is a kind of system development method.
The following steps show the HDT process (see Figure 1).

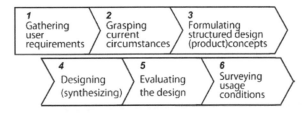

Figure 1. illustlation of the HDT process.

70 design items are included in HDT and composed as follows.
(a) Usability and user interface design items (29 items)
   < Construction of user-oriented UI system>
   Flexibility, Customization for different user levels, User protection
   Universal design, Application to different cultures.
   < Encouragement of the user's motivation >
   Encouragement of the user's motivation, Provision of user enjoyment,
   Provision of sense of accomplishment , The user's leadership, Reliability.
   <Construction of effective interaction>
   Clue, Simplicity, Ease of information retrieval, At a glance interface, Mapping ,
   Identification , Consistency, Mental model , Presentation of various information ,
   Term/Message , Minimization of users' memory load, Minimization of physical
   load, Sense of operation, Efficiency of operation, Emphasis, Affordance,
   Metaphor, System Structure, Feedback, Help.
(b) Kansei(sensitivity) design items (9 items)
   Color, Fit, Shape, Functionality and covinience, Sense of material, Design
  images, Ambiance, New combinations, Unexpected application.
(c) Universal design items (9 items)
   Adjustability, Redundancy, Understanding function and feature at a glance,
   Feedback, Error tolerance, Acquisition of information,
   Understanding and judgment of information, Operation,
   Continuity of information and operation
(d) Product liability (PL) design items (6 items)
   Elimination of risk, Fool proof, Tamper proof, Guard, Interlock, Warning label
(e) Robust design items (5 items)
   Strong material, Examining shape, Strong structure , Design reduced or avoided
stress, Design for unconscious behavior
(f) Maintenance items (2 items)
   Keeping space, Easy operation
(g) Ecological design items (5 items)
   Durability, Recycling, Very few materials, Most suitable materials, Flexible
   design.

(h) Other (5 items)

Physical aspect ,Information aspect ,Temporal aspect, Environmental aspect, Organizational aspect

2.2   The service design framework (yamaoka,2010)

The structure of service design consists of the following 5 aspects. The overview of the service design framework is illustrated in Figure 2.

1) Service organization system

As the policy of service organization system influences the service design concept, the policy is very important especially.

2) Service design concept

The service concept is structured. The concept items on the second layer of service design concept are weighted for making the service design concept concrete and clear.

3) Interaction between customers and service employees / machine.

4) Produced good service quality

As good service is imitated sometimes and customers are tired of the service, a new service should be developed.

5) Increased service productivity

6) Service design

The service design is done based on above-mentioned items. If customers' evaluation for service is better than their expectation, they feel satisfaction with the service.

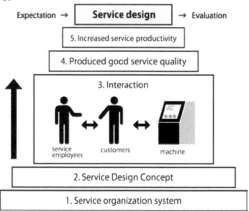

Figure 2. illustlation of the service design framework

2.3   Service design items in B to C:

The service design items are used in the relationship between customers and service employees.

(a) Careful consideration, (b) correspondence to the situation and (c) caring attitude are 3 service design items for designing and evaluating the relationship between customer and service employee.

(a) The careful consideration is related to grasp situation and feel sympathy.

(b) The correspondence to the situation means to correspond quickly, flexibly, precisely, equally and at ease, and confirmation such as an order in a restaurant and so on is also important in interaction between user and employee.

(c) The caring attitude means making a good impression, getting sympathy, getting trust and being tolerant of mistakes.

2.4 Structured service design concept

The structured service design concept is constructed based on 70 design items and 3 service design items. The service is visualized according to the structured service design concept.

3 AN EXAMPLE USING THIS METHOD:

"Web printing service" using this method was designed. Firstly the policy in this service was made. The policy is "customer centered service policy".

Next, according to the policy, the requirements such as "can print anywhere" and so on were extracted. Based on these requirements, the main concept is "Everyone prints anytime anywhere". Sub concepts are "usable (60%)" and "easy to operation (40%)".Each sub concept is weighted and visualized based on 70 design items in HDT items and 3 service design items. Figure 2 is the part of shown structured service design concept.

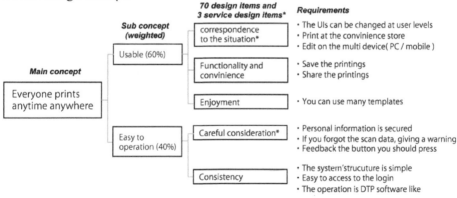

Figure 2. The service design concept of a web printing service.

The system design was made based on this concept (see Figure 2). This service is to print at the convenience stores by the mobile phone or PCs. The image of this system is shown in figure 3.

274

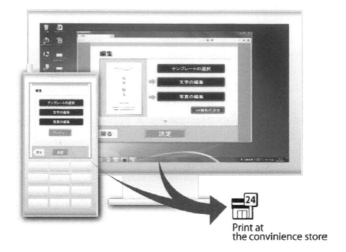

Print at
the convinience store

Figure 3. .The image of proposed web printing service

3.1   Evaluation

The proposed method was evaluated in order to check the validation.   The GUI design created using this method was compared with the conventional one. The participants of the study are three university students (21 years old).
An example was evaluated by GUI Checklist, SUM (Simple Usability Test) and Usability Task Analysis (Yamaoka,2007). The usability task analysis carried out two times (conventional / using proposed method). Its result is shown in Table 1.

Table 1 A Part of usability task analysis's results.

Screen	Conventional	Using proposed method
Home	**Good:** The user's task is clear. **Bad:** The relationship between items is complex. **Score: 1**	**Good:** The user's task is clear. **Bad:** none **Score: 3**
Login	**Good:** After inputting user ID, OK button is shown. **Bad:** none **Score: 2**	**Good:** After inputting user ID, OK button is shown / Simple Design **Bad:**none **Score: 3**
Edit	**Good:** Mapping information. **Bad:** Technical terms, Back button is not found. **Score: 1**	**Good:** Easy to recognize. **Bad:** The description is noisy **Score: 1**

Print	Good: none Bad: No confirmation. Picture only. Score: 1	Good: Confirmation screen is good. Bad: The screen is noisy Score: 2
...	...	...

4   CONCLUSIONS

According to the result of the evaluation, the example using this method was evaluated highly. The logical design method causes the design proposal to be an original design efficiently.

REFERENCES

Yamaoka,T.2003. Introduction to Human design technology, Morikita publishing.
Toshiki Yamaoka 2010, A study on service design method based on Human Design Technology, Proceedings of the 2nd International service Inovation design Conference, 107-112.
Toshiki Yamaoka 2007, Japan Ergonomics Society, The section meeting of Ergo-Design, Task analysis for usability evaluation.

CHAPTER 28

Facilitating Service Development in a Multi-agent Environment: Different Roles of Consultancy

Inka Lappalainen, Katri Kallio and Marja Toivonen

VTT Technical Research Centre of Finland
Espoo, Finland
inka.lappalanen@vtt.fi, katri.kallio@vtt.fi and marja.toivonen@vtt.fi

ABSTRACT

The topical issue of the co-development of services in multi-agent environment is approached with the concepts of social and value chain innovations as well as prerequisites for networked activity. In this paper the aims, practices, and challenges of co-development between municipal organizations, local SMEs and citizens are studied. The specific focus is on the roles of the external consultant in the co-development. Two empirical case studies, carried out in Finland, deepen understanding on the dynamics of service co-development, the importance of external facilitation as well as multifaceted innovation activity. Promising, tentative results are presented.

Keywords: Co-development, innovation, multi-agent network, KIBS

1.   INTRODUCTION

Western economies are referred as service economies (Vargo & Lusch 2008). However, public services face critical and complex demographic, economic, and structural challenges. This study illustrates the situation in Finland, whose municipal sector is under pressure to find more efficient, customer-oriented ways to organize, produce, and develop services. Citizens have shown growing interest in participating in this development, while digitalization and social media enable new kinds of services and their co-development. In addition, trends such as new public management and public–private partnerships have opened new opportunities for co-operation between municipalities and local companies. However, innovative thinking and novel participatory practices are required. (See Kivisaari & Saari 2012; Hartley 2005.)

Studies of service innovation and new service development (NSD) have focused on the provider–customer dyad and less on co-development in a multi-agent environment (e.g. Smith & Fischbacher 2005; Syson & Perks 2004). Recently, adoption of new concepts has broadened the perspective. For the public context, the concepts of value-chain innovation and social innovation provide relevant frames, which we present as theoretical background. They help to deepen understanding of multi-agent co-development (Sundbo 2011; Harrison et al. 2010). This paper examines the aims, practices, and challenges of co-development in the public (particularly municipal) context.

If we are to achieve goal-oriented, efficient, and useful co-development from the viewpoint of all parties involved, the issue of process facilitation is urgent. However, in-depth studies of knowledge-intensive business services (KIBS) companies' facilitation practices in multi-agent environments are still rare (cf. Miles 1999; den Hertog 2002). The present study aims to narrow this research gap: it systematically analyses the roles of a consultant in co-development.

Our empirical material is based on two case studies carried out in Finland. Both featured a facilitator and representatives of a municipal organization, citizens, and companies. The primary aim for the municipality was to construct a general model for the co-development process. This model was defined, tested, and improved collaboratively in two multi-agent contexts. We suggest as our main conclusion that, from an innovation perspective, the cases provided a basis for both a social innovation and a service innovation.

2. THEORETICAL BACKGROUND

2.1 Social and value-chain innovations facilitated by KIBS

With efficiency and renewal pressures, the roles of various players in public services are changing. Municipal and private organizations find it challenging to understand the profound changes needed in their core tasks and how they should manage new service development with their partners and citizens.

Recent innovation literature uses the concept of social innovation to refer to collaborative innovation processes addressing complex economic and social problems. Social innovations can be created at three interlinked levels: grassroots level, among individual citizens; the intra- or inter-organizational level by private, public, and third-sector organizations; and the societal and policy level, in the form of radical systemic changes. (Harrison et al. 2010.) The concept of value-chain innovation (Sundbo 2011) has in common with social innovation by focusing on the innovation process, which consists of dynamic interaction and participatory practices among varied actors. They link social and economic elements, claiming that goal-oriented co-development contributes to better services, welfare, and new business opportunities while boosting empowerment, and collective responsibility (Harrison et al. 2010; cf. Sundbo 2011). Because social innovation is a shared phenomenon with differing targets and timelines, often without a clear leader, there is a danger of co-development not progressing effectively. On the other hand, it opens possibilities for new actors to be involved (cf. Kivisaari & Saari 2012). Thus, new, manifold, and changing needs for the facilitation of innovation processes are emerging.

Today's KIBS companies rely heavily on professional knowledge and provide intermediate services to be used for the creation of end products and services. Their customers are other businesses and the public sector. They are typically leading users of information technology (IT) as a means to support customers' activities and their own (Miles 1999). The figure below illustrates the framework of our study: the interaction of different stakeholder groups and a KIBS company in social innovation.

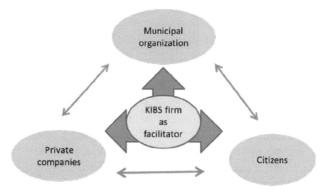

Figure 1. KIBS as facilitator of multiple actors' co-developed social innovation.

2.2 Multi-agent networks as forums for social innovation: Aims, challenges, and collaboration practices

In addition to value chains, social innovations can be facilitated and studied as multi-agent networks, typically with diverse innovation sources and goals. Participation interests may range from political to professional and business to voluntary.

A municipal organization in co-development. Public organizations have often been regarded as problematic from the innovation standpoint: legislation, local politics, communal decision-making culture, and organizational inertia are typically seen as the main inhibitors (e.g. Hartley 2005; cf. Sundbo 2011). Innovation activity is rendered complex by differing, even conflicting, drivers related to policy, professional and managerial issues. Traditionally, innovations are seen as top-down activities initiated by policymakers rather than bottom-up activities (Hartley 2005). Recently, growing interest has been shown in bottom-up activities: combining employee- and user-driven approaches (Hasu et al. 2011). However, this calls for commitment from the top management. Smith and Fischbacher (2005) point out a strategic role of new service development: granting legitimate power for an innovation activity. Co-development calls for new competencies in involving and managing complex and multilevel processes.

Involvement of local SMEs. For local small and medium-sized enterprises (SMEs), multi-agent networks may serve as important arenas for renewal. As Syson and Perks (2004) claim, innovation networks should be seen and utilized more as a flexible resource pool, which may also create a strategic competitive edge, difficult to imitate. The main interest in co-development from SMEs' viewpoint lies in short- and long-term business potential, which networking, benchmarking, and exercise influence on local politics, may create. Sundbo (2011) stresses entrepreneurship as the key driver of value-chain innovations. However, practice oriented SMEs hesitate to commit themselves into processes whose benefits cannot be seen clearly.

Citizens as end users and participation via social media. Even though co-development with customers and users has been discussed a great deal in literature, participation seems challenging in public service development in practice (e.g., Brabham 2009; Hartley 2005; Smith & Fischbacher 2005). However, citizens now have more interest in societal issues and willingness to take active part in decisions related to their day-to-day life and environment. While e-government solutions present top down services that support current political processes, social media may create new possibilities for bottom-up innovation (Bäck et al. 2012). Co-development via social media may provide so-called 'wisdom of crowds' and practical involvement of citizens, as active participants in ideation and piloting of new solutions (Brabham 2009).

In summary, the following issues seem to be essential considerations when organizing co-development in the public–private–citizens context:

- Cross-sector collaboration and combination of complementary competencies and personalities
- Commitment and strategic orientation of the top management
- Dialogue between top-down and bottom-up activities
- Clear ownership in the innovation process
- Entrepreneurial spirit and fit between parties
- Attention to power relations and balance among the various interests in iterative and knowledge-intensive processes
- Integration of material, service, and experiential elements
- Technological infrastructure and competencies
- Enabling mechanisms within organizational parties and between interest groups
- Understanding of the multifaceted nature of innovation.

2.3   KIBS as facilitators of social innovation

Studies of the activities of consultants (and, more broadly, of KIBS) have indicated that an innovating organization can gain fundamental benefit from using an external facilitator (den Hertog, 2002). Three roles of KIBS in innovation are identified: 1) facilitators (supporting the innovation process), 2) carriers (transferring the innovations emerging), and sources (creating innovations themselves) of innovation. The facilitating function of KIBS is our focus in this paper and may include the following (Miles, 1999):

- Performing diagnosis and problem-clarification
- Synthesizing general expert knowledge with context-specific issues
- Experience-sharing and benchmarking
- Brokering and promoting the interaction of actors of different types
- Acting as an agent of change

In addition to the apparent utility of these various functions of KIBS, the companies are often highly innovative in their own right (den Hertog 2002). They seek to renew their services and create service innovations constantly, for differentiation. In technology-oriented KIBS, one of the main trends is a move toward customer- and value-orientation (cf. Vargo & Lusch 2008); here, human, process, and organizational knowledge are increasingly important.

3.   PROCEDURE FOR THE EMPIRICAL STUDY

Empirical material has been collected from two case studies conducted in multi-agent network settings. Case study A (co-designing a meeting place at a market square) was conducted between August 2010 and February 2011. Case study B (co-developing collaborative modes between a 'Youth workshop services' and local SMEs) started in August 2011 and is, at the time of writing, still in progress, to be finished in March 2012. The parties involved in both were a medium-sized municipality (approx. 50,000 inhabitants), a few SMEs, and citizens. A technology-oriented KIBS company interested in offering new kinds of services to municipalities acted as a facilitator. The first two authors of this paper have been involved in both cases from their planning to their current state.

The main approach applied in the study was participatory action research (McIntyre 2008) that included participatory activities, observation, and the use of service design tools. Our main dataset consists of the memos from the meetings and collaborative workshops. We also interviewed the key persons involved, from the municipality (in case A) and the KIBS company (in both cases). End-user data were collected from the collaborative workshops; discussions via the Web platform; and, in case A, the feedback in evaluation of the process. Qualitative analysis based on theming and categorization was developed from the theoretical background described.

4.   MAIN RESULTS

The aims, collaborative practices, and challenges of the networked co-development are summarized in tables 1 and 2, below, from the viewpoint of different actors. Table 3 presents the main roles of KIBS.

Table 1 Aims and expected results of the co-development

	Citizens	(SME) companies	Municipal level	KIBS
Case A	-Personal interests -Communal decision-making regarding the local environment	-A designer (not local) -Provision of broad service solutions for a renewing marketplace -Keeping the customer satisfied; remaining the main service provider	-Co-designing a meeting place at the market square -Co-designing a collaborative model to enhance service innovations and boost civic participation	-Develop and pilot a new service concept, to create new service business in the future
Case B	-Personal interests -Rehabilitation -Practising collab. -Employment opportunity	-Recruitment and training channel -New, flexible ways of utilizing resources -Social responsibility	-Renew collaboration modes between youth workshops and local SMEs -Apply and diffuse the co-development approach	-Test and validate the new service concept and explore the business potential

Table 1 shows the multi agent networks' multiplicity of interests, discussed in the next section. In more detail, in both cases the co-development process included three sub-processes: 1) enabling the participation of the end-user groups (citizens and local SMEs), 2) utilizing collaborative technologies to support interaction, and 3) planning and managing the tasks of the back-office consortium (municipal and KIBS). Since service development is highly context-dependent, we found that the processes had to be rebuilt in practice in both cases; however, having general features in common. In Table 2, the main features of networked co-development and related challenges are described, based on the literature review.

Table 2 The features and challenges of the co-development

Criteria	Case A	Case B (on-going, tentative)
Cross-sector competencies and personalities in the core group	-Complimentarity -Inspiration and solution-orientation -Municipal employees' exploration of renewal/change in expert roles	-Complimentarity **Challenges**: Heterogeneity and temporary work contracts
Commitment / dialogue between top-down and	-Top mgmt. commitment to the pilot (municipal EE's chosen target) **Challenge**: Scarcity of	-Need for facilitation from top mgmt. (core group chose the focus) -Dev. resources' allocation:

bottom-up activities/interest	resources; internal confusion of authorities (legitimacy)	strong political and strategic message (legitimacy) **Chall.**: Internal tension
Ownership of the innovation process	**Chall.**: Confusion between the facilitator and the municipal (risk of organisational inertia)	-Clearer definition **Chall.**: Project-based dev. team as an owner on behalf of the youth workshop (risk of org. inertia and complexity)
Entrepreneurship and fit	-Creativity and networking -Fit with top mgmt. of municipalities **Chall.**: Local SMEs (invited by trade assoc.) not being interested in co-dev. (a designer was involved too late)	-Mutual benefit and entrepreneurial spirit stressed but not yet fully reached **Chall.**: Means of involving local SMEs and trade assoc. (initial collab. interest but involvement limited by resources)
Power relations and balance of interests	-Experience of community spirit and chance to exert influence -Contrib. of dialogue and respect for each other's competence/ experience	-Continuous combination of interests **Chall.**: Collab. with local SMEs as main interest (however, left in a rather minor role amidst the complex renewal)
Integration of material, service, and experiential elements	-Inspiring and purposeful acts serving several process aims and preferences of interest groups -Service design methods (prototype exhibit.): great user experience /feedback **Chall.**: Resource requirements	-Application -Service design methods (solution-based role-playing) to test and evaluate the new collaborative modes between SMEs, municipal employees, and youth quickly
Technological infrastructure and competencies	-Facilitator who provided a tailored, open Web platform -Positive user/municipal experience -Recognition of the importance of the local Web moderator	-Facilitator recommendation of an open Web-based platform (with the municipal org. in operators' roles) **Chall.**: Utilization of social media (means, competence and resources)
Enabling mechanisms within and between interest groups	-Regular virtual core-group meetings -Flexible process model, roles, schedule, and risk analysis -Dynamic multi-channel comm. **Chall.**: Incomplete process view (citizens, some municipal employees)	-Regular virtual core-group meetings -Adoption of process-orientation -Pursuit of multi-channel communication **Chall.**: Use of few facilitation tools Communication and co-creative spirit
Understanding of the multifaceted nature of innovations/results	-Co-dev. model for the municipality -Service innovation embryo for the facilitator **Chall.**: Primary/generic co-dev. goal left unclear partic. for citizens: end result was criticized	- Co-dev. process as on-going, aimed at many types of renewals, by different interest groups **Chall.**: Service apparently applicable for the facilitator yet requiring a lot of resources and tailoring

Table 2 shows how, despite their shared general features, the cases differed in their basis and challenges; hence, we assume that they reflect different facilitation needs. The main roles of KIBS in cases A and B are compared to the classification employed by Miles (1999) and summarized in Table 3.

Table 3 The main roles of KIBS in the co-development

Roles	Case A	Case B (tentative)
Performing diagnosis and clarification of problems	Decisions regarding the development target and ad-hoc problem clarification paralleling the process	Stressed at the start: making decisions regarding the development target
Combining general expertise with context-specific issues	Introduction and development of the co-development process	Applying and tailoring the co-development process
Experiencing sharing and benchmarking	Knowledge about collaborative technologies and platforms: offering the bespoke platform and guidance	Making recommendations for usable collaborative technologies and platforms
Brokering and promoting interaction of different actors	Action as a continuous contact point and clarification of core-group roles Encouragement of dynamic outside communication	At the start, promoting roles' clarification in the core group (challenges); encouraging external communication (later)
Acting as an agent of change	Introducing the new way of co-developing to the municipal organization via face-to-face workshops and collaborative technologies	Tailoring the new way of co-developing in the multi-agent network to a more complex development target and authorities

5. DISCUSSION AND CONCLUSIONS

The empirical results showed diverse *aims* and interests in the multi-agent networks: citizens' personal and SMEs' business interests were combined with social engagement and the contribution of 'the common good'. The municipality's aims seemed twofold: in addition to the practical end results, creation and testing of a more general process of co-development was pursued. In addition, the KIBS aimed at creating new business. However, some tensions arose between long-term interests of the municipality's top management and practical, project-based interests of municipal employees, SMEs, and citizens (cf. Bäck et al. 2012).

Comparison of the cases showed that the development targets varied dramatically, from concrete and rather neutral to complex, politically charged issues. Therefore, the bases for establishing and managing *co-development practices* were more challenging in case B. The results thus illustrate the context-dependence of service development and common features and challenges related to networked co-development. The results support recent studies of growing mutual interest in bottom-up and user-driven innovations via social participation and combining local resources for better services (e.g., Bäck et al. 2012; Hasu et al. 2011). Furthermore, they revealed internal challenges within the municipality in movement toward empowerment and co-development (cf. Hartley 2005). Similarly, local SMEs and citizens showed ambiguity; the processes began with enthusiasm but were challenged by the new methods, scarcity of resources, and expected results (cf. Sundbo 2011).

Analysis of the external *facilitator's roles* revealed that the categories of Miles (1999) fit well but received differing emphasis: in case A, diagnosis, problem-clarification, and continuous communication alongside collaboration were emphasized. Also, perceiving the entirety in a target-oriented manner was important. In case B, the customer sought process consultation but challenges emerged because of the poor definition of the needs for facilitation, complicated organization, and questions of management and ownership. Therefore, the role of the consultant culminated in clarification and direction of the co-development. Our theoretical classification regarding the prerequisites for the co-development seemed to be useful. Coupling empirical results with the classification enables deeper understanding of the dynamics of co-development, particularly in the municipal context, providing a framework for management and facilitation.

In addition, as a conclusion our study revealed that co-development in a multi-agent environment can create *two types of innovations* at the same time: *social and service-related.* As a social innovation in case A, the co-development process (a new way of interacting and making decisions) was created, tested, and evaluated. Later, in case B, it was applied, further developed, and diffused for different needs and stakeholders (cf. Harrison et. al, 2010; Hartley 2005). From the results, it seems that such a relatively systematic and multilevel co-development process requires considerable resources and is more applicable (however challenging) to development of targets with strategic importance, where the process of participation is as important as the end results are.

Likewise, the study showed how the co-development enabled the KIBS facilitator to develop a new service concept, to test and apply it in different contexts. Thus the outcome of the social innovation emerged as a service innovation for the facilitator and benefited participants from the communities involved (cf. Harrison et al. 2010). Hence, it seems that KIBS types of organizations may assist in facilitating and creating innovations in public–private contexts (cf. den Hertog 2002; Saari & Kivisaari 2012). In practice, however, it seems that this kind of end-user-driven co-development approach is still a fairly project-based activity rather than a legitimated activity integrated into public decision-making. In terms of validity, in participatory action research the community members participate more as partners in the research than as external members and are acting for social change (McIntyre 2008). The validity improved also by triangulation of the results; nonetheless, the results are still tentative and need further study. More interestingly, what kinds of co-development practices and organizational innovations are required within the participating organizations if the practices are to be reformed in the longer term?

ACKNOWLEDGMENTS

The authors wish to thank the case organizations for their fruitful co-operation. The research has been funded by TEKES (the Finnish Funding Agency for Technology and Innovation), participating organizations, VTT and University of Lapland, Finland.

6. REFERENCES

Bäck, A., Näkki, P., Ropponen, T., Harju, A., & Hintikka, K.A. 2012. From design participation to civic participation – participatory design of a social media service. International Journal of Social and Humanistic Computing, forthcoming.

Brabham, D.C. 2009. Crowdsourcing the public participation process for planning projects. Planning Theory, 8(3), 242–262.

den Hertog, P. 2002. Co-producers of innovation: On the role of knowledge-intensive business services in innovation. In Gadrey, J. & Gallouj, F. (eds), Productivity, Innovation and Knowledge in Services: New Economic and Socio-economic Approaches. Cheltenham & Northampton: Edward Elgar.

Harrison, D., Klein, J.-L., & Browne, P.L. 2010. Social innovation, social enterprise and services. In: Gallouj, F. & Djellal, F. (eds), The Handbook of Innovation and Services. Cheltenham & Northampton: Edward Elgar.

Hartley, J. 2005. Innovation in governance and public services: Past and present. Public Money & Management, 25(1), 27–34.

Hasu, M., Saari, E., & Mattelmäki, T. 2011. Bringing the employee back in. Integrating user-driven and employee-driven innovation in the public sector. In: Sundbo, J. & Toivonen, M. (eds), User-based Innovation in Services. Cheltenham, UK: Edward Elgar.

Kivisaari, S., Saari, E., Lehto, J., Kokkinen, L. & Saranummi, N. 2012. System innovations in the making: Hydbrid actors and the challenge of up-scaling. Technology Analysis & Strategic Management. (Accepted to be published).

McIntyre, A. 2008. Participatory Action Research. Qualitative Research Methods Series, vol. 52. Thousand Oaks, CA: Sage Publications.

Miles, I. 1999. Services in national innovation systems: From traditional services to knowledge intensive business services. In: Kuusi, O & Shienstock, G. (eds), Transformation Towards a Learning Economy: The Challenge for the Finnish Innovation System. Sitra 213. Helsinki, 57–98.

Smith, A.M. & Fischerbacher, M. 2005. New service development: A stakeholder perspective. European Journal of Management, 39(9–10), 1025–1048.

Sundbo, J. 2011. Value chain innovation: An actor-network-theory based model for innovation at the interface between service and other economic sectors. In: Sundbo, J. & Toivonen, M. (eds), User-based Innovation in Services. Cheltenham & Northampton: Edward Elgar.

Syson, F. & Perks, H. 2004. New service development: A network perspective. Journal of Services Marketing, 18(4), 255–266.

Vargo, S.L. & Lusch, R.F. 2008. From goods to service(s): Divergences and convergences of logics. Industrial Marketing Management, 37, 254–259.

CHAPTER 29

Three Approaches to Co-creating Services with Users

Eija Kaasinen, Kaisa Koskela-Huotari, Veikko Ikonen, Marketta Niemelä and Pirjo Näkki

VTT Technical Research Centre of Finland
Espoo, Finland
firstname.lastname@vtt.fi

ABSTRACT

The role of users in service design is changing from passive research subjects to active co-designers and content creators. This new direction can be supported with inspiring physical or virtual spaces where users, designers and other actors can meet informally and participate in service design as equals. In this paper we describe three different approaches to co-creation spaces: web-based Owela, physical showroom Ihme, and Living Labs that combine both physical and web elements. We compare these approaches based on the innovation phase they are most suitable for, the methods as well as the strengths and challenges of the approaches. All the three co-creation spaces manage to bring co-creation close to the users' everyday life. Participation is quite independent of time and in Owela independent of place as well. Users can select their level of contribution, varying from short comments to long-term participation in development projects. Direct designer/user interaction supports turning the designer's mind-set from technical features to user experience. This facilitates the design of services that are accepted by and interesting to users.

Keywords: co-creation, user participation, service design, Owela, Ihme

1 INTRODUCTION

The role of users in service design is changing. Instead of passive research subjects, they are seen as active co-designers and content creators. Users are the best experts in their everyday lives and therefore have great potential as sources of innovation. User participation can affect the success of services directly by better quality, fit to needs and innovation speed. The effects can also be indirect such as more customer-centered image, customer-driven organizational culture and increased motivation of employees.

Today, human-centered design is quite an established practice for designing products and services so that forthcoming users are represented in the design process (ISO, 2010). Human-centered design starts once the decision to design a certain kind of service has been made. To increase the users' role in design and innovation, we should increasingly involve them in deciding what is needed and what kinds of services should be designed for them and with them. Kanstrup and Christiansen (2006) describe this change as changing the user's role in design from a victim who needs support to a valuable source of inspiration.

Co-creation stresses the collective creativity of all stakeholders including end-users (Sanders and Stappers, 2008; Prahalad and Ramaswamy, 2000). A crucial factor for the success of service development is the performance in the early stages of the development process, that is, the 'fuzzy front-end' in which the targeted service has not yet been decided (Khurana and Rosenthal, 1998). User participation could be especially useful at this stage due to its high level of uncertainty and low formalization (Alam, 2006). In addition to ideation, user participation 'at the moment of decision' is attracting increasing interest (Sanders and Stappers, 2008).

In traditional human-centered design, only small numbers of users have been involved in the design activities. New methods are needed to reach the masses of potential innovators. Computer-supported methods for co-creation are promising (Sawhney, Verona and Prandelli, 2005; Schumacher and Feurstein, 2007) but also new kinds of face-to-face collaboration methods are needed.

Co-creation in service design is usually referred to as *value co-creation* (e.g., Lusch and Vargo, 2006; Vargo and Lusch, 2008), which is an integral part of the paradigm called service-dominant logic. In service-dominant logic, value is always determined by the beneficiary (e.g. customer) of the service. This means that companies can only offer value propositions to their customers and actual value is created collaboratively, making customers co-creators of value. Lusch and Vargo (2006) also acknowledge the customer participation in the development of the core offering itself and view it as a component of value co-creation; however they call this co-production. Kristensson et al. (2007) also suggest that co-creation includes two ways of collaboration: value co-creation and co-production, the latter also leading to value-in-use but in a more indirect way. In this paper we use the term co-creation similar to Kristensson et al. when referring to user-involving approach in the innovation of services. We use the term co-creation instead of co-production to emphasize the creative nature of end users' participation.

Co-creation of new services requires approaches that support collective creativity. As design has shifted from work to leisure and pleasurable engagements (Björgvinsson et al., 2010), easy and effortless participation has become increasingly important. The design should happen close to the use context in order to give the users a familiar context to act and the stakeholders a real life experience of use context (Buur and Bødker, 2000). According to Ainasoja et al. (2011), co-creation of services requires clear communication of the goals, open and informal atmosphere, high quality of inspirational and background materials as a basis for innovation, concretizing of the service in situ, documentation and sharing of ideas and notes as well as feedback of the follow-up process of the user-generated ideas. The participants should have alternative ways to contribute depending on their individual interests and competencies. Informal and equal interaction between different actors encourages contributions (Ainasoja et al., 2011).

Our vision is that co-creation can be supported with inspiring physical or virtual spaces where users, designers and other actors can meet informally and as equals. Based on the above described findings from earlier research we have set requirements for our co-creation spaces regarding context, participants, motivation and activities as described in Table 1. The table also describes the requirements we set for data analysis.

Table 1 Requirements for the co-creation spaces

Context	Close to use context, intertwined with everyday life
Participants	Low threshold to participate, for anyone
Motivation	Brings value to all stakeholders, fits personal goals, is fun
Activities	Alternative ways to contribute, depending on participants' interests, time limits and capabilities Encourages creativity and informal interaction
Analysis	Agile gathering and analyzing of data with restricted time and resources Continuous applying of results, iterative development

In the following sections we describe three different approaches to co-creation spaces that we have been developing and using. Open Web Lab (Owela) utilizes social media as co-creation space. Ihme innovation showroom facilitates co-creation in public everyday spaces. Living Labs combine both physical and web elements and interweave design and use. We describe our experiences of co-creation activities in the spaces. Our main focus is on user participation but we also touch the viewpoints of other co-creation actors. We compare the co-creation spaces according to their suitability to different innovation activities, the co-creation methods as well as their strengths and challenges. Finally we conclude with suggestions on the suitability of each co-creation space for different innovation activities.

2   SOCIAL MEDIA AS CO-CREATION SPACE: OWELA

Open Web Lab (Owela, http://owela.vtt.fi/) is an online platform designed by VTT for co-creation between end-users, customers, developers and other stakeholders (Figure 1). Owela is built on social media-type interaction and thus enables user participation regardless of time and place. Owela provides tools and methods for understanding users' needs and experiences as well as innovating and designing new products and services together (Näkki and Antikainen, 2008).

Figure 1 The Owela online co-creation platform

Over 40 different kinds of co-creation cases have been carried out in Owela. In most of the cases, ordinary consumers and citizens have had the chance to interact with companies and researchers in order to create new products and services. Most of the cases have been related to the early phases of the innovation process such as gathering information on needs, generating ideas and evaluating new product and service concepts. There have also been encouraging experiments to involve end-users in the later stages of new product and service development, especially in the software context. Table 2 illustrates some of the studies that have been carried out in Owela.

Table 2 Selected cases of Owela co-creation

Name of the study, length	Participants	Topic	Phase of the innovation process
Mobideas, 6 months	33 users, 4 developers, 2 researchers	Social media service	Idea generation, concept design, development, testing
Monimos, 1 year	70 users, 5 researchers, 1 designer, 1 developer	Multicultural social media service	Idea generation, concept design, testing phase
City Adventure, 1 month	36 users, 1 researcher	City adventure service	Need capturing, ideas, concept evaluation
Events, 1 month	4 users, 3 developers, 3 researchers	Mobile event management service	Prototype testing

Users can participate in Owela studies from their own environment, whether it is at home or on the go. They only need access to internet and basic skills on social software. Owela makes participating in co-creation activities easy for users, regardless of the time and place. Owela encourages users to make micro-contributions, and thus enables contributions also from people who would not have the time to participate otherwise. Users are empowered to act as innovators, design partners and decision-makers as they are continuously connected in the innovation process. Open and transparent design processes have been achieved through Owela. With Owela, designers and developers can reach large numbers of users quickly and cost-efficiently. Owela has enabled companies to establish long-term interaction relationships with the users. The flexibility of the online co-creation platform has enabled ad hoc changes in the implementation of intensive development projects.

Owela enables different levels of participation based on users' own interest. Since most participants will not read long and complicated instructions online, Owela tasks are as short and simple as possible yet contain all the necessary information. Assigned tasks contain possibilities for micro-contributions. Most active users spend multiple hours per week in Owela co-creation and they contribute also to tasks that require more intensive participation (e.g. idea chats). The Owela tool itself does not guarantee success but experienced facilitators are needed. The goals and tasks must be defined beforehand and clearly communicated to the participants. Most of the communication in Owela is text-based, and this has to be taken into account when analyzing users' ideas and comments, as the text may lack some crucial information or be subject to misunderstanding for some other reason. The advantage of web-based co-creation is that all developers have real time access to user feedback without intermediates and are able to ask more questions directly from the users. This helps developers to better understand the users and vice versa.

3   CO-CREATION IN PUBLIC EVERYDAY SPACES: IHME

VTT's Ihme innovation showroom concept (Figure 2) was launched to test and further develop the idea of an open public co-creation environment. Ihme aims to fulfill the existing gap between laboratory research and a living lab approach. Ihme is an open, low threshold environment where ordinary people can visit easily according to their own schedules. People can experience, see and try physical proofs of concepts and other tangible illustrations of new technology and services. Visitors can freely just look and try pilot services, participate in co-creation sessions or just leave their ideas and comments. Presented technology and service pilots are designed so that they are entertaining and fun, e.g. set up in the form of a game. Ihme emphasizes direct designer/user interaction and encourages designers to come and introduce their ideas and discuss of them with potential users. Direct interaction enables agile, iterative development.

VTT's first Ihme environment was set up in the Ideapark shopping centre (Lempäälä, Finland) in a 61 square meter facility in summer 2010 for two months. Besides the Ideapark Ihme Innovation Showroom, more temporary Ihme innovation showrooms have been set up in the contexts of fairs and exhibitions.

Figure 2 The Ihme innovation showroom

In Ihme, the ideation theme has to be such that it will tempt passers-by to take a closer look. Each user should be able to devote as much (or little) time to the ideation as (s)he happens to have. In Ihme, we have studied, e.g. a virtual travel service, games, augmented reality applications, Internet of things and mobile consumer services. As data gathering methods, we have used interviews as well as posters on which users can put their ideas and comments as post-it notes. The interviews and ideation sessions typically last from twenty minutes to one hour for each individual or group.

In Ihme the value proposals have to be presented in such ways that ordinary, less technology-oriented users can quickly understand them. Direct user/designer interaction has been fruitful and has produced concrete ideas. The dialogue not only foments ideas but also makes the designer understand the user's world.

The Ihme innovation showroom at the Ideapark shopping center reached a large number of visitors during the first opening period, summer 2010 (approx. 2500 visitors). In a visitor survey, interactivity, entertainment factor, innovative visual representation, presence of sound feedback, possibilities for further development and broad applicability were mentioned as reasons that made presented applications appealing. Of the survey respondents, 69% showed a positive response towards participating in the design of new technologies and services. Participation was seen as useful and important but also fun and interesting. Visiting the Ihme space was reported as a positive experience by all the survey respondents. The main positive aspects were the opportunity to participate, experiencing new technology trends, the public appearance of the research institute, the opportunity to meet experts, an easily approachable location and a low threshold to participate.

4   INTERWEAVING DESIGN AND USE: LIVING LABS

Living Labs are open innovation ecosystems that engage users in the co-creative process of new services, products and societal infrastructures in real-life settings (European Commission, 2010). A Living Lab offers services which enable the users to take active part in research and innovation as part of their everyday lives. As a development and innovation environment, a Living Lab is more participatory than traditional social pilot studies and ethnographic research, which focus on observing rather than interacting. Living Labs can provide reliable information about the market behavior of users, further contributing to reduced risks for new business and technology (Näkki and Antikainen, 2008).

The Living Lab co-creation approach is based on our experiences of applying and developing co-creation methods in different field pilots. It is already an established practice to organize field tests in pilot services before they are launched on the market. Prototype services that are reliable enough for long-term use and include content valid for actual use are good candidates for field tests. Long-term field evaluations give user feedback beyond first impressions. Very often, user attitudes are only established after the first few weeks of a 'honeymoon' period.

The Living Lab approach changes the setting of a traditional field test so that in addition to actively gathering user feedback, users are also encouraged to propose improvement ideas. The best ideas are put into practice right away and thus the users can see immediately how their feedback influences the service. We call this 'design-in-use'. This motivates additional comments and development ideas and gradually creates a positive spin of continuous improvements based on everyone's contributions. Another motivating factor is positive experiences of the participation: to belong a Living Lab user community and give feedback can be simply fun.

As Living Labs require long term commitment, registration should be simple but still inform clearly the participants of the expectations. The Living Lab should provide different co-creation activities such as online contribution in Owela, focus groups, individual evaluations and face-to-face interviews. The possibility to choose the ways to participate motivates the users. The interaction between the users, researchers as well as other stakeholders should be continuous and informal to allow the design-in-use be part of everyday life and to encourage creativity of the users. The design-in-use idea actualizes best when methods to collect and analyze user feedback data are agile and carried out in appropriate intervals. Online methods complement field evaluation methods and these can be used in parallel.

Our most recent Living Lab case was focused on the development of postal services. We applied co-creation efforts in several phases of the service development. We used different methods from household and individual interviews and questionnaires to co-design sessions with group dialogical methods. Owela was also in use as a method for collecting feedback and ideas and for informing participants. The co-creation process produced plenty of user ideas and opinions also on related services. The Living Lab community, part of a small village in southern Finland, benefited from the social agenda and the technical platforms set by the Living Lab, as the participants could meet each other and collaborate for the common goals of the Living Lab.

5   COMPARISON OF THE CO-CREATION APPROACHES

In Table 3 we compare the co-creation spaces regarding setup, methods, participation as well as strengths and challenges.

Table 3 Comparison of the three co-creation approaches

	Owela	Ihme	Living Lab
Innovation phase	All (needs, ideas, concepts, prototypes)	All (needs, ideas, concepts, prototypes, even market research)	Design-in-use, ideating complementary services
Typical illustration material	Text, images, videos, slideshows	Scenarios, demonstrators, proofs of concepts	Actual or pilot service
Data gathering	Online discussion, ideas, polls, surveys, votes, ratings	Interviews, observation, questionnaires	Household interviews and evaluations, group discussions, Owela
Participating users	Internet users (i.e. almost anyone)	Visitors at the place where Ihme is set up, e.g. shopping mall	A focused user group of the service
Form of user interaction	Mostly text-based commenting, rating, voting, chatting	Face to face	Online, face-to face, group meetings, phone discussions
Role of service provider /designer	The role can vary from active participant to observer	Presenter, interviewer	Leader and motivator of the Living Lab, observer
Strengths	Easy to reach users; enables micro-contributions of masses; long-term collaboration	Open to all in public space – low threshold to participate	Design feedback and ideas based on long-term actual use; user empowerment
Challenges	No face-to-face contact; requires continuous facilitation; mainly text-based communication	Quite resource intensive; data gathering is challenging in ad-hoc face to face meetings	The service should be technically reliable and have real content; motivated and engaged long-term users may be hard to find and keep

The main differences between the co-creation spaces are in participation space and participation role. Owela is an online approach and Ihme is a physical world approach whereas Living Labs can have both elements depending on the co-creation activity. Owela participation focuses on reflecting ideas and developing them further whereas Living Lab and Ihme focus on people experiencing themselves, and giving feedback based on actual experiences.

Owela is at its best in early ideation and concept design. Ihme is at its best when people can look, feel and experience physical demonstrations or proofs of concepts. Living Labs are needed when the service already exists as a working pilot or when an existing service is further developed during use. These boundaries are not fixed however and the co-creation spaces can be used in parallel.

6   DISCUSSION AND CONCLUSIONS

Owela, Ihme and Living Labs are all co-creation spaces that manage to go close to user's world and thus are able to reach masses and intertwine with the everyday lives of people. The threshold to participate is especially low in Ihme and Owela. All the co-creation spaces provide various ways to participate, and let the users choose the ways that they personally like. In Owela, users can participate in a web-based innovation community independent of time and space. Users can select their level of contribution, varying from short comments to long-term participation in development projects. In Ihme, designers can meet and interact with users in a physical environment that has been designed to encourage ideation. Living Lab environments enable 'design-in-use': long-term service development with users in parallel with using the service. Owela is at its best in early ideation, especially when the ideation theme is such that it tempts people to create ideas and comment on them based on their own experiences. Ihme is at its best when designing new interaction concepts or other tangible experiences. Future usage possibilities of Ihme include long term company specific Ihme spaces and short term pop-up Ihme spaces at public places. Living Labs enter the picture when co-creation extends to the actual use, and services are continuously improved in parallel with their use. At its best, a Living Lab enables firm and continuous connections to actual users and co-creation based on actual everyday experiences. Living Labs can be complemented with Owela and Ihme to facilitate online and face to face interaction between participants.

The co-creation spaces can also have indirect impacts on better company image and they can assist in marketing. With all three co-creation spaces, direct designer/user interaction supports turning the designer's mind-set from technical features to user experience. This facilitates the design of services that are better accepted by users and more interesting to them. All three approaches produce a lot of material to be analysed. Agile ways to analyse the feedback are needed and this is our main research focus for the future.

ACKNOWLEDGEMENTS

The co-creation spaces and related user-driven innovation methods have been developed in research activities funded by Tekes, the Academy of Finland, cooperating companies and public sector organizations, and VTT. The authors wish to thank all the cooperating organizations, as well as their colleagues who have contributed to the studies.

REFERENCES

Ainasoja, M, E. Vulli, E. Reunanen, R. Hautala, et.al. 2011. User Involvement in Service Innovations - Four Case Studies. MCPC 2011 Conference. San Francisco.

Alam, I. 2006. Removing the fuzziness from the fuzzy front-end of service innovations through customer interactions. *Industrial Marketing Management,* 35: 468-480.

Björgvinsson, E., P. Ehn, and P-A. Hillgren. 2010. Participatory design and "democratizing innovation". PDC '10. ACM Press.

Buur, J. and S. Bødker. 2000. From usability lab to 'design collaboratorium': reframing usability - practice. Conference on Designing Interactive Systems, New York, USA.

European Commission. 2010. Advancing and applying Living Lab methodologies. An update on Living Labs for user-driven open innovation in the ICT domain. Accessed February 20, 2012. http://ec.europa.eu/information_society/activities/ livinglabs/docs/pdf/newwebpdf/living-lab-brochure2010_en.pdf

ISO. 2010. ISO 9241-210:2010. Ergonomics of human-system interaction -- part 210: Human-centred design for interactive systems.International standard. International Organization for Standardization. Geneva.

Kanstrup, A. M. and E. Christiansen. 2006. Selecting and evoking innovators: Combining democracy and creativity. 4th Nordic conference on Human-computer interaction: changing roles.

Khurana, A. and S. R. Rosenthal. 1998. Towards holistic 'front-end' in new product development. *The Journal of Product Innovation Management,* 15(1): 57-74.

Kristensson, P., J. Matthing and N. Johansson. 2008. Key strategies for the successful involvement of customers in the co-creation of new technology-based services. *International Journal of Service Industry Management,* 19(4): 474-491.

Lusch, R. F. and S.L. Vargo. 2006. The service-dominant logic: reaction, reflections and refinements. *Marketing Theory,* 6 (3): 281 288.

Näkki, P. and M. Antikainen. 2008. Online Tools for Co-design: User Involvement through the Innovation Process. Workshop: 'New Approaches to Requirements Elicitation & How Can HCI Improve Social Media Development?' NordiCHI, Lund, Sweden.

Prahalad, C.K. and V. Ramaswamy. 2000. Co-Opting Customer Competence. *Harvard Business Review,* January 2000.

Sanders, E. B. and P. Stappers. 2008. Co-creation and the new landscapes of design. *CoDesign,* 4: 1: 5-18.

Sawhney, M., G. Verona and E. Prandelli. 2005. Collaborating to create: The Internet as a platform for customer engagement in product innovation. *Journal of Interactive Marketing,* 19 (4): 4-17.

Schumacher, J. and K. Feurstein. 2007. Living Labs – the user as co-creator. ICE 2007, 13th International Conference on Concurrent Enterprising, Sophia Antipolis, France.

Vargo, S. L. and R. F. Lusch. 2008. Service-dominant logic: continuing the evolution. *Journal of Academic Marketing Science,* 36: 1-10.

CHAPTER 30

Interactive Design Method Based on Structured-Scenario

Koki KUSANO, Momoko NAKATANI, Takehiko OHNO

NTT Cyber Solutions Laboratories
1-1 Hikarinooka Yokosuka-shi, Kanagawa 239-0847, Japan
kusano.kouki@lab.ntt.co.jp

ABSTRACT

To design useful User Interfaces (UIs), the designer must consider not only beauty but also the users' situation; who will use the UI, why/how they will use it, and so on. However, most designers fail to adopt a well-structured approach to UI design. This paper proposes a Scenario-based Interactive System Design and tool. Our proposed tool enforces the integration of many details through its three processes. 1. Structured scenarios are explicitly written by the designer using a hierarchy structure. 2. Nodes of the hierarchy are examined to identify the functions needed, the attributes to be handled, and the order in which the functions are invoked. 3. At the request of the designer, the tool can visualize the functions and attributes in the form of conceptual layouts. With this tool, a designer can create well-formed comprehensive UIs smoothly and easily.

Keywords: user interface, design method, scenario, visualization

1   INTRODUCTION

The well-known approach is the scenario-based design process, which can yield user-friendly UIs (Cooper et al., 2003)(Carroll, 2000). In this process, the designer must consider the users' situation; who will use the UI, why/how and when/where will they use it. However, most designers fail to adopt a well-structured approach to UI design and so have difficulty in extracting the information needed and utilizing it (Tanahashi, 2008).

This paper introduces a scenario-based interface design method and tool. Our proposed method, called Scenario-based Interactive System Design, avoids these difficulties. We embed Structuring Scenario Process and Extracting and Visualizing

Design Constraints into the process (Fig.1). With this method and tool, a designer can write and analyze scenarios smoothly, and connect scenarios and UIs easily. Our proposed method is expected to reduce the cognitive workload through its visualization. As a result, the designer can create well-formed comprehensive UIs easily and smoothly.

The next section describes current scenario-based UI design process. Section 3 points out key weakness of the current scenario-based design method. Section 4 details Scenario-based Interactive Design Process. Section 5 introduces about experiments of a prototype that is implemented part of our proposed tool, and section 6 shows the results. Section 7 discusses about effectiveness of our proposed method with results. Section 8 concludes the paper with future work.

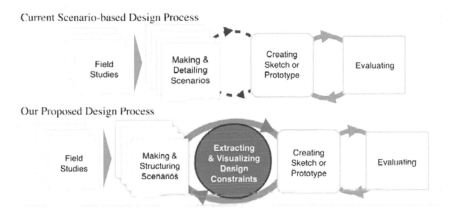

Figure 1 Illustration of current scenario-based design process (top) and our proposed design process (bottom)

2 CURRENT SCENARIO-BASED UI DESIGN PROCESS

Most of current scenario-based UI design schemes contain 4 steps; field studies, making & detailing scenarios, creating sketches or prototypes, and evaluations.

First step is Field Studies. The designer investigates the target user and user needs. The designer observes real users and their activities to find user needs.

Second step is Making and Detailing Scenarios. The designer writes a detailed scenario in this step with reference to the field studies data. Scenarios are useful in imaging how the user will employ the service. According to the method of goal-directed design (Cooper, 2003), the designer can create a scenario with target users as imagined from the field-studies data. Next, the designer extracts functions from the scenario, and details the scenarios using the extracted functions.

Third step is Creating Sketch or Prototype. First, a designer roughly conceptualizes the forms of functions and decides form layouts (function-layout). The designer also considers relationship to other functions, run time, execution frequency and so on from scenarios. Second, the designer sketches a rough UI that contains controls (ex. button, icons, input form). By externalizing scenarios as UIs, the designer can clearly perceive how users will employ each UI. Therefore, the designer readily finds the omission of functions and/or controls and adds the functions needed to UI. This step should be iterated in parallel with the step of making and detailing scenarios. Because that the designer should check necessity of functions and controls, whether the functions and the controls are needed not only to UIs, but also to scenarios. If the designer does not do this step, UI has too many functions and controls, which is not needed by anybody.

Fourth step is Evaluation. After creating sketches or a prototype, the designer evaluates them to find usability problems. At first, the designer evaluates low-fidelity UIs with paper prototyping (Snyder, 2003) or uses the heuristic-evaluation (Nielsen, 1994) method. By evaluating low-fidelity prototypes, we can find usability problems at an early design stage. Furthermore, the designer can review usability problems with the scenarios, refine them, and recreate the sketch or prototype. The result of evaluation is used to refine and detail scenarios and UI, which are reevaluated. These steps must be iterated (Nielsen, 1993)(ISO 9241-210).

3 LIMITATION OF CURRENT UI DESIGN PROCESS

Previous work pointed out that extracting information from scenarios and using this information to design UIs is difficult for most designers (Tanahashi, 2008). It means that the current process has a "wall" between "2.2 Making and Detailing Scenario" step and "2.3 Creating sketch or prototype step". The scenarios and UI prototypes are quite different in form and most designers cannot translate them smoothly. Therefore, we propose the "Scenario-based Interactive System Design" to solve this problem.

There are some related works in this research area. For example, one method defines writing scenario rules to ensure that scenarios can be organized easily (Yanagida et al., 2009). Another one uses workflow-diagrams to visualize scenarios (Tanahashi, 2008). Unfortunately, these methods cannot connect scenarios and UIs directly. Designer can use various prototyping-tools that support making and sketching UIs (Lin et al., 2000)(OmniGroup, 2011). some prototyping-tools adopt storyboarding (Justinmind, 2011)(Li et al., 2008). The storyboard helps the designer to create UI screens but provides no support in determining the form of the functions needed and their order.

4 SCENARIO-BASED INTERACTIVE SYSTEM DESIGN

Our proposed method avoids these difficulties. We embed "Making & Structuring Scenario Process" and "Extracting and Visualizing Design Constraints" into the current process (Fig.1). We detail these processes in Section 4.1 and 4.2.

4.1 Making and Structuring Scenario

First, the designer creates a hierarchical scenario from field studies data, and nodes of the hierarchy are examined to identify the functions needed. The designer extracts and tags sentences as functions from the hierarchical scenario. This step clearly links the original scenario to the extracted functions. In our experimental prototype, Scenarios are created in HTML, and embedded hyper links are set between the scenario and the functions. In addition, this experimental prototype provides outlines, they summarize scenarios that contain functions.

This process clearly elucidates the scenario structure, and provides high traceability between scenarios and functions. Therefore, the designer can organize and extract information more easily than is possible with current scenarios. In addition, since the designer is relieved from the need to continually track the relationship between the scenario and extracted functions, more time is available to deeply focus on the image of the target user and target users' behavior.

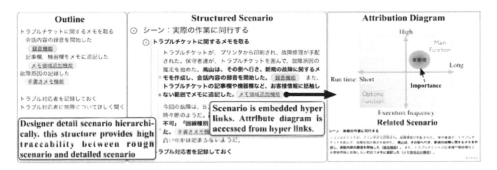

Figure 2 Example of structured scenario and attribute-diagram and related-scenario.

4.2 Extracting and Visualizing design constraints

We define five design constraints. The designer analyzes the scenario to extract design constraints that represent, for each function, 1) importance, 2) relationship to other functions, 3) run time, 4) execution frequency, 5) order. The design constraints provide viewpoints that ensure the efficient analysis of functions and scenarios.

1. **Importance:** This constraint reflects how important the function is in the system. By defining importance, the designer finds it easier to invoke function priority.
2. **Relationship to other functions:** By defining the relationship, the designer can understand that which functions should be grouped and which function should occupy the center of the group.
3. **Run time:** This constraint shows the execution time. The designers can utilize this to consider the functions' appearance (ex. window, frame, pop-up, button).

4. **Execution frequency:** This constraint addresses the functions' execution frequency. The designer can utilize this to consider about functions' appearance.
5. **Order:** This constraint shows the order in which the functions are invoked. The designer can utilize this to understand when functions should be shown.

The designer analyzes each scenario and sets up all design constraints for each function in our proposed method. At the same time, our tool records the link between design constraints and scenarios.

Needless to say, visualizing information is effective in various fields; therefore we also provide visualization rules for design constraints. Fig.3 shows a visualization example, and a sketch based on it. The designer can create a UI by directly manipulating visual objects representing functions.

Our experimental prototype uses the boxes to represent functions and lines indicate the relationship between functions. The visualization rules provided by our tool allow the designers to visualize a function's design constraints and attributes in the form of conceptual layouts easily.

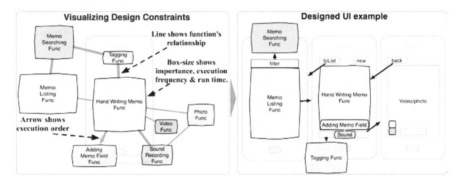

Figure 3 Example of visualization of design constraints (left) and sketch (right).

A line between two boxes shows the relationship between them and its thickness shows the strength of the relationship. An arrow indicates invoking-order. Box attributes represent three constraints; Importance, Run time and Execution frequency (in the prototype screen of Fig. 3, box-size represents the simple summation of these constraints). These constraints are entered/edited in an attribute-diagram (see Fig.2). In this diagram, the cross axle shows Run time, the vertical axis shows Execution frequency, and Point size shows Importance.

This process yields well-formed comprehensive UIs smoothly and easily. In addition, this process provides high traceability between scenarios and UI.

5 EXPERIMENT

We create a prototype that is implemented part of our proposed tool (Fig.4). We evaluated it focusing on the structured scenario and the visualization of design constraints.

5.1 Environment and Participants

We implemented the experimental prototype with OmniGraffle (Omnigroup 2011) and HTML, see Fig.4. We set up the visualized design constraints and scenarios manually. We arranged all functions showing their importance, run time and execution frequency by box-size, as well as line connections and line thickness. We implemented structured scenarios with HTML with embedded hyperlinks. Therefore, participants could access scenarios and attribute diagram from a function with just two mouse-clicks.

Nine participants (ID1~ID9, 20~30 years old, 6 men and 3 women) designed two UIs in 25 minutes. All participants are information engineers, but have never studied UI design. They have various backgrounds in software development.

In the experiment, they used two scenarios, written for i) telecommunication service, ii) digital memo tool for personal use. There were three conditions: 1) With structured scenario, design constraints and function list (Proposal), 2) With scenario and function list (Scenario), 3) With function list only (Function list). Five participants used 1) and 2), four used 1) and 3).

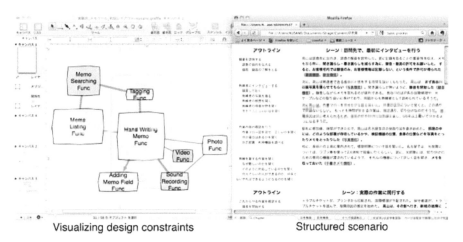

Visualizing design constraints   Structured scenario

Figure 4 Experimental environment for proposal condition.

5.2 Procedure

First, we introduced the scenarios and systems to the participants. Next, they designed UIs in a training session in 20 minutes. There were some questions; experimenter answered these questions at that time.

After the training session, participants created 2 UIs with each scenario. They had 25 minutes to design each UI. If they couldn't finish, they were given an extra 5 minutes. On the other hand, if they could finish within 25 minutes, the experimenter ordered them to review the UI until they could not find any further points to revise.

After the experiment, the participants answered a questionnaire and were interviewed during which the participants watched a video of their design activities.

302

6 RESULTS

Seven HCI researchers evaluated the 18 UIs heuristically and ranked them (Rank 1~9 for each kind of UI). They used a usability check-sheet that was modified from the design-guideline of About Face 3, see Tab.1.

Table 1 Evaluation Criteria.

Criteria (1. Strong Disagree, 2. Disagree, 3. Not sure, 4. Agree, 5. Strong Agree)
1. The UI fits the device; no screen has too many functions.
2. Number of screens (form, dialogues, page, window etc.) is suitable.
3. The UI has consistency and can be used easily and smoothly.
4. The UI has good navigation; user can understand what he is doing.
5. Use can expect UI feedback.
6. User can understand which is main function and which is not.
7. Optional functions are less obvious than main functions.
8. User can invoke main function with fewer steps.
9. Hierarchy suits the number of functions.

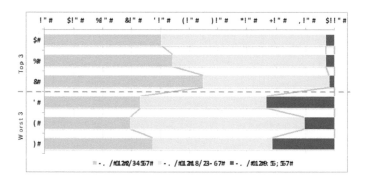

Figure 5 Top 3 and worst 3 participants' design activities with Proposal condition.

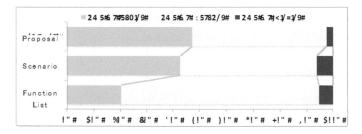

Figure 6 Average of top 2 participants' design activities.

The results of UI evaluation showed that there were a few people who got high scores and utilized the proposed method quite well (ID2, ID5), but no significant differences were seen in averages of all participants. We also analyzed the videos of the participant's design activities. We focused on three activities; 1) time for reading, 2) time for operating, 3) time for thinking. Time for reading contains reading scenario, related-scenario includes scenarios that are related to target function, and requirement-list. Time for operating contains control function-layout, grouping, modifying and creating other screen. Time for thinking does not include reference to external information. In addition, we excluded "time for questioning" and "time spent in confusion" from analyzing-target.

We found differences in design activity between the top 3 participants and the worst 3 participants (Fig.5). Participants who could utilize the prototype and could design effective UIs, took less time to think; they found it easy to plan and design UIs. They also spent more time reading documents and tended to extract much more information, see Fig.5. Furthermore, Fig.6 suggests that the proposed prototype better supports these activities than other conditions.

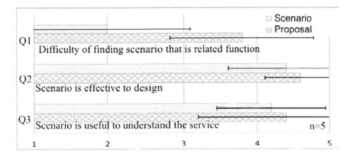

Figure 7 Results of questionnaire about scenario (Proposal vs. Scenario condition).

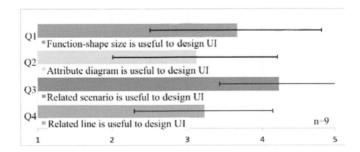

Figure 8 Results of questionnaire about Proposal condition.

6.1 Featured activities

In this experiment, experimenter ordered participants to read the scenario first. Participants imagined the target user to decide the design-concept. The results of Fig.7 Q2, Q3 indicated that the scenario was useful for imagining the target-user clearly. For instance, ID8 said *"This service will be used by elderly people and small children, so I thought that I should make this service plain and simple"*. This result supports previous works, which reported the effectiveness of scenarios.

The participants utilized not only decided concepts but also the scenario for reviewing designed UI and deciding next action. Concretely, the participants read those parts of the scenario related to functions (ID1, ID5, ID6, ID8), outlines (ID2, ID9) and picked up colored words (ID7). The results of Fig.8 Q1 and Fig.9 Q3 indicate the importance of traceability between UI and related scenario parts.

Participants utilized our proposed visualization rules that represented importance, run time and execution frequency. We got several positive comments; *"I thought that big functions are important"* (ID1, ID4), *"In reviewing my video I noticed that I judged big functions as important."* (ID6) and so on. In addition, the result of Fig.9 Q1 indicates that box-size is well utilized for designing UI. On the other hand, we found that the prototype failed to provide strong support for UI refinement. Because participants directly dragged and dropped function-objects, the prototype lost the association to the original box-size.

Attribute diagram was utilized by ID5 and ID9 for reviewing main functions. For example, ID5 commented, *"I read execution frequency from attribute diagram when deciding the number of screens."*

ID5, ID8 utilized related line for creating function groups in the first half of design. We also got some hints for improving implementation; *"I cannot see line well because it is too pale"* (ID4), *"When I deleted some function and then undid it, related line was also deleted but I cannot undo it"* (ID8). These effects appear in the result of Q4 in Fig.8.

7 DISCUSSION

We discuss the effects of using structured scenarios and visualizing design constraints for designing UI using the analyzed activity data, interviews, and questionnaire responses.

The results of Section 6.1.1 suggest that if the designer utilizes a structured scenario, he/she finds it easy to find the information needed for UI design from the scenario smoothly and easily. Interviews and questionnaires indicate that our proposed method was utilized for deciding the concept of UI design and reviewing a designed UI. Since these activities are important and frequent, our proposed method allows a designer to design UIs much more efficiently. On the other hand, some participants pointed out problems with the implementation. In particular, participants felt troubled when accessing the scenario, because two mouse clicks were needed in the system tested. Namely, we should design a UI that can access the related scenario part automatically as in mouse-in links.

The results of interviews and questionnaires that were written in section 6.1.2 suggest the effectiveness of using box-size to visualize importance, run time and execution frequency for setting function-layout. They also revealed that visualizing relationships with lines and line-thickness was useful for creating groups of related functions in the first half of design. In addition, time for thinking was reduced when participants used our proposed method. Therefore, these results indicate that if a designer utilizes information that was extracted from structured scenarios and properly visualized, he can design UIs easily and smoothly.

To better support the second design step, we need new visualization rules. For example, to visualize interactively only the information associated with selected and related functions. Furthermore, we found that we should avoid using indirect visualization; box-size currently uses one parameter to summarize three constraints. Namely, three visualization parameter are needed to better express the three constraints.

The relatively naïve implementation of the prototype hindered its smooth use by the subjects and further work is needed to fully demonstrate its effectiveness in the key design step of extracting functions from scenarios and placing them in a UI.

8 CONCLUSION

We identified a key weakness of the current scenario-based design method, the designer has difficulty in extracting information from the scenario and linking UI and scenario. To solve these problems, we proposed and evaluated a method offering Scenario-based Interactive System Design.

Evaluation results showed the effectiveness of setting traceability between scenario and UI; visualizing the design constraints helps a designer to design a UI from a scenario. However we found some problems with the experimental prototype, which should be resolved. In the future, we will create more practical prototypes, and evaluate their effectiveness in case studies.

REFERENCES

Carroll, J.M. 2000. Making Use: Scenario-Based Design of Human-Computer Interactions, The MIT Press.

Cooper, A., Reimann, R. and Cronin, D. 2007. About face 3: the essentials of interaction design, John Wiley & Sons, Inc., New York, NY, US.

ISO 9241-210 2010. Ergonomics of human-system interaction -- Part 210: Human-centred design for interactive systems.

Justinmind 2011. Justinmind Prototyper (2011). http://www.justinmind.com/.

Li, Y. and Landay, J.A. 2008. Activity-based proto- typing of ubicomp applications for long-lived, everyday human activities, Proceeding of the twenty-sixth annual SIGCHI conference on Hu- man factors in computing systems, CHI '08, New York, NY, USA, ACM, pp. 1303–1312.

Lin, J., Newman, M.W., Hong, J.I. and Lan- day, J.A. 2000. DENIM: finding a tighter fit between tools and practice for Web site design, Proceedings of the SIGCHI conference on Human fac- tors in computing systems, CHI '00, New York, NY, USA, ACM, pp.510–517.

Nielsen, J. 1994. Heuristic evaluation, pp.25-62, John Wiley & Sons, Inc.

Nielsen, J. 1993. Iterative user-interface design, Computer, Vol.26, No.11, pp.32-41.

OmniGroup 2011. OmniGraffle.
    http://www.omnigroup.com/products/omnigrafflelash.

Snyder, C. 2003. Paper Prototyping: The Fast and Easy Way to Design and Refine User Interfaces (Interactive Technologies), Morgan Kaufmann.

Tanahashi. H. 2008. Persona tsukutte sorekara dousuruno? User tyuushin de tsukuru website. [What should I do after creating a persona? – Design the website with user-centered design]", Softbank creative.

Yanagida, K., Ueda, Y., Go, K., Takahashi, K., Hayakawa, S. and Yamazaki, K. 2009. Structured Scenario-Based Design Method, Human Certered Design (Kurosu, M., ed.), Lecture Notes in Computer Science, Vol.5619, Springer Berlin / Heidelberg, pp.374–380.

CHAPTER 31

Black Female Voices: Designing an HIV Health Information Artifact

Fay Cobb Payton*, James Kiwanuka-Tondo*, and Lynette Kvasny**

North Carolina State University*
Raleigh, NC, USA
Fay_payton@ncsu.edu
Pennsylvania State University**
University Park, PA, USA

ABSTRACT

HIV online information and services targeting African American females are of particular interest given the impact of this infectious disease in this underserved community. That is, data from the Centers of Disease Control and Prevention show that African Americans are the racial/ethnic group most affected by HIV. In 2009, Black women accounted for 30% of the estimated new HIV infections among all Blacks. The estimated rate of new HIV infections for Black women was more than 15 times as high as the rate for White women, and more than three times as high as that of Latina women (CDC Data, 2009). Numerous technology-based tools and applications have been designed for intended populations and users based on age, gender, ethnicity and medical conditions, including smoking cessation, breast cancer, Alzheimer's Disease and heart disease. Prior research (Payton, 2009; Payton and Kiwanuka-Tondo, 2009; Payton, et al., 2011) evaluated health information technology applications relative to people, processes and patients as well as the HIV prevalence among Black women. As articulated in Payton, et al. (2011), patient-centered care has its focus in community involvement, participation and formulation, such as social networks or support groups, thus enabling social change. According to Healthy People 2020, the criticality of health communication and health information technology lends itself to a myriad of topics, including building health skills and knowledge, supporting community and home care, facilitating clinical and consumer decision-making and improving the public health infrastructure. Further, Healthy People 2020 articulates the need for the application of evidence-based best practices in user-centered design approaches to dissemination health information across underserved and under-represented populations to better understand health disparities and health outcomes. Given this context, interdisciplinary approaches can enable improved design and awareness while creating a service experience to address the health issues affecting and infecting a vulnerable population, such as Black female college students. Aforementioned research studies provide insight to social models that can drive the technology design and experiences among Black females.. In an effort to reach underserved populations, engagement, design research, communication and human computer interaction frameworks collectively provide the theoretical bases to better engage and design service systems.

In this paper, we will provide preliminary findings from a study of African American female college students and their use of online HIV health information. We will also discusses the use of interdisciplinary approaches to address the design and creation of a dynamic IT artifact intended for the population under investigation.

Keywords: Black Females, HIV, Technology, Social Media, Service Design

1   MOTIVATION AND SOCIAL NETWORKS FRAMEWORKS

Perisse and Nery (2007, p S362) argue that a "social network consists of a series of individuals or groups connected by links that represent some kind of relationship (e.g., friendship) or interaction (e.g., sexual). The investigation of social networks permits an evaluation of the influence that the connections between people have in the transmission of a given disease". We adopt this definition as we look to more recent work with a focus on HIV in ethnic minority communities and information dissemination in a social network. Kimbrough, Fisher, Jones, et al (2009), created the Social Networks Demonstration Project which provides a theoretical foundation for our research.   These scholars evaluated the use of social networks to reach persons with undiagnosed HIV infection in ethnic minority communities and linked them to medical care and HIV prevention services.

According to the Centers for Disease Control (2009), African-Americans represent 14 percent of the U.S. population, but account for 44 percent of new HIV infections in 2009.   The estimated rate of new HIV infections among African-American women is 15 times that of white women and over three times that of Latina women.   Earlier research (Chong, Kvasny and Payton, 2006) indicated that patients are becoming more empowered and educated via the Internet as electronic health information sources proliferate.   As reported in Payton (2009), health information seekers, according to the Pew Internet and American Life Project (2006)   release, are "Internet users who search online for information on health topics" whether they take on the role of consumers, caregivers, or e-patients. Given its ubiquitous nature, the Internet continues to be a key source of health information among health seekers. Specifically, the Pew Internet data indicates that eight of ten

Internet users, or roughly 113 million adults, have sought after health information online. The question is: what type of health information do these Internet users want or need? In 2006, sixty-six percent (66%) of all health seekers searched for information on a specific disease or medical problem for themselves, family members, or friends. Based on this metric, the Pew study committee concludes that health searches have reached parity with common uses of the "Net," such as electronic bill paying, blogging, or directory lookups for addresses and telephone numbers. Enabled by Internet-enabled applications, social media tools will play a significant role in emerging forms and definitions of social networks.

Keckley and Hoffmann (2010) of the Deloitte Center for Health Solutions explained further the concepts enabling social networks in health care. They offered that "social networks transmit media such as video, web logs (blogs), ratings and reviews, podcasts and audio among a group of people who are linked by a common characteristic, such as friends and family, profession, school, residence and even likes and dislikes. Figure 1 also shows the reach (in terms of usage) of social-media by prescribed delivery, such as video and text feeds.

Figure 1: Social network description, usage and applications

Network	Description	Usage	Business Applications
Twitter	140-character news feed	105 million users	• Posting press release-like announcements which, in some cases, can supplant traditional news formats
Facebook	Recreational peer-to-peer social network	400 million users	• Building fan pages for specific causes, organizations or products • Sharing recreation-oriented campaigns
YouTube	Video	6.5 billion views	• Posting educational videos and testimonials
Blogs	Internet web diary	112 million blogs	• Discussing happenings in an organization (e.g., product launches, executive changes)
LinkedIn	Professional peer-to-peer networking	60 million users	• Recruiting talent, announcing staff news

Figure 1. Social Network Description, Usage and Applications. (From P.H. Keckley and M. Hoffmann, 2010. From Issue Brief: Social Networks in Health Care: Communication, collaboration and insights, Deloitte Center for Health Solutions).

The growth of social-media applications in health is growing exponentially. For instance, the YouTube Health Care Team has documented the use of videos in health dissemination and reports that[1]:

---

[1] (http://blog.ogilvypr.com/2010/08/looking-to-reach-consumers-in-the-healthcare-space-consider-tuning-into-youtube/?utm_source=feedburner&utm_medium=feed&utm_campaign=Feed%3A+360DI+%28Ogilvy+PR+360+Digital+Influence+Blog%29_)

- Of YouTube's 180 million viewers, 32% watch health videos - more than food or celebrity; Of those viewers, 79% of health consumers have watched videos about their specific health condition
- 93% take action after viewing health information; 69% conduct further online research as a result of the video they watched

Research also shows that young African-Americans are among the fastest-growing users of social media networks (Pew Research Center, 2009). Thus, social media tools are promising channels for the dissemination of culturally salient prevention messages and services. Socially inclusive eHealth applications can, therefore, play an important role in reducing health disparities. Yet, web content draws heavily on the myopic norms, ideologies and power structures which are often translated into daily social interactions (Brock, Kvasny and Hales, 2010). Others assert that there is minimal web content that reflects cultural competency (Lazarus and Lipper, 2000).

"While information and communication technologies (ICT) were thought to be the "magic bullet" to health care information education and dissemination, structural features are the specific types of rules and resources, or capabilities offered by the application or technology. Spirit is the general intent with regard to values and goals underlying a given set of structural features. We contend that technology designers convey both spirit and cultural competency in their designs of advanced information technologies. Thus, the spirit in question is not that of the users. To understand spirit, or the general intent of an advanced technology, one cannot ignore cultural competency, as this discourse calls for the inclusion of a targeted population's cultural values. To ignore cultural relevance would be the equivalent of user-centered design devoid of the user (Klein and Hirschheim, 2001). Others (Nakumara, 2002) go further and describe this phenomenon as the "architecture of belief" to signify how designers, through their choice of keywords, images, and use of language, create interfaces which represent the identities of some idealized user population(s). Thus, the interface reflects the cultural imagination of the designer, and performs familiar versions of race, gender, and class (Payton and Kiwanuka-Tondo, 2009, 193-194).

Hence and as reported in Payton and Kiwanuka-Tondo (2009, p. 194), the lack of cultural competencies in the communication of HIV/AIDS prevention information between Blacks, in general, and the provider/research community plays a significant role in the high infection rates (National Minority AIDS Education and Training Center, 2002; Scott, Gilliam and Braxton, 2005; U.S. Department of Health and Human Services, 2005; Prather et al., 2006). Cultural competency is defined as the demonstrated awareness and integration of three population specific issues: health-related and cultural values, disease incidence and prevalence, and treatment efficacy (LaVeist, 2005). Although online health information is available from multiple sources, evidence shows that racial and ethnic disparities and lack of cultural competency still exist.

2 DESIGN PRINCIPLES – RECOGNIZING THE VOICES FOR ARTIFACT CREATION

Design science lends itself to the concept of development with emphasis on the ICT artifact (Hevner, et al, 2004). The design of ICTs can be informed by models, foundational constructs and evaluative metrics. As articulated by Lau (1997) and demonstrated in Payton and Kiwanuka-Tondo (2009), action research has also been used by IS scholars in their discussion of systems analysis, design, implementation, usability and human computer interaction. In addition, Payton and Kiwanuka-Tondo (2009) offered justification for the inclusion of cultural competency in designing ICT artifacts:

> *The (un)intended spirit inscribed on ICT may be (ex)inclusive with*
> *regard to the population or society it intends to serve. Hence, grassroots*
> *approaches and audiences can, in fact, offer effective insight into user-*
> *centered designs focused on educational and prevention content among*
> *those most affected and infected by chronic diseases, such as HIV (p 1).*
> *Health policy researchers (Schulz, Freudenberg and Daniels, 2006)*
> *document the voluminous new evidence of racial, socio-economic, and*
> *gender inequalities and disparities in the U.S. The triangulation of race,*
> *class, and gender exacerbates the complexities of service delivery, in*
> *general, and HIV/AIDS care, in particular. The epidemic is too often*
> *met with societal stigma and the antiquated moral debate. Despite these*
> *and other tribulations, an awareness and inclusion of social, cultural and*
> *political context is critical. By (un)consciously neglecting these social*
> *constructions and merely bounding our focus strictly to organizational*
> *entities, we reduce our understanding of IT artifacts while omitting*
> *substantive impacts (Galliers, 2003). Further, web designers,*
> *government agencies, and other entities must not preface our results by*
> *concluding that nonuse of the ICT described, herein, is abnormal.*
> *People, however, can have carefully crafted rationale as to why they do*
> *not engage with a technology, and the social context often facilitates our*
> *knowledge leading to these outcomes* (Cushman and Klecun, 2006).

Based on the Theory for Design and Action, a myriad of conditions enable ICT artifacts in their quest to engender domain-specific design and service utilization. Gregor (2006) outlined the conditions such as the utility to a community of users and the novelty of the artifact and effectiveness, as well as completeness, simplicity, consistency, ease of use and quality. Each of these conditions impact both design and utilization of ICTs. We further contend that cultural competency should be included in this discourse. More recently, scholars (Sein, Henfridsson, Purao, Rossi and Lindgren, 2011) have posited that designing and building ICT artifacts is only a part of the argument. Rather, Sein, et al. (2011) advocate action design research (ADR) which incorporates user perspectives in a given context while remaining attentive of the problem formulation.

Hence, the voices of Black females are critical to the design, building and sustainability of any IT artifact intended to target the population, in general, and HIV health information object, in particular.

3 DATA COLLECTION: MOVE FROM RECOGNITION TO HEARING THE VOICES

Using the Kimbrough, Fisher, Jones, et al (2009) social network framework, we sought to collect the voices of Black female college students - as they have a greater propensity to be socially connected, and as pointed out by the Pew Research Center, are significant users of social media tools. *The questions guiding our research are:*

- What do you think HIV/AIDS prevention information targeted to African American women should be focused on in order to improve educational outcomes?
- Should this information focus on prevention, individual behaviors, or both? Explain.
- Do you find the HIV/AIDS information on existing websites to be culturally compelling for African American women? Explain.

3.1 Data Collection

Starting in October 2011, we conducted focus group with 40 Black female college students who matriculated at North Carolina State University. We obtained the informed consent of each participant and entered each woman into a raffle for a gift card. We used a series of focus groups to solicit the input of these young adults in order to address our guiding questions noted above. We constructed an interview protocol based on prior studies (a number of which are listed in the earlier sections of this manuscript) of the conceptualization of health care, HIV and cultural identity.

Of the 40 female participants, all were Black and ranged between 18 and 24 years old. The women reported a mean family size of 4 (based on parents, themselves and siblings) and between six and ten years of Internet experience. This indicates that they have been "connected" a significant portion of their lives and illustrates the concept of digital natives. Thirty-eight of the 40 were undergraduate students with two (2) graduate students. Twenty-five (25) or roughly 63 percent reported studying in science, technology, engineering and mathematics (STEM) fields, such as computer science and various engineering disciplines. The women are socially connected through their involvement in various collegian academic and social organizations. Thirty (30) of the participants partake in at least one university-affiliated association while twenty-four (24) are involved in two student organizations.

4 PRELIMIARY FINDINGS

Our initial findings, herein, are taken from the prior six months. The female participants were consistent in their responses on the issues of cultural competency and/or compelling online HIV health information. The participants expressed critical themes in our design and build initiative:

1) There is a clear need to explain the difference between the clinical meaning of HIV and AIDS. As several participants pointed out, " Believe me...not everyone understands the difference and how to effectively protect themselves."

2) Multimedia is critical for this population. Overuse of linear text with technical medical jargon is not effect in communicating prevention messages and services. Accordingly, we heard the following voices:
"The language on some health sites is full of clinical information which is fine for me as a biology student, but no everyone knows how to interpret the data even as educated people." "The medical jargon works for those in the health area not those of us as patients." 'We need to consider literacy rates of the general population in the design process."

3) Social media messages at key points would be beneficial. The women expressed the need for the push messaging using social media and mentioned ideas on blogging, Facebook and Twitter. One participant summed the social media inclusion well by stating, "This is what we do."
The women want to "see" images that look like them and which they can relate to. However, images of blackness need to communicate (i.e. message) positive trends as well. According to a number of participants, "Despite the data, not all of us have HIV or AIDS, so don't make the information suggestive that we do...don't assume we have the disease but tell us how to prevent getting it." To this end, several participants noted: "Everything that is black need not be negative even in the discussion of HIV and AIDS in our community".

4) Geo-and graphical depiction of meaning HIV and health statistics. This theme offers the opportunity to provide a deep drive into the numbers with heat maps, geo-service offerings and health resources and dynamic updates

5 CONCLUSIONS

The voices of Black women are vital, yet often unheard, in the design and development of an ICT artifact intended to communicate HIV prevention information. As implied by action science design theory, social networks of PEOPLE and grassroot channels can strengthen health dissemination and improved messaging among the target population. Our initial findings have provided with a themed approach in our design initiative which will be an iterative process to better ensure content is both culturally competent and accurately represents the social identities of Black women. Information support will be valuable among members of the members of the social network. This support will come in the form of dissemination of personal experiences, meaningful data, and meaning representation of these data. At the same time, the data should be absence of punitive or presumptuous messaging often associated with HIV.

ACKNOWLEDGMENTS

The authors would like to acknowledge the National Science Foundation for its support of this research (Grant # IIS-1144327) and a team of consultants, undergraduate and graduate students.

REFERENCES

Brock, A., Kvasny, L. and K. Hales. 2010. Cultural appropriations of technical capital: Black women, weblogs, and the digital divide. *Information, Communication and Society*, 13, 7: 1040-1062.

Centers for Disease Control. 2009. HIV among African Americans. Accessed on February 15, 2012, http://www.cdc.gov/hiv/topics/aa/.

Chong, J., Kvasny, L. and F. C. Payton. 2006. Lacking cultural competency: Online AIDS/HIV health information among minority women,.IFIP 8.2 Conference, Limerick, Ireland.

Cushman, M. & Klecun, E. 2006. How can non-users engage with technology: Bringing in digitally excluded. In IFIP International Federation for Information Processing, 208, *Social inclusion: Societal and organizational implications for information systems*. Trauth, E. Howcroft, D., Butler, T., Fitzgerald, B. and J. DeGross (eds), Boston: Springer, 347-364.

Galliers, R.D. 2003. Change as crisis or growth? Toward a trans-disciplinary view of information systems as a field of study: A response to Benbasat and Zmud's call for returning to the IT artifact. *Journal of the Association of Information Systems 4, 6*: 337-351.

Gregor, S. 2006. The nature of theory in information systems. *MIS Quarterly*. 30, 3: 611-642.

Health People 2020. 2011. Accessed February 1, 2012, http://www.healthypeople.gov/2020/topicsobjectives2020/overview.aspx?topicid=18.

Hevner, A., March, S., Park, J., and S. Ram. 2004. Design science in information systems research. *MIS Quarterly*. 28,1: 75-105.

Keckley, P. H. and Hoffmann, M. 2010. *Issue brief: Social networks in health care: Communication, collaboration and insights*. Deloitte Center for Health Solutions.

Kimbrough, LW., Fisher, HE., Jones, KT, Johnson, W, Thadiparthi, S. and S. Dooley. 2009. Accessing Social Networks with High Rates of Undiagnosed HIV infection: The social networks demonstration project. *American Journal of Public Health*, 99, 6: 1093-1099.

Lau, F. 1997. A review on the use of action research in information systems Studies, In *Information Systems and Qualitative Research*, L.A. Liebenau and J. DeGross (eds.), London: Chapman & Hall, 31-68.

Laviest, TA. 2005. *Minority populations and health*. San Francisco, CA: Jossey-Bass.

Lazarus ,W. and Lipper, L. 2000. *Online content for low-income and underserved Americans: the digital divide's new frontier.* Santa Monica, CA: The Children's Partnership,

Nakamura, L. 2002 *Cybertypes: Race, ethnicity, and identity on the Internet.* New York: Routledge.

National Minority AIDS Education and Training Center. 2002. *Be safe: A cultural competence model for African Americans.* Washington, DC: Howard University. Accessed on January 5, 2012, www.nmaetc.org.

Payton, F.C. and Kiwanuka-Tondo, J. 2009. Contemplating Public Policy in AIDS/HIV Online Content, Then Where is the Technology Spirit?, *European Journal of Information Systems,* Vol. 18, 192-204.

Payton, F.C. 2009. Online health information seeking: Access and digital equity considerations. In Tan, J. with FC. Payton. *Adaptive Health Care Management Information Systems,* Third Edition, Jones & Bartlett Publishing.

Perisse, A.R.S. and Nery, J.A. da Costa. 2007. The relevance of social networks analysis on the epidemiology and prevention of sexually transmitted diseases. 23, Sup 3: S361-S369.

Pew Internet and American Life Project. 2006. *Online health search.* Accessed on June 10, 2011, http://www.pewinternet.org/pdfs/PIP_Online_Health_2006.pdf

Prather, C., TR. Fuller, W. King, M. Brown, M. Moaring, S. Little and K. Phillips. 2006. Diffusing an HIV prevention intervention for African American women: Integrating Afrocentric components into the SISTA diffusion strategy. *AIDS Education and Prevention.* 18, A: 149–160.

Scott KD., A. Gilliam, and K. Braxton. 2005. *Culturally competent HIV prevention strategies for women of color in the United States.* Healthcare for Women International. 26:1 /–45.

Sein, M. K., O. Henfridsson, S. Purao, M. Rossi. and Lindgren, R. 2011. Action design research. *MIS Quarterly.* 35, 1: 37-56.

U.S. Department of Health and Human Services. 2005. *Transforming the face of health professions through cultural and linguistic competence education: The role of HSRA Centers of Excellence.* Washington, DC: U.S. Department of Health and Human Services.

Section VI

Organizations and Change

CHAPTER 32

Smart Governance to Mediate Human Expectations and Systems Context Interactions

P. Piciocchi [1], J. Spohrer [2], C. Bassano[3], A. Giuiusa[4]

[1] University of Salerno (ITALY)
p.piciocchi@unisa.it
[2] IBM Almaden Research Center ,CA (USA)
spohrer@us.ibm.com
[3] "Parthenope" University of Naples (ITALY)
clara.bassano@uniparthcnope.it
[4] "Tor Vergata" University of Rome (ITALY)
alessio.giuiusa@uniroma2.it

ABSTRACT

Today we need to reframe our modes of thinking about governance as an issue to reformulate and introduce a new logic – the art of governmentality (Foucault, 1977, 1978) – to illustrate how decision makers assess the role of human resources for Service Systems viability.

Our perspective, based on an integrated approach, examines the governance of human expectations, an analysis that implies focusing on processes by means of which government (i.e. decision makers or policy makers) acknowledge, evaluate, measure and graduate the expectations and aims of an organization's stakeholders.

From a Service System based view ($S_E S_Y bv$) – derived from Service Science Management and Engineering + Design (SSME+D) and the Viable Systems Approach (VSA), our conceptual hypotheses are:

Hp1)Whether governance can be considered as the result of a continuous transformation of governance structure and practices according to the new business and societal liberalism;

320

and

Hp2) Whether Service Systems are: a) a dynamic configuration of resources; b) a set of value co-creation mechanism between suitable entities; c) an application of competencies-skills-knowledge any person(s) in job or stakeholder roles; d) an adaptive internal organization responding to the dynamic external environment; d) learning and feedback to ensure mutual benefits or value co-creation outcomes; e) a complex viable structure enable to respond to environmental change, where environmental change is mostly generated by other viable systems;

Then

Th) Smart Mechanism of Governance (SMG) to assess skills and mediate human expectations need a conceptual re-interpretation of government as an institution (structure) as well as practice (system) according to the new deal of governance in service systems: governmentality.

In fact, smart governance, based on the new art of governmentality, has to be read as the result of political and economic and technological processes characterized by network cooperation and collaboration at every level of organizations and/or society. In other words, in searching for models to adopt for smart governance, it is important to reject the top-down logic, based on despotic and individual power, and to embrace the bottom-up logic – of networks of governance (Triantafillou, 2004; Piciocchi and Bassano, 2009) –, based on shared knowledge and trust. This change of perspective, is coherent with an ($S_E S_Y bv$) assumption: that the viability of service systems depends on the capability of its government to create and develop mechanisms of value co-creation and to guarantee systemic equifinality, based on a continuous process of mediation of stakeholder expectations (customers, citizens, etc...). According to $S_E S_Y bv$, in service systems, value co-creation and equifinality imply networking cooperation and collaboration, i.e. the new deal of governance (governmentality): multilevel governance systems (Rosenau and Czempiel, 1992).

Keywords: Human side of Governance, Service Systems based view ($S_E S_Y bv$), Smart Mechanisms of Governance, Governance, Governmentality

1 AN INTRODUCTION: GOVERNMENTALITY AS A NEW DEAL OF GOVERNANCE IN SERVICE SYSTEMS

In this study we propose a Smart Governance Model (SGM) on the basis of current societal and business needs to re-think managerial practice for the viability and survival of Service Systems.

First of all, this requires clarifying the differences between concepts of Government and Governance. Both refer to decisional processes for ensuring the viability of organizational systems, rule oriented in the explanation of activities and objectives. While government implies a formal authority for ensuring the correct implementation of certain policies, governance refers to supportive activities aimed at common goals without necessarily relying on the powers that be for dealing with disloyalty or for achieving compliance. In this sense, governance is a wider aspect of government, which embraces governmental institutions, but also subsumes informal and non-governmental mechanisms. Nowadays, governance is the norm in new public management, corporate governance, urban governance and security governance. In general, governance is any strategy, process and plan for controlling, directing and managing problems in any field: global, national, local, organizational; it is used as a *catch word* (Smouts, 1998:81), or as a *buzz-word* (Jessop, 1998:29) to indicate any coordination of interdependent activities.

Referring to our Hp1), building on Golinelli (2010), and Barile (2009), the concept of governance refers not only to a mere problem solving approach but also to the need to comprehend the subjective features and relational means qualifying the dynamics of an articulated common decision-making process to achieve systems viability. From a S_FS_Ybv perspective, this means a new deal of governance – governmentality – in which the prominent role of governance is smarter, based on common objectives, practice and actions and on synergic cooperation and the collaboration of each service systems component (people, universities, local institutions, cities, regions, nations, etc…).

Governmentality implies creating the conditions for the viability of Whole Service Systems (Spohrer, Piciocchi and Bassano, 2011), ensuring the collaboration and equally shared aims of each component. In other words, the new deal of governance is able to answer an historical question: *How can we be smarter in governance?* In fact, the smart logic of governmentality refers to governance processes that make a significant change within the government's role as a structure: from summary transcendent satisfaction of guided components to regulator of common interests, since it is in the nature of expectations and interests, in the practice of governance and its objects, that its utility can arise. Governmentality is here conceived rather as an internal function of modern governmental rationality that does not ask about the conditions of its enhancement through the exercise of government, but moves by the company asking why we need government and what kind of relationship it should have in society to justify its existence (Foucault, 2004:256).

On the basis of our Hp2), the concept and practice of smarter governmentality enables the transition from the structure to the system of government, in fact, the viability of service needs to be recognized in the collaborative and shared method of governance and not in the power structure of its government.

2   THE METHODOLOGICAL APPROACH FOR MULTILEVEL GOVERNMENTALITY IN SERVICE SYSTEMS

Today, the focus on governmentality is attributed to the invention and the assembly of a series of expertise, knowledge and technologies that link up the strategies developed in the different decisional nodes scattered throughout the

322

territory to strengthen the ability of governments to manage economic life and the health and habits of the population.

Governmentality Studies translate this approach into *steering governance from a distance* (Kickert, 1995). Government does not act directly on interests and power relations, but indirectly by linking a variety of more or less independent organs in order to direct the outcome of individual and collective conduct. Steering at a distance in the government, therefore, means establishing flexible relationships between separate entities that exist in space and time and in formally separate and independent areas. In other words, governance represents a process capable of using new technical and collaborative guideline.

Relevant is the definition of governance given by the Organization for Economic Cooperation and Development (OECD, 2001), in which the obsolescence of traditional government concepts is confirmed compared to the new way of organizing and administering economic and social service systems.

«*Government is no longer an appropriate definition of the way in which populations and territories are organized and administered. In a world where the participation of business and civil society is increasingly the norm, the term 'governance' better defines the process by which we collectively solve our problems and meet our society's needs, while government is rather the instrument we use*» (OECD, 2001)

The semantic difference between government and governance and governability, isn't simply conceptual. Indeed, while government implies a network of few formal authorities, governability refers to a spreading network of informal relationship able to express the steering governance from the distance.

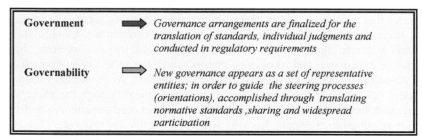

Figure 1 – Semantic differences between government and governability.

If traditionally governance meant a complex set of mechanisms and procedures designed to streamline and harmonize the interests of investors and at the same time to ensure the governability of the organization, currently, in the age of service systems, governmentality characterizes self-regulation processes – therefore independent of the rules and control and verification procedures – by which it becomes more efficient, more transparent and more secure for the benefit of the stakeholders involved (Arienzo, 2004). In other words, since the organization is a collective entity operating in a changing environment and to achieve conflicting goals, smarter governance is needed to uphold a framework of shared responsibility between service management, the board, the shareholders and other stakeholders (i.e. between the organization, ownership, shareholders, investors, customers, suppliers, etc...).

Therefore, although governance develops in a horizontal manner, as networks of collective actors in which players may be involved in the negotiation process through mediation and consensus procedures, such procedures seem designed to acquire legitimacy and compliance rather than participation. As concerns services, this means that the adoption of the new deal of governance – governmentality – needs a change of perspective: from the prescriptive complexity of government to the descriptive complication of governance. For this reason, we can affirm that nowadays, the governmentality is intrinsic to de-vertical modern government, which goes in the direction of intelligent governance of complex services. There is a reconfiguring of sovereign power that expresses the paradox of the concentration of the power without centralization: *multilevel governmentality of services systems* in which competences, skills, collaboration, sharing and competitiveness are favored by a regulatory interest and widespread stakeholder expectations.

Consequently, $S_E S_Y bv$, a methodological approach merging the Service Science and Viable Systems Approach, in order to qualify this multilevel governance in services, illustrates and interprets how governance mechanisms have to be smart – in universities, cities, regions and nations – to ensure a good quality of life and competitiveness mediating interactions among the players in a holistic context of service for value co-creation (Spohrer and Maglio, 2010).

New smarter governance, to enhance the competitiveness and consonance of the service systems as opposed to the dimensions of the context entities, has to be based on attitudes and predispositions (such as rationale, creativity and loyalty) and on the knowledge each individual working in the organization has at his/her disposal. Collaboration, knowledge sharing aimed at generating new knowledge and skills has to be stimulated. Therefore, mediation is needed between the variety of interests involved, guaranteeing reciprocal satisfaction and as a result, systemically equal aims (value co-creation) in terms of the ability of a system to reach the same target starting from different conditions and/or along different pathways.

VSA based on assumptions relative to concepts of "consonance" (structural compatibility) and resonance (systems viability) explains how in a complex service system, the process of value co-creation by holders of specific and subjective interests can be ensured by multilevel governmentality able to guarantee systemic objectives as a whole and the consequent satisfaction for all involved. On the contrary, in our integrated perspective (SSME+D and VSA), consonance sums up the necessary condition for guaranteeing the ability, for instance, of governments to make decisions having a collective (i.e. national) relevance being ideally at least, sensitive to stakeholder needs (citizens, universities, firms and other organizations, employees, investors; institution). In other words, any decision by government has to be made taking into account the mediation effect of the diverse aims (the "why" and not the "how or by what means") expressed by the players interacting and involved in the process of service value co-creation.

Furthermore, if resonance refers to the governmental capability to define decisional context, allocation of resources, strategies and compliance with bottom-up demands, then assessing the role of human resources for service systems governmentality represents the process by means of which smart government acknowledges, evaluates, and measures the expectations and aims of stakeholders and/or of the other service systems. This implies governmentality not necessarily of an optimum kind, but certainly capable of covering the diverse areas of needs (aims and desires) to plan complex service systems according to efficient value co-creation policies (Spohrer, Piciocchi and Bassano, 2011).

The need for greater cooperation involving numerous players (suppliers, research groups, competitors, clients, etc), symptomatic of a greater extent of openness, not to mention the need to acquire, keep and increase consent concerning strategic and operational initiatives, has led to a careful consideration of the needs, not only economic, of the various stakeholders the system even when their contracting ability is vastly inferior to that of the governing individuals (Golinelli and Bassano, 2012).

3   THE HUMAN ASPECT OF GOVERNANCE IN A SERVICE SYSTEM BASED VIEW ($S_ES_YB_V$)

Service systems for value co-creation require analyzing in an interactional perspective whereby the systems context prevails over the subjectivity of the individual stakeholder (single expectations). In other words, the representation of services based on "single expectations" does not fully explain a phenomenon and its implications; services need to be integrated using a systemic interpretation – the so called "Whole Service" (WS) perspective – which highlights shared aims (consonance) within a service system in order to generate value (Spohrer et al., 2011).

From a Service System based view ($S_ES_Y$bv) service systems have to be interpreted as context dependent/related and co-created from a dynamic viewpoint which in other words, means that the emphasis is not on single elements (parts) but on the service process (whole) which contributes to creating value in a specific relational context. The ontological meaning of the innovative concept of Whole Service (WS) integrates the value proposition of Holistic Service Systems (HSS) (Spohrer, 2010) by means of multilevel governance contributing to the growth of the overall value co-creation (Spohrer et al., 2011). This implies that the creation and the development of services depends on the capability of governmentality to read and interpret the context from a multilevel but consonant viewpoint. If we assume value co-creation as the mediation of the expectations of diverse, but collaborative, entities then governmentality involves the interpretation, representation and coordination of the various stakeholder interests ("the reason why") to define and promote a participative strategy useful for value co-creation. In $S_ES_Y$bv, this means:

- searching for levels of consonance within the observed holistic service system which involves members of the same epistemic community or service system context having similar backgrounds in terms of values, formal training and job experience (T-Shape people) with a widet variety of interests;
- focusing on symbolic approaches to understanding service system entities and value co-creation mechanisms (Spohrer and Maglio, 2010).

Each entities (public or private, local, national or international) needs to have not merely a specific role – I-Shaped for playing specialized activity of value – but rather one of sufficient variety – T-Shape for participating in value co-creation – (Macaulay et al, 2010) cooperating and collaborating according to a win-win logic for Whole Service global improvement (Gummesson, 2008). This assumption, in line with $S_E S_Y bv$; service systems, requires an interactional view whereby systems dynamics prevail over structural components with a shift in focus from conventional governance mechanisms to smarter shared and collaborative governance processes. The network synthesis of multiple entities consonance has to be guarantee by a smart governance role – in the new way described – mediating different stakeholder expectations in a cohesively designed, network structure capable of changing in changing environments. A smart governance role is fundamental for the growth and improvement of service systems. As governmentality is the ability to interpret the environment, to describe the context, to dialogue with other system, to reconcile the different expectations of the entities involved directly and/or indirectly in the whole system, we can assume that people, involved in governmentality require a multi-skilled professional profile (T-&-Π Shaped and Wedgies).

$S_E S_Y bv$ underlines the relevance of harmonizing different interests; on one hand, the need of a multi-skilled profile for taking on a governing role is emphasized and, on the other, the capability of focusing global governance on Whole Service to arrange and to assess the various conditions of compatibility and the expectations of the leading pressure groups (stakeholders). Multilevel governance can be conceived as governmentality characterized by *evolving and directional processes* (Spohrer et al, 2011).

In fact, directional and evolving processes ensure both external and internal relational interaction planning as well as improving shared competences and innovation. This means, that in complex organizations, or Whole Services, the bottom-up process for value co-creation is responsive to "evolutionary governance" (March and Simon, 1958), characterized by three logical inter-connected roles:

   a) *entrepreneurial governance*, devising new value proposition creation mechanisms;
   b) *brokerage governance*, creating new relationships and alignment mechanisms;
   c) *funding governance*, creating new shared risk and reward mechanisms.

Shared smart governance favours the limited "dispersion" of individual strategic guidelines (directional processes) and leads to modular development for improving specialization and integration (evolving process). Decision making, not necessarily of an optimum kind, but certainly capable of covering the diverse areas of needs (aims and desires) for designing and administrating complex service systems is consequently, ensured on the basis of value co-creation in Whole Service (Spohrer et al, 2011). In this context, we argue that the capability to co-create service value depends on the capability of multilevel governmentality to interpret, mediate and synthesize the context in a $S_E S_Y bv$ perspective.

As mentioned previously, if value depends on the interacting of multiple systems (Spohrer and Maglio, 2010), then value co-creation, in a Whole Service perspective – as well as in the Smarter Planet Project –, depends on the consonance with the different levels of networked holistic service systems oriented towards the common objective of sustainable and competitive growth. Consonance also underlines the aspect of trust, necessary to achieve cohesion in relations and interactions in service systems. In service systems the philosophy of balance and equal aims underpinning governmentality, depend on the consistency of relations and interaction based on a win-win logic.

In this perspective, Whole Service can ensure global value co-creation only if service system interactions guarantee shared growth as represented in Figure 2. As the figure shows, in each kind of trust relation, HSS share different conditions of consonance (structural adequacy for service systems) and different degrees of resonance:

1) *Calculus-based trust* = relations are based on contingent consonance schemes for achieving opportunistic simple exchange resonance (win-lose or lose-win logic). In this way, Whole Service can be compared to an *in embryo system*;

2) *Knowledge-based trust* = relations are based on indeterminate consonance to achieve opportunistic and also collaborative resonance (win-lose versus win-win logic). In this case, Whole Service can be compared to an *evolving system*;

3) *Identification-based trust* = relations are based on established consonance to achieve integrated and synergic resonance (win-win logic). In this case, Whole Service can be compared to a *viable system*.

Whole Services, established on knowledge-based trust, or rather, on identification-based trust, are able to generate a context of value co-creation in which each component is a participant in the development of the supra-system characterized by shared aims and no constraints in a win-win logic.

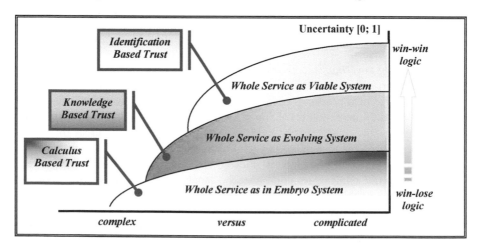

Figure 2 The bottom-up process for building Whole Services based on relations of trust.

4 LIMITATIONS & IMPLICATIONS FOR SERVICE RESEARCH

According to $S_ES_Y$bv, governmentality roles help to explain "how" Whole Services can be built, seeking the development and growth of each holistic service system on the basis of shared planning. The implications are that future research could focus on Whole Service as a polycentric network where holistic service systems collaborate in promoting multilevel governmentality – as defined – to ascertain and interpret cohesion between governance roles and stakeholder expectations (Ostrom, 2009).

In this perspective, the role of human resources in value co-creation process and in Whole Service multilevel governmentality is fundamental in terms of human talent. In fact, in a Service Science perspective, the "shape of a professional's" means the capabilities that professionals can apply when they are involved in problem solving and when they are communicating with other professionals as part of a project team (Spohrer et al, 2010:10). If governance refers to the government activity aimed at guaranteeing the viability of the systems, then T-Shaped professionals are individuals with talents for producing global value co-creation and Whole Services. In effect, T-Shaped professionals manage decisional uncertainty in a complex scenario; this means in other words, that T-Shaped professionals know "what to do" – targets and ends – and do "what they know" – i.e. their competences and means – (Spohrer et al, 2010:12).

Future research is needed to ascertain how to implement smart mechanisms of governance and how the former relate to an "improved governance framework in order to guide policy makers in achieving long-term, regional, quality of life" (Fleming and Spohrer, 2011:1).

REFERENCES

Arienzo, A. 2004. "Dalla corporate governance alla categoria politica di governance". In. *Governance. Controdiscorsi I*, eds. Dante & Descartes, Napoli.

Barile, S. 2009. *Management sistemico vitale*. Giappichelli.

Fleming, M. and Spohrer, J. 2011. "Co-evolution of Future Technologies and Regional Skill-Job-Career Landscapes: Connecting Frameworks, Theories, and Models". *CFTSJC*.

Foucault, M. 2004. *Sécurité, Territoire, Popolation. Coirs au Collège de France, 1977-1978*. Michel Senellart, Gallimard-Seuil, Paris.

Golinelli, G.M. 2010. *Viable Systems Approach (VSA). Governing Business Dynamics*. Cedam, Kluwer, 2010.

Golinelli, G.M., Spohrer, J., Barile, S., Bassano, C. 2010. The evolving dynamics of service co-creation in a viable systems perspective. *Proceedings of the 13th Toulon-Verona International Conference* in Coimbra, Portugal, 2-4 September.

Golinelli, G.M. and Bassano C. 2012. "Human resources for governing business dynamics. The Viable Systems Approach". *Proceedings of the 4th International Conference on Applied Human Factors and Ergonomics (Ahfe)*, San Francisco, July, 21-25.

328

Gummesson, E. 2008b. Quality, service-dominant logic and many-to-many marketing. *The TQM Journal* Vol. 20, No. 2: 143-153.

Jessop, B. 1998. The rise of governance and the risk of failure. *International Social Science*, (50) 155:29-45.

Kickert, W. 1995. Steering at a Distance: A New Paradigm of Public Governance in Dutch Higher Education. *Governance*, Vol. 8, issue 1:135-157.

Latour, B. 1987. *Science in action*. Milton Keynes: Open University Press.

Macaulay, L. Moxham, C., Jones, B., Miles, I. 2010. "Innovation and skills. Future Service Science Edication". In. *Handbook of Service Science*. eds. P.P., Maglio, C., Kieliszewski, and J. Spohrer. Springer.

Maglio, P., Kieliszewski, C., Spohrer, J. 2020. eds. *Handbook of Service Science*. Springer.

March, J. G. and Simon, H. 1958. *Organizations*. Wiley.

OECD. 2011. *Governance in the 21st century*. OECD publications, Paris.

Ostrom, E. 2009. "Beyond Markets and States: Polycentric Governance of Complex Economic Systems". Prize Lecture, December 8, Workshop in Political Theory and Policy Analysis, Indiana University, Bloomington, IN 47408, and Center for the Study of Institutional Diversity, Arizona State University, Tempe, AZ, U.S.A.

Piciocchi, P. and Bassano, C. 2009. "Governance and viability of franchising networks from a Viable Systems Approach (VSA)". Proceedings of the 2009 Naples Forum on Service. Service Dominant Logic, Service Science and Network Theory, Capri, June 16-19, Giannini Editore, Napoli.

Rose, N. 1999. *Power of Freedom. Refraiming political thought.* Cambridge: Cambridge University Press.

Rosenau, J.N. and Czempiel, E.O. 1992. *Governance without Government*. Cambridge UP. Cambridge.

Smouts, M.D. 1998. The proper use of governance in international relations. *International Social Science Journal*, (50) 155.

Spohrer, J., Golinelli, G.M., Piciocchi, P., Bassano, C. 2010. An integrated SS-VSA analysis of changing jobs. *Service Science*, Vol. 2, No. (1/2): 1-20.

Spohrer, J. and Maglio, P. 2010. Service Science: Toward a Smarter Planet. In Service Engineering. eds. Karwowski and Salvendy. Wiley, New York, NY.

Spohrer, J. 2010. "Working together to build a Smart Planet". *ICSOC*, San Francisco, December 8, available at: http://www.slideshare.net/spohrer/icsos-20101208-v2.

Spohrer, J., Piciocchi, P., Bassano, C. 2011. Three frameworks for service research: exploring multilevel governance in nested, networked systems. *Proceedings of the 2011 Naples Forum on Service. Service Dominant Logic, Service Science and Network Theory: integrating three perspectives for a new service agenda,* Capri, Italy, June 14-17, Giannini Editore, Napoli.

Triantafillou, P. 2004. "Conceiving "Network Governance". The Potential of the Concepts of Governmentality and Normalization". Working paper 2004/4. *Centre for Democratic Network Governance*. Roskilde, May.

Vaccaro, S. 2008. *Governance e governo della vita*. In. *Biopolitica, bioeconomia e processi di soggettivizzazione*. eds. A. Amendola, et al. Quodlibet, Macerata.

Vargo, S.L. and Lusch, R.F. Service-dominant logic: continuing the evolution. *Journal of the Academy of Marketing Science*, Vol. 36, No. 1, pp. 1-10, 2008.

Williamson, O. 1996. *The mechanisms of governance*. Oxford University Press, Oxford.

CHAPTER 33

Governing Human Relations to Promote Local Service Systems in Processes of Internationalization

I. Orlando[1], A.Siniscalchi[2], P. Piciocchi[3], C. Bassano[4], A. Amendola [5]

[1]Intertrade – Special Agency CCIAA of Salerno (ITALY) - orlando@intertarde.sa.it
[2]University of Salerno (ITALY)   asiniscalchi@unisa.it
[3] University of Salerno (ITALY) – p.piciocchi@unisa.it
[4]Parthenope" University of Naples (ITALY) – clara.bassano@uniparthenope.it
[5]University of Salerno (ITALY) – alamendola@unisa.it

ABSTRACT

Currently in public-private organizations, governance has changed in nature, in intensity, as well as in the variety of stakeholder expectations.

Our study aims to qualify the relations created between the strategic actors of a territory and the developing of a local network to facilitate the internationalization process. In particular, the role is analyzed of *Intertrade* – the Special Agency of the Chamber of Commerce (CCIAA) of Salerno for the internationalization of SMEs – seeing as it is the Institution that mediates, coordinates and promotes local service systems in international scenarios. The main constraints to the development of internationalized SMEs are represented by the gap relative to cognitive and managerial opportunities in international markets and the lack of an adequate competitive, economic structure in global contexts. For this reason, a smart entity for governing the mediation process of stakeholder expectations would be desirable. Therefore, Intertrade is analyzed on the basis of its coordinating role. The institutional mission of the Agency is the promotion of SMEs in the Salerno area on international markets. Two main lines of action characterize the role of Intertrade

(Globus, 2010): a) training and information to consolidate and support existing know-how; b) the creation of contacts abroad, in close connection with the Chamber of Commerce Offices, supporting enterprises at fairs, during commercial missions and business meetings in Campania in order to accelerate the achieving of positive economic results deriving from International business. In other words, this means Intertrade as a governing body capable of playing a prominent role (Piciocchi and Bassano, 2009) in international-local networks to support the recognition and visibility of a local service system (the Salerno area), coordinating *economic capabilities*, *pressures*, *resources*, and the *socio-technical system* and *relationships*. In this perspective, a distinctive identity for competitiveness on international markets (Pilotti, 2011) is favoured.

Keywords: Multilevel Governance, Internationalization Process, Service Science Management Engineering+Design, Viable Systems Approach

1   INTRODUCTION

From a government point of view, structural and systemic change reflects the relevance of synergic relationships between local institutions and private actors based on collaborative and cooperative projects (Borgonovi, 2000).

At the present time, global competitiveness proposes that for development, each organization should share its business with similar entities through participation in collaborative ventures for value co-creation. In other words, for competitive processes of internationalization, resonant interaction in network systems is required whereby each partner shares informative units, interpretative schemes and category values (Barile, 2009).

The role of multilevel governance is extremely relevant for competitiveness, for coordinating strategies and actions and mediating public and private expectations according to a bottom-up process of participative planning. In the service era, people and organizations cannot express their business competitiveness if they are strictly autopoietic, because stand-alone policies do not enable them to manage the wide variability of local and global contexts.

This implies that organizations of local services have to participate in virtuous processes of promotion and strategic planning, shared by the local productive and social system and sustainable by global markets. Multilevel governance (Locke and Schweiger, 1979; Ostrom, 2009) refers to the relationship between local actors and to their economic cohesion or consonance (Golinelli, 2010), stressing the relevance of multilevel policies as well as value co-creation (resonance) for the whole local service system. Multilevel governance in effect, seeks to provide a critical understanding of local cohesion as an economic policy and aim for territorial systems, interpreting governance dynamics within the transformation frame of government activities and the emergence of complex and collaborative decision-making processes.

Starting from a review of the literature on the transformation of the business logic with regard to new governance based on a collaborative business approach, the study offers an analysis of local joint governance, stressing the relationship between no-profit governance and territorial consonance in accordance with collective agreements as instruments of local network contracts.

2 INTEGRATED METHODOLOGIES FOR MULTILEVEL GOVERNANCE IN LOCAL SERVICE SYSTEMS

"Multi-level governance can be defined as an arrangement for making binding decisions that engages a multiplicity of politically independent but otherwise interdependent actors – private and public – at different levels of territorial aggregation in more-or-less continuous negotiation/deliberation/implementation, and that does not assign exclusive policy competence or assert a stable hierarchy of political authority to any of these levels" (Schmitter, 2004: 49).

In recent years, the main focus of territorial policy has been on sustaining growth to make local service systems more competitive. This is complex because each territory is different from the others in terms of resources, capacities, competences, social characteristics which imply specific policy and investment needs. In this perspective, the competitiveness of local economies through multilevel governance mechanisms (Williamson, 1996; Spohrer et al, 2011) on which the implementation of development strategies is based, needs to be addressed.

Multilevel governance, seen as the exercise of authority and the various dimensions of relations across levels of government, has changed. In particular, decentralization has delegated to local governments more autonomy in decision making and has increased their capacity to formulate and deliver local policy of development. Local governments, by virtue of their economies are increasingly exposed to global competition, and have to guide - in a collaborative way – positive impact strategies for improving the competitiveness of entire local service systems. Governance in other words, has become both more complex and more demanding, involving multiple public and private actors and requiring a rethinking of how to govern participation and cooperation. Therefore, the enhancement of the role of regional no-profit organizations such as Intertrade and the reformulation of international projects is imperative and justified by consolidated theories such as *Public Governance* (PG), the *Network Theory* (NT), the *Resource Based View* (RBV), *Service Science Management Engineering+Design* (SSME+D) and the *Viable System Approach* (VSA).

PG (Kooiman and Van Vliet, 1993; Kickert, 1997) through the composition of formal and informal relationships, vertical and horizontal, integrates the perspective of a single organization with networks. This change of perspective suggests the widespread adoption of the NT logic (Granovetter, 1985; Killworth et al., 1990) that verifies the weight and definition of strategic roles of the players in a territorial perspective of appropriate services to enterprises (e.g. Intertrade activities) for internationalization. Obviously, it is important maximize collaborative synergies implemented by the prominent role in governance networking.

In other words, local networks create collaborative competitiveness, in relation to propositions of the SSME+D (Spohrer and Maglio, 2010; Vargo and Lusch, 2004) within service systems. In fact, the SSME+D approach considers that performance and competitiveness depend on service systems, that are dynamic configurations of resources.

The proper functioning of the local network for internationalization is connected to knowledge sharing (RBV) (Moingeon et al. 1998; Siano, 2001) as a source of competitive advantage, sustainable and difficult to replicate being based on the structured knowledge and the peculiarities and needs of stakeholders in the local system.

In the VSA view (Golinelli, 2000, 2010) the interpretation of the entrepreneurial fabric as an organized cohesive entity (evolving system) suggests smart governance is needed to address and coordinate the expectations of stakeholders in a manner consistent with the expectations of the market.

In this perspective, Intertrade, represents both a responsive and proactive system of governance able to define the cognitive approach for supporting SMEs in the internationalization process, but also capable of coordinating participative structures suitable for the emergence of a competitive territorial system. Joint governance in the systems approach, aims to qualify above all, a collective identity in order to ensure equal aims and competitiveness for every individual node in the internationalization process of territorial systems.

2.1   Governance of local networks and services to support the internationalization of SMEs in Salerno: the role of Intertrade

On the basis of integrated methodology, Intertrade appears a strategic component of local government oriented to promoting internationalization processes, in an evolutionary perspective. This means, however, that Intertrade has to acquire a more central role and identity in the sense of proactive governance institution for supporting and promoting the productive system of Salerno on global markets.

The SMEs in Salerno which also include some cases of excellence and best practices, do not possess the vocation to confront in an organized and collaborative way, the challenges of globalization. By virtue of this deficit in management, Intertrade effectively supports SMEs in their approach to foreign markets but these promotional activities do not appear to be consonant or consistent with real enterprise needs. On the basis of this gap, Intertrade has changed its planning policies, focusing on two main objectives: 1) the international promotion of integrated local service system (Piciocchi et al, 2011); 2) supporting and promotion of local service sectors on an international basis, of the whole territorial system.

The above mentioned objectives have been defined according to a bottom-up logic; in other words, starting from sector best practices, Intertrade tends to generate a virtuous circle of syntropy among local competences to develop a smart whole local service system, highly integrated in global scenarios thanks to the political relationships with foreign players, as figure 1 shows.

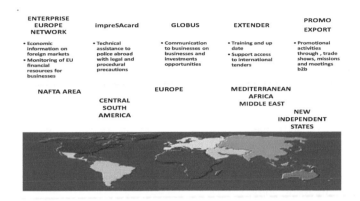

ENTERPRISE EUROPE NETWORK	impreSAcard	GLOBUS	EXTENDER	PROMO EXPORT
• Economic information on foreign markets • Monitoring of EU financial resources for businesses	• Technical assistance to police abroad with legal and procedural precautions	• Communication to businesses on businesses and investments opportunities	• Training and up date • Support access to international tenders	• Promotional activities through , trade shows, missions and meetings b2b
NAFTA AREA	CENTRAL SOUTH AMERICA	EUROPE	MEDITERRANEAN AFRICA MIDDLE EAST	NEW INDEPENDENT STATES

Figure 1: *Intertrade Salerno's* "Regional" activities

In a SSME+D&VSA perspective, this means that Intertrade Salerno is a strategic node of the local multilevel governance system in establishing consonance conditions (drivers) for coagulate the different stakeholder viewpoints relative to a shared process of the entire internationalization system and in defining a series of actions to improve the local service system reputation on global markets. This support and promotion on the part of Intertrade Salerno, makes them a prominent actor in the coordination of internationalization strategies.

3   GOVERNANCE FOR THE COMPETITIVENESS AND REPUTATION OF LOCAL SERVICE SYSTEM

Intertrade Salerno is an entity that can be qualified as a territorial "star" node for local service system internationalization. Such a role combines: i) identity as organizer and co-ordinator of international processes, but lacking a propositional identity in terms of establishing guidelines/planning of participation policies; ii) timely concerted action to promote specific international cooperation programs (Evans, 2004; Stiglitz, 2002). Participatory/collaborative/communicative planning of local governance is built on the search for collaborative synergies that can generate virtuous circuits of co-creation value in the entire local service system.

The focus on governance on the part of Intertrade Salerno certainly intercepts and consolidates the approval of a wider structure of relationships between local stakeholders in a "variable geometry" logic in which the local, national and international intertwine and create a series of coordinated actions to ensure effectiveness and efficiency in the processes of internationalization. This implies more efficacy in Intertrade Salerno role in the contemporary reinterpretation of the territory as a whole system to promote the competitiveness of which is the result of collaborative synergies (Piciocchi and Bassano, 2009) co-directed by public and private actors. Such a systemic view of the territory implies, therefore, the promotion of local networks whose international competitiveness derives not only from their endowment but also from their structural dynamic behavior (Paz, 1997).

From a SSME+D viewpoint, international service systems are dynamic configurations of resources capable of creating and delivering value for stakeholders. In such a perspective, Intertrade Salerno is classified as a strategic hub of governance, a generator of professional network services that support and exhort internationalization. This is the result of an internal process of value co-creation that seeks to establish a destination brand consistent with local conditions; but also through an external process that tends to improve brand reputation (e.g. a territorial and productive brand) mediating the satisfaction of the interacting entities (Gronross, 2008). The definition of a brand destination implies the design and implementation of a service structure, that ensures the sharing of resources in order to produce greater benefits for all the stakeholders interacting on the basis of a win-win logic (Gummesson, 2008b).

Figure 1: Multi-level governance to create value proposition for LSS competitiveness

Figure 2 shows how the territory can be represented by varying degrees of cohesion and cooperation among stakeholders through local multilevel governance.

If the territory is considered a "set of resources", competitive advantage is based on the performance of companies involved in the international market; network components interact for instance, without a common planning policy, aims are single as opposed to common, links are fragmentary and shared governance able to provide guidelines and rules is entirely lacking. In an integrated SSME+D&VSA perspective, the local network is an in embryo system that has no clear cut common identity within and outside the territory.

If the territory is to be considered as a "product to promote" its competitive advantage is based on territorial identity characterized by all the visual elements through which it represents itself in the context of a global system (Siano, 2001). In this case, competitive strength depends on the ability to read the context and on a definition of a policy of communication in line with stakeholders' expectations. However, the focus remains that of tangible goods offered by SMEs in Salerno even though the service component is finally beginning to emerge.

In other words, if the territory is conceived as a generic "scenario" then competitive advantage is based on image defined as a non-standardized subjective perception that stakeholders have about a given territory at a given time (Siano, 2001), constituting, consequently, a complete system in which government provides guidelines (constraints), defines roles (rules) and exercises strict control. A limit of such a system is the lack of stability resulting from the variability of decision makers based on different interpretative patterns and value-categories (Barile, 2009) and the consequential creation of a reputational gap. In SSME+D terms, the delivery of valuable services is framed, but this value has a very short life cycle, leaving no sedimenting in the social structure of the territory.

Consequently, if the territory is conceived as a "system", competitive advantage stems from reputation or socially shared judgment based on the enduring capability to create value for stakeholders (Siano, 2001). In this context and in a SSME+D perspective, territory is described as a viable system or stable service system the competitive advantage of which is triggered by the whole local service system; in this perspective, Intertrade Salerno plays the role of coordinator and proposer thus enhancing place reputation (Siano et al., 2009) and, therefore, increasing competitive advantage. Consequently, in the transition from Local Area to Local Stable Service System, territory is no longer the object of action rather the subject, capable of expressing, by means of rationale and smart governance, competitiveness (consonance and reputation) in the international arena. Figure 3 illustrates how the competitiveness of LSS depends on the sharing of global value propositions through multi-level governance (Sphorer et al., 2011) representative of local components. Consequently, Intertrade Salerno has to govern, together with the other levels of governance, two drivers:

1. *consonance*: the evaluation of structural conditions for joint value propositions at international level, in order to define a brand destination;
2. *reputation*: the evaluation of systemic conditions to generate trust capital in order to enhance the place reputation of the whole local system.

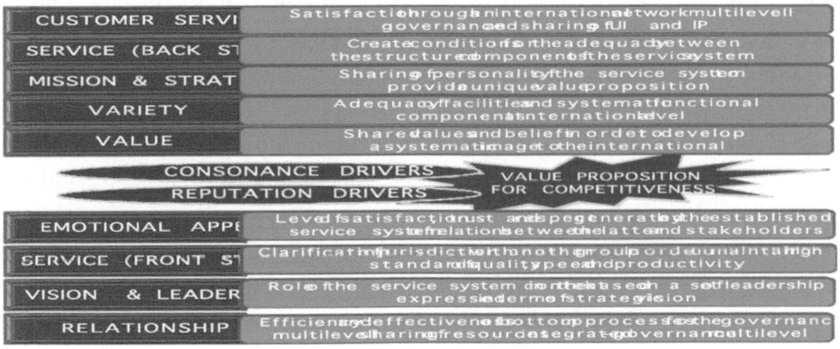

Figure 2: Consonance and reputation drivers for LSS competitiveness

To identify the drivers, the attribution of specific weights to each items should be pondered in order to define the impact of the drivers for value co-creation in the international network. This goal will be pursued in future research.

4 CONTRACTUAL AGREEMENTS FOR MULTILEVEL GOVERNANCE OF INTERNATIONAL TERRITORIAL NETWORKS

In the present day's internationalized setting, territories have difficulty in surviving and developing, due to their lack of systemic organization and the widespread idea of local area as an object rather than a system of competition.
As a result, local smart governance can be conceived in terms of :

* *guiding strategic decision making;*
* *identifying main factors for the competitiveness of a Local Service System (LSS);*
* *mediating different interests;*
* *co-finalizing the efforts of local players/stakeholders on the basis of joint projects;*
* *monitoring the adoption and assimilation of the value co-creation perspective in the local service system.*

With reference to the experience in Salerno, that of improving forms of cooperation to carry out joint projects for the international development of the territory as a system, the network contract instrument (Law dated 9 April 2009, no. 33) has a predominant part. Network contracts offer the opportunity of creating extended networks – locally and globally –, capable of supporting the international development of the whole local service system. The main limitation of this contractual agreement lies in the need for smart multilevel governance for the coordination and dissemination of knowledge and to avoid systemic entropy.

In general, network contracts have their main objective in the definition of norms through which enterprises, while remaining independent, can participate in common international projects, in particular to increase competitiveness in specific SMEs, but even in the entire local service system. Network contracts are also an opportunity for companies that do not have sufficient weight in local and/or global markets and that are virtually under protected, especially in supply chain relationships. In particular, these contracts are a network tool that allows small companies characterized by insufficient financial resources and expertise, to undertake a participatory form of development and to be guided in internationalized scenarios. Therefore, networks need to be constructed in terms of internal resonance in order to link up and integrate them into international contexts, thus creating sustainable competitive advantage and a stable and strong reputation. Through the network contracts instrument/tool, Intertrade Salerno could generate synergies between producers, distributors, suppliers of goods and services, research laboratories, institutions and other stakeholders in order to implement virtuous processes of internationalization of SMEs in Salerno. From an integrated SSME+D-VSA perspective, Intertrade Salerno could devise a specific structure that underpins the potential expressed by an expanded framework which governs the international network throughout its life cycle.

5 ORIGINALITY VALUE AND CONCLUSION

The value of this work lies in the qualification of the territory as an evolving system building reputation capital for the Local Service System. Sharing information units, interpretive schemes and value-categories, in other words, multilevel governance, enables the building of a strong local identity and, therefore, a value proposition in local and international contexts. LSS can be vital for the territory if it equipped with conditions of survival that ensure equal goals for each actor. In such a scenario, the opportunity to strengthen the cohesive role of Intertrade Salerno in promoting and monitoring local networks to support internationalization processes, represents a milestone for smart multilevel governance of Salerno system. In conclusion, the territorial organization of the Local Administrative Authorities is becoming increasingly detached from the logic of hierarchical power. In such a perspective, government cannot be defined as a set of techniques and procedures regulating the conduct of individuals throughout the course of their lives, subjecting them to the authority of a responsible but powerful leadership. On the contrary, smart governance encourages the trend for creating partnerships with the multitude of local actors/stakeholders, based on a policy of collaborative and cooperative initiatives.

REFERENCES

Barile, S. 2009. *Management sistemico vitale*. Giappichelli.

Borgonovi, E. 2000. L'organizzazione a rete nelle amministrazioni pubbliche. *Azienda Pubblica*, N.4 .

Evans, P. 2004. *Development as Institutional Change: The Pitfalls of Monocropping and the Potentials of Deliberation*, Studies in Comparative International Development. Vol. 38, N.4.

Globus. 2010. Salerno Internazionale. *La Tua Impresa nel Mondo*, Globus Special Number 81, December.

Golinelli, G.M. 2000. *L'approccio sistemico al governo dell'impresa*. Cedam, Padova.

Golinelli, G.M. 2010. *Viable Systems Approach (VSA). Governing Business Dynamics*. Cedam, Kluwer.

Granovetter, M. 1985. Economic action and social structure: the problem of embeddedness. *American Journal of Sociology*, Vol. 91, N.3.

Gronross, C. 2008. *Adopting a Service Business Logic in Relational Business-to-Business Marketing: Value reation, Interaction and Joint Value Co-Creation*. Otago Forum.

Gummesson, E. 2008b. Quality, service-dominant logic and many-to-many marketing. *The TQM Journal*, Vol. 20, N.2.

Kickert, W.J.M. 1997. Public governance in the Netherlands: an alternative to Anglo-American 'managerialism'. *Public Administration Review*, N.4.

Killworth, P.D., Bernard, H.R., McCarty, C., Shelley, G.A., Robinson, S. 1990. Comparing four different methods for measuring personal social networks. *Social Networks*, N.12.

Kooiman, J. and Van Vliet, M. 1993. *Governance and public management*. In. *Managing public organization: lessons from contemporary European experience*, eds. El Jassen, K.A. and Kooiman, J. Sage, Londra.

Locke, E. A. and Schweiger, D. M. 1979. Participation in decision-making: One more look. *Research in Organizational Behavior*, Vol. 1: 265-340.

Moingeon, B., Ramanantsoa, B., Métais, E. and Orton, D. 1998. Another Look at Strategy-Structure Relationships: the Resource-Based View. *European Management Journal*, Vol. 16, N. 3.

Ostrom, E. 2009. "Beyond Markets and States: Polycentric Governance of Complex Economic Systems". Prize Lecture, December 8, Workshop in Political Theory and Policy Analysis, Indiana University, Bloomington, IN 47408, and Center for the Study of Institutional Diversity, Arizona State University, Tempe, AZ, U.S.A.

Paz, M., 1997. *Structural dynamics: theory and computation*, Springer.

Piciocchi, P. and Bassano, C. 2009. "Governance and viability of franchising networks from a Viable Systems Approach (VSA)". *Proceedings of the 2009 Naples Forum on Service. Service Dominant Logic, Service Science and Network Theory*, Capri, June 16-19, Giannini Editore, Napoli.

Piciocchi, P., Siano A., Confetto, M.G. and Paduano E. 2011. "Driving Co-Created Value Through Local Tourism Service Systems (LTSS) in Tourism Sector". *Proceedings of the 2011 Naples Forum on Service - Service-Dominant Logic, Network & System Theory and Service Science: integrating three perspectives for a new service agenda*, Capri, Giannini Editore, Napoli.

Pilotti, L. 2011. *Creatività, innovazione e territorio – Ecosistemi del valore per la competizione globale*. Il Mulino, Bologna.

Schmitter, P. 2004. "Neo-functionalism". In. *European Integration Theory*. eds, A. Wiener and T. Diez. Oxford: Oxford University Press: 45-74.

Siano, A. 2001. *Competenze e comunicazione del sistema d'impresa. Il vantaggio competitivo tra ambiguità e trasparenza*. Giuffrè, Milano.

Siano, A., Confetto, M.G., Siglioccolo, M. 2009, "Place reputation management and leverage points. Rethinking cultural marketing for weak areas". *Proceedings of the 8th International Marketing Trends Congress*, Paris.

Spohrer, J., Piciocchi, P., Bassano, C. 2011. "Three frameworks for service research: exploring multilevel governance in nested, networked systems". *Proceedings of the 2011 Naples Forum on Service. Service Dominant Logic, Service Science and Network Theory: integrating three perspectives for a new service agenda*, Capri, Italy, June 14-17, Giannini Editore, Napoli.

Spohrer, J. and Maglio, P. 2010. Service Science: Toward a Smarter Planet. In. *Service Engineering*, eds. Karwowski and Salvendy. Wiley, New York, NY.

Spohrer, J., Golinelli, G.M., Piciocchi, P., Bassano, C. 2010. An integrated SS-VSA analysis of changing jobs. *Service Science*, Vol. 2, No. (1/2): 1-20.

Stiglitz, J.E. 2002. Participation and Development: Perspectives from the Comprehensive Development Paradigm. *Review of Development Economics*, N.6(2).

Vargo, S. L. and Lusch, R. F. 2004. Evolving towards a new dominant logic for marketing. *Journal of Marketing*, Vol. 68.

CHAPTER 34

A VSA Communication Model for Service Systems Governance

Alfonso Siano[1], Paolo Piciocchi[2], Maddalena della Volpe[3],

Maria Giovanna Confetto[4], Agostino Vollero[5], and Mario Siglioccolo[6]

[1] University of Salerno - Fisciano (Italy), e-mail: sianoalf@unisa.it
[2] University of Salerno - Fisciano (Italy), e-mail: p.piciocchi@unisa.it
[3] University "Suor Orsola Benincasa" - Naples (Italy), e-mail: maddalena.dellavolpe@unisob.na.it
[4] University of Salerno - Fisciano (Italy), e-mail: mconfetto@unisa.it
[5] University of Salerno - Fisciano (Italy), e-mail: avollero@unisa.it
[6] University of Salerno - Fisciano (Italy), e-mail: msiglioccolo@unisa.it

ABSTRACT

This paper proposes a conceptual framework for governance and management of the decision making process for corporate communication. The aim is to address a gap in the literature in that to date, by adopting some basic assumptions from Viable Systems Approach, integrated with Service Science perspective, in order to identify communication resources, the nature (static and dynamic) of the corporate communication activities, and "engineering" communication process. Parsons' sociological approach, with reference to the classification of organisational decisions (policy, allocation and coordination) results fundamental to contextualise this approach to the decision-making within corporate communication function/department.

Key words: corporate communication management, communication resources, communication decision-making, communication strategy.

1. INTRODUCTION AND LITERATURE REVIEW

In recent years, the growing significance of managing corporate communication has lead both pratictioners and academics to develop different conceptual models for the management of the corporate image (Dowling, 1986) and identity (Markwick and Fill, 1997). These models provide a broad variety of practices for the management of corporate communication, but do not examine corporate communication strategy, in terms of the types of decision that are entailed, and the link between the strategic and tactical decisions. Indeed, the idea that corporate communication requires strategic thinking has received very little attention in the literature (Steyn, 2003), since "communication is often still seen as a largely tactical activity" (Kitchen and Schultz, 2000) with practitioners acting as 'technicians' (Cornelissen, 2008: 99).

Nevertheless, it has been suggested that the management of communication needs to move from being tactical to strategic (Holm, 2006), and that communication strategy must contribute to the corporate strategy (Cornelissen, 2008). Adopting the central corporate perspective means regarding communication strategy as a functional strategy formulated by corporate communication departments that operate at the highest corporate level (van Riel and Fombrun, 2007). Corporate communication strategy is the outcome of a strategic way of thinking and of a decision-making process involving different parties (Steyn, 2003).

In an era of stakeholder management (Freeman, 1984), corporate communication strategy should be inseparable from the strategic management of a firm's relationships with its stakeholders (Steyn, 2003), and aim to build, protect, and consolidate a favourable corporate reputation (van Riel, 1995).

As briefly described, significant contributions to the debate on communication strategy may be found in the literature, but a conceptual framework that offers a deeper perspective on communication decisions within corporate communication management still needs to be devised.

2. AIMS OF THE PAPER

Decision-making related to the use of resources is one of the main issues arising from earlier studies of corporate strategy (Ansoff, 1965), and, particularly, by the resource-based view (RBV) (Penrose, 1959; Barney, 1991), which considers resources at the general level of money, human resources, and time.

Nevertheless, resources are a key factor in corporate communication as well: a strong, well-managed corporate brand, for example, meets the criteria posited by RBV, since it is rare, imperfectly imitable (Balmer and Gray, 2003), and durable (Grant, 1991).

Apart from considerations about brand, in the literature on the management of corporate communication there is little mention of how resources are identified or of the relevant decisions required to create and use them. The present paper aims to address this gap and to provide an in-depth examination of the issues neglected in the past. Therefore, the study presents a conceptual framework for government and management of corporate communication based on Viable Systems Approach (vSa) (Golinelli, 2010) integrated with Service Science approach (SS) (Spohrer et al., 2010; Spohrer et al., 2011; Spohrer and Kwan, 2008; Maglio et al., 2010; Vargo and Lusch 2004), short for "Service Science, Management, and Engineering" (also known as SSME), a recent stream of studies which offers the chance to develop further interesting interpretation about corporate communication management.

3. A FRAMEWORK BASED ON THE STRUCTURE-SYSTEM PARADIGM FOR GOVERNANCE AND MANAGEMENT OF CORPORATE COMMUNICATION

In line with basic assumptions made by the RBV, communication resources can be categorised as firm-specific or non-firm-specific (Siano et al., 2011).

Firm-specific communication resources express the specific capabilities, core values, and/or historical references relevant to a particular organisation, are rooted its identity, allowing it and its products to be recognised by stakeholders. Therefore, firm-specific communication resources are valuable and unique, and support the creation and maintenance of competitive advantage. They are the outcome of strategic decisions, and have long-term implications.

In contrast, non-firm-specific communication resources are typically acquired through market transactions, not based on a firm's identity, and have short-term implications. Typical non-firm-specific communication resources are the means of communication (media) and the services/expertise of communication (copywriters, art directors, etc.).

3.1. The main firm-specific communication resources

The main firm-specific communication resources are: basic key words, such as common starting points (van Riel, 1995), themed messages (Cornelissen, 2008), and brand mantras (Keller, 1999); distinctive short messages derived from the basic key words (typically slogans/taglines); unique symbols and sounds, i.e. elements of corporate visual identity and corporate jingles (Bernstein, 1986); codices (sets of rules or heuristics) related to the management and expression of corporate brand (Tilley, 1999); organisational storytelling derived from corporate stories (Brown et al., 2004).

In order to identify a decision-making framework, Parsonsian three types of organisational decisions (1956) are applied to corporate communication management. Policy decisions are those that commit the organisation as a whole, concern the creation of the resources required by an organisation, or lead to significant changes in the resources that belong to it. Allocation decisions concern the utilisation of the resources available to the organisation. Finally, coordination decisions are aimed at achieving the integration of the organisation as a whole system.

3.2. Policy and coordination decisions of communication at the strategic level (governance)

Policy decision-making at the strategic level concerns the firm-specific communication resources, has long-term implications, and involves a number of different parties (communication managers, CEOs, managers of other functions or departments, and practitioners). Firm-specific resources are made coherent, harmonised, and synergised by decisions related to the coordination of communication, which occurs both at a strategic and a tactical-operational level (Schultz and Kitchen, 2000).

In this respect, firm-specific communication resources should be designed via policy and coordination decisions to support the communication requirements to build a strong reputation, e.g. visibility, distinctiveness, authenticity, transparency, and consistency (Fombrun and Van Riel, 2004).

Decisions on policy and coordination are a part of the communication strategy, which is aligned with corporate strategy and corporate culture and considered to be a 'dual' decision, related to: (1) the strategic intent of an organisation, concerning its desired position in terms of corporate reputation (Cornelissen, 2008); and (2) the creation and coordination of firm-specific communication resources. The strategic intent involves a change or consolidation of the organisation's reputation.

Policy decisions are governance decisions and refer to planning, organization and coordination of firm-specific communication resources, which represent communication structure. The concept of structure is derived from Viable Systems Approach (VSA) (Golinelli, 2010; Barile and Saviano, 2011) and indicates the set of communication resources that organization has got at a given time. This set of resources may undergo various types of changes and adjustments over time. VSA paradigmatic distinction between structure and system (Golinelli, 2010) enables to grasp the difference between decisions about creating communication resources and decisions of utilization of these structural elements. The integration of VSA with Service Science makes it possible to qualify viable systems in terms of "service systems", which, according to SS view, are dynamic configurations of resources (people, technologies, organizations and shared information) that co-create and deliver value.

3.3. Allocation and coordination decisions of communication at the tactical-operational level (management)

Decisions on allocation and coordination pertain to choices on the utilisation of communication resources (firm-specific and non-firm-specific) in short-term

campaigns made to carry out the strategic intent, and concern:
- the utilisation of firm-specific communication resources for conveying messages (advertisements) ;
- the selection and utilisation of the communication mix (the techniques and means/channels of communication);
- the integration, harmonisation, and realisation of coherent, concrete messages (internal and external communications) and the combination of promotional elements of the communication mix (advertising, sales promotion, public relations, digital media, etc.) to obtain the synergistic effects of integrated communications (van Riel, 1995; Schultz and Kitchen, 2000);
- concrete formulation of short-term programmes and campaigns (annual/infra-annual communication plans) in line with the available budget.

Decisions on allocation and coordination are taken by communication managers and practitioners/consultants, and refer to the stages of planning, organisation, and coordination at the tactical-operational level.

According to Viable Systems Approach, allocation decisions are management decisions, and refer to the use of firm-specific and non firm-specific communication resources. Their use enables to generate corporate communication flows, and, in this way, create a service system of corporate communication of the firm, according to the structure-system paradigm (Barile and Saviano, 2011), and the SS perspective (Maglio et al., 2010).

3.4. Policy, allocation and coordination decision-making within the circular process of corporate communication management

Following the implementation of a communication strategy, in the subsequent stage of communication control a gap analysis procedure is carried out to ascertain how the organisation is seen by different stakeholders (its corporate reputation) (van Riel and Fombrun, 2007) and its desired position (Cornelissen, 2008). This analysis provides guidance for further decision-making on policy, allocation and coordination. An overview of the process of corporate communication management is illustrated in figure 1.

The SS perspective teaches us that only the interaction between provider (in VSA, organisation) and customers (in VSA, stakeholders), makes it possible to co-create value, which is lastly represented by corporate reputation. In this sense, the allocation and coordination decisions of communication require a sense-adapt-respond approach (Schultz and Kitchen, 2004), in which communication directors and practitioners must constantly listen to the various stakeholder groups (see figure 2).

344

Figure 1 – The decision making process for corporate communication: a framework based on decisions on the creation and utilisation of communication resources

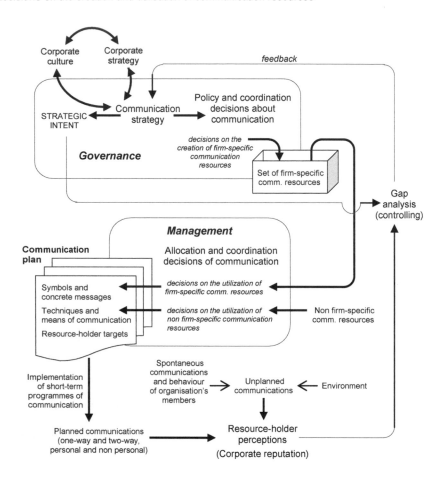

Source: Adapted from Siano, Vollero, Confetto, and Siglioccolo (2011).

Beyond tools provided from the web 2.0 (corporate blogs, forums, wikis, etc.) (della Volpe, 2007, 2011), many organizations are considering their stakeholders in the creation and/or change of firm-specific communication resources (eg. only to name one, think of the recent example offered by GAP: the famous US clothing retailer got back to its initial logo a few days after launching a new one, after the negative comments and reactions left by many users on GAP fan page on Facebook).

Figure 2 – The process of creation and change of firm-specific communication resources

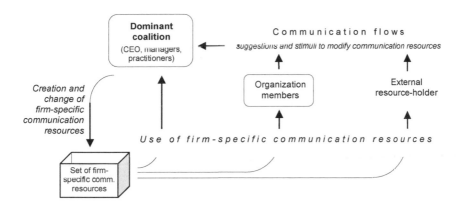

Source: Adapted from Siano, Vollero, Confetto, and Siglioccolo (2011).

Any changes that arise from adjustments to decisions are aimed at improving planned corporate communication, at changing corporate reputation, and, lastly, to enable service system co-creating value with all parties involved in corporate communication management process.

The proposed conceptual framework focuses on planned (organisation-controlled or intentional) communications (Hartley and Pickton, 1999), implemented through corporate identity namely symbolism (corporate visual identity), planned forms of public relations and sponsorship, publicity, advertising and sales promotion and representational forms of behaviour (e.g. the behaviour of store employees or a firm's call centre staff) (van Riel and Balmer, 1997; Cornelissen and Elving, 2003). Stakeholders also perceive organisational behaviour from the point of view of organisational performance. Moreover, socially responsible behaviour can be seen as planned communications.

Nevertheless, the signals conveyed by an organisation may also be unplanned; in this category are accidental cues or unintended messages (Markwick and Fill, 1997), and a firm's environment may also be a source of uncontrolled and uncontrollable communications (e.g. comparative advertising, online/offline word-of-mouth etc.) (Grönroos, 2000).

4. PRACTICAL IMPLICATIONS

The conceptual framework represents a means of support for managers of corporate communication, with a comprehensive view of the different types of communication decisions taken in corporate life. Policy and coordination decisions may regard, for example

- to redefine common starting points/themed messages, thereby adapting them to express core themes around which an organisation can focus its actions and distinctiveness;
- to reformulate the slogan/tagline in order to increase the consistency between an organisation's actions and brand characteristics;
- to modify one or more elements of corporate symbolism to strengthen emotional appeal and to enable the organisation to be perceived as being more transparent and coherent (via the alignment of core organisational purpose, values, and beliefs with employees, managers, and CEO behaviour);
- to rethink the means of story-telling to render the organisation's statements more credible, in situations of insufficient transparency in the conducting of its affairs.

At the same time, practitioners frequently focus on decisions of allocation and coordination in order to make the best use of a firm's communication resources. Such decisions may be useful in a variety of different ways to increase the visibility and distinctiveness of communication initiatives. A more focused and selective utilisation of a firm's means of communication can help to achieve better visibility through exposure to stakeholders. This involves a rather different use of firm-specific communication resources. For instance, an unusual or alternative way of using a corporate blog may be an appropriate sort of allocation decision.

5. CONCLUSIONS

The framework for decision-making within corporate communication management that is proposed herein has the overall aim of contributing to the further development of corporate communication as a separate area of management. To this end, we have focused on communication resources and decisions concerning the creation and utilisation of these resources, two topics that have been somewhat neglected in the existing literature.

The Parsonian view of organisational decision-making and the integration of VSA and SS approaches enables to understand the dynamic nature of corporate communication process, and the role played by all parties involved in the co-creation of value.

The setting of communication strategy as a dual decision also contributes to the view that corporate communication is a separate area of management, and encourages to suppose that any strategic intent may be pursued only if strategic communication resources are created and then managed at a tactical-operational level.

Ultimately, the insights provided by the study may be considered as elements that lend further arguments to support the consideration of corporate communication as an autonomous area of management, and 'stimuli' to encourage a change in the perception of corporate communication as a management function based on professional knowledge and competencies, rather than as a peripheral area made up of the set of technical skills possessed by practitioners.

The analysis of firm-specific/identity-based communication resources represents a first attempt at addressing a topic that requires in-depth examination. Such investigation should stimulate vigorous debate among researchers from a variety of different backgrounds, for example, corporate communication, branding, and corporate reputation.

REFERENCES

ANSOFF, HI. 1965. Corporate strategy: an analytic approach to business policy for growth and expansion, New York: McGraw-Hill.

BALMER, J.M.T., and GRAY, E.R. 2003. "Corporate brands: what are they? What of them?", European Journal of Marketing 37, nos. 7/8.

BARILE, S., and SAVIANO, M. 2011. Foundations of systems thinking: the Structure-System paradigm, in: VV. AA., *Contributions to theoretical and practical advances in management. A Viable Systems Approach (VSA)*, Avellino: International Printing Editore.

BARNEY, J.B. 1991. "Firm resources and sustained competitive advantage", Journal of Management 17.

BERNSTEIN, D. 1986. Company image & reality. a critique of corporate communications, Eastbourne: Holt, Rinehart and Winston.

BROWN, J.S., DENNING, S., GROH, K., and PRUSAK, L. 2004. Storytelling in organizations: why storytelling is transforming 21st century organizations and management, Boston: Dutterworth Heinemann.

CORNELISSEN, J. 2008. Corporate communication. A guide to theory and practice, London: Sage.

CORNELISSEN, J.P., and ELVING, W.J.L. 2003. "Managing corporate identity: an integrative framework of dimensions and determinants", Corporate Communications: An International Journal 8, no. 2.

DELLA VOLPE M. (2007), "Nuove tecnologie e Comunicazione d'impresa", in: ELIA, A., LANDI, A., eds., La testualità. Testo Materia Forme, Roma: Carocci.

DELLA VOLPE M. (in press), "Enterprise 2.0: l'impresa senza confini", Paper - University of Naples "Suor Orsola Benincasa" 2011.

DOWLING, G.R. 1986. "Managing your corporate image", Industrial Marketing Management 15.

FOMBRUN, C.J., and VAN RIEL, C.B.M. 2004. Fame & fortune. How successful companies build winning reputations, New Jersey: Prentice-Hall.

FREEMAN, R.E. 1984. Strategic management: a stakeholder approach, Boston: Pitman.

GOLINELLI, G.M. 2010. Viable System Approach (VSA). Governing Business Dynamics, Padova: Cedam (Kluwer Italia).

GRANT, R.M. 1991. "The resource-based theory of competitive advantage: implications for strategy formulation", California Management Review 33, no. 3.

GRÖNROOS, C. 2000. "Creating a relationship dialogue: communication, interaction, value", Marketing Review 1, no. 1.

348

HARTLEY, B., and PICKTON, D. 1999. "Integrated marketing communications requires a new way of thinking", Journal of Marketing Communications 5.

HATCH, M.J., and SCHULTZ, M. 1997. "Relations between organizational culture, identity and image", European Journal of Marketing 31, no. 5.

HOLM, O. 2006. "Integrated marketing communication: from tactics to strategy", Corporate Communications: An International Journal 11, no. 1.

KELLER, K.L. 1999. "Brand mantras: rationale, criteria and examples", Journal of Marketing Management 15, nos. 1-3.

KITCHEN, P.J., and SCHULTZ, D.E. 2000. "A response to theoretical concept or management fashion", Journal of Advertising Research 40, no. 5.

MAGLIO, P., KIELISZEWSKI, C., and SPOHRER, J., eds. 2010. Handbook of Service Science, London: Springer.

MARKWICK, N., and FILL, C. 1997. "Towards a framework for managing corporate identity", European Journal of Marketing 31, nos. 5/6.

PARSONS, T. 1956. "Suggestions for a sociological approach to the theory of organizations", Administrative Science Quarterly 1, no. 1.

PENROSE, E.T. 1959. The theory of the growth of the firm, New York: John Wiley & Sons.

SCHULTZ, D.E., and KITCHEN, P.J. 2000. Communicating globally: an integrated marketing approach, New York: Palgrave.

SCHULTZ, D.E., and KITCHEN, P.J. 2004. "Managing the changes in corporate branding and communication: closing and re-opening the corporate umbrella", Corporate Reputation Review 6, no. 4.

SIANO, A., VOLLERO, A., CONFETTO, M.G., and SIGLIOCCOLO, M. 2011. "Corporate communication management: A framework based on decision making with reference to communication resources", Journal of Marketing Communications, DOI: 10.1080/13527266.2011.581301.

SPOHRER, J., and KWAN, S.W. 2008. "Service Science, Management, Engineering and Design (SSMED): Outline and References", in: SPATH, D. and GANZ, W., eds., The Future of Services: Trends and Perspectives, Munich: Hanser.

SPOHRER, J., GOLINELLI, G. M., PICIOCCHI, P., and BASSANO, C. 2010, "An integrated SS-VSA analysis of changing jobs", Service Science 2, no. 1/2.

SPOHRER, J., PICOCCHI, P., BASSANO, C. 2011. "Three frameworks for service research: exploring multilevel governance in nested, networked systems", Naples Forum on Service. Service Dominant Logic, Service Science and Network Theory: integrating three perspectives for a new service agenda, Capri, Italy, June 14-17.

STEYN, B. 2003. "From strategy to corporate communication strategy: a conceptualisation", Journal of Communication Management 8, no. 2.

TILLEY, C. 1999. "Built-in branding: how to engineer a leadership brand", Journal of Marketing Management 15, nos. 1-3.

van Riel, C.B.M. 1995. Principles of corporate communication, Hemel Hempstead: Prentice-Hall.

van Riel, C.B.M., and Balmer, J.M.T. 1997. "Corporate identity: the concept, its measurement and management", European Journal of Marketing 31, nos. 5/6.

van Riel, C.B.M., and Fombrun C.J. 2007. Essentials of corporate communication, Abingdon: Routledge.

Vargo, S.L. and Lusch, R.F. 2004. "Evolving to a new dominant logic for marketing", Journal of Marketing 68.

CHAPTER 35

The ARISTOTELE Project for Governing Human Capital Intangible Assets: A Service Science and Viable Systems Perspective

S. Salerno[1], P. Piciocchi[2], C. Bassano[3], P. Ritrovato[4], G. Santoro[4]

[1] Research Centre in Pure and Applied Mathematics (CRMPA) – University of Salerno (ITALY) – salerno@unisa.it
[2] University of Salerno (ITALY) – p.piciocchi@unisa.it
[3] "Parthenope" University of Naples (ITALY) – clara.bassano@uniparthenope.it
[4] MOMA SpA – (ITALY) ritrovato@momanet.it - (ITALY) santoro@momanet.it

ABSTRACT

In the modern so called "knowledge society" characterized by a profound crisis the organization are more and more realizing the relevance to shift from tangible assets to intangible one for increasing their competitiveness. Information technology alone cannot provide an answer. It is important to enlarge the vision considering complex environments that integrate models, processes and technologies with organizational aspects in a more systemic approach. In this paper we present a complex system under development in the frame of the ARISTOTELE research project and analyze its results under the lens of Service Science Management Engineering+Design (SSME+D) and Viable Systems Approach (VSA) demonstrating its viability in modern knowledge intensive organizations.

Keywords: Human Capital Management, Knowledge Management, Semantic Modeling, Adaptive Learning Systems, Service Science, Viable Systems Approach

1 INTRODUCTION

Organizations increasingly have recognized that intangible resources (e.g., knowledge, skills, innovation processes) enable an organization, when managed accordingly, to increase effectiveness, efficiency, and competitiveness (Schultze and Leidner, 2002). Hence, organizations need a constant flow of ideas they can draw upon to foster innovation. Innovation processes do not occur in a vacuum, within solitary geniuses (Hargadon and Sutton, 2000), but are rather part of complex systems that might span even across organizations and institutions, related through networks. Creativity and innovation inside organizations are directly related to the way people learn (informally and/or formally), collaborate, share ideas, knowledge, and organizational goals. Innovation may happen in several ways. However, knowledge-sharing within networked communities faces several barriers such as outsourcing and employment insecurity which hinder people from interacting with others and sharing their knowledge (Adler and Heckscher, 2006). At the same time, communities provide a way for managing knowledge as an asset by allowing free and creative flow of knowledge and experiences and hence have positive effects on the overall organizational performance (Wenger and Snyder, 2000).

Enhancing the learning capability of an organization is indeed key to achieve the above mentioned integration and merging of intangibles but still today learning capabilities of an organization are mainly devoted to allow acquisition of personal knowledge, skills and competencies that are very difficult to share with peers.

One of the main problems is that a great part of the learning processes that occur in organizations are neither self-planned nor derived by the fine-grained analysis of the specific activities forming a business process but basically pre-designed for specific targets and confined to a small number of reserved time slots.

We can summarize the main identified lacks as follows:

- Separation and missing interconnection between enterprise learning pathways and real contingent workers' needs.
- Missing links between learning strategies oriented to different working and organisational contexts.
- Difficulty to capture and reuse formal and informal knowledge as a basis for organisational learning. Once the needed knowledge is acquired, all information on how this knowledge was built disappears.
- High-level fragmentation of data, information, tools and environment used by workers to operate in their working life (studying, carrying out process activities, discussing, collaborating, etc.).
- Difficulty to assess, exploit, share and reuse learning experiences in terms of approach, contents and knowledge both at personal and enterprise level.
- Missing of any form of Collective Intelligence.

Unfortunately technologies don't help to overcome these issues.

Current Technology Enhanced Learning solutions are not a solution to the issues outlined above since the suffer from the traditional L(C)MS vision. They are centered on the content with very limited personalization capabilities and embed pedagogies in the content itself. In the working environment, this type of rigid TEL solutions are integrated with the general Enterprise learning objectives in a very simple (and often simplistic) way, e.g. through definition of pre-determined learning paths (courses) for class of employees. On the other side Competency management systems often operate in isolated manner providing a simple inventory that it is updated few times per year and with long and even costly processes. Moreover, even as we speak, the adoption of Web 2.0 approach (even in the enterprise) emphasized the central role of the user and, more in general of user-generated contents that become the main source for discovering intangibles.

The ARISTOTELE integrated project, a 6M€ European Commission co-funded research and development project (2010-2013), recognizes this flow of problems and challenges that come along with the management and exploitation of intangible assets within an organizational context. ARISTOTELE aims to provide an answer to these issues providing sets of models, methodologies and tools integrated in an enterprise software platform leveraging on existing technologies, namely the IWT – Intelligent Web Teacher (Albano et al., 2007) a learning and knowledge management platform and Microsoft Share Point Server, in order to speed-up the development and facilitate the adoption, acceptance and assimilation within the organizations.

2 THE ARISTOTELE PROJECT VISION AND APPROACH

2.1 The Vision

In organizational contexts, three main kind of processes are traditionally present and, generally, managed in an isolated fashion: organizational processes (such as marketing& communication, human resources management, business processes), learning processes (such as group training sessions), and social collaboration processes (such as processes involving the spontaneous formation of group inside the organization).

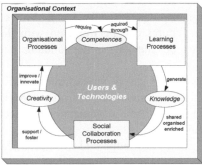

Figure 1: ARISTOTELE Vision

The Figure 1shows how ARISTOTELE conceives the synergies among these processes in an organizational context where, in a sort of virtuous circle, intangible values of an organization (graphically shown as ovals in the figure) are created by

352

processes in order to be exploited (or to improve/innovate) another process generating other intangible value, and so on. The figure shows also another key aspect of the ARISTOTELE vision: the centrality of the users and the "enabled" role played by the technologies.

According to our vision, to raise competitiveness of an organization, its key processes should be build, organized and managed taking into account either the strategic goals and the human factors (such as motivation, acceptance etc.) of the users.

2.2   The Approach

To achieve the above vision, the ARISTOTELE project will investigate, design and develop capabilities and tools to enable the building blocks in the middle of the following picture.

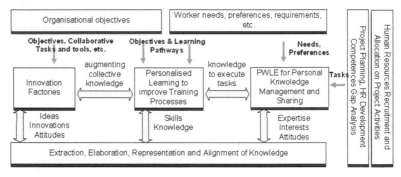

Figure 2: The three main differentiators enabled by ARISTOTELE

Basically, at the top we have the enabling features of a successful organizational learning process: the organizational objectives and the worker preferences, needs and requirements. These features can be modeled with the adoption of semantic technologies in order to be machine-understandable (i.e. the ARISTOTELE models).

In the middle we have three key processes supported by ARISTOTELE. The central one is the building of personalized learning experiences on the basis of a learning pathway defined starting from organizational objectives and worker needs, preferences and requirements.

As part of this process new knowledge and two needs can arise.

The need to further elaborate this knowledge with other workers to boost innovation can arise and be managed by fostering an innovation factory. A guidance process support the creation of this environment taking into consideration organizational goals, heterogeneity of competencies of the members of a group, use of suitable tools for creative activities.

The second need that can arise is to re-use the knowledge to execute independent tasks relating to the business process of an organization. On the other hand it may happen that during the execution of creative activities or daily business processes activities, personalized learning experiences have to be created and executed on demand.

In any of the previously mentioned cases, the outputs of the above three processes are knowledge, ideas, innovation, expertise and skill improvements that will be extracted, elaborated, formalized and aligned in the knowledge base of the organization and re-used for several organizational purposes (highlighted in Figure 2 as vertical blocks) such as project planning, competence analysis, HR allocation.

2.3 The key elements of the ARISTOTELE project: Models, methodologies, tools and the Integrated Platform

According to the proposed approach the ARISTOTELE project have defined a set of models, namely the Knowledge Model, Competence Model, Worker Model, Learning Experience Model described using standards and specification of the semantic web and used to store relevant information providing the knowledge base for supporting methodologies to reasoning on this information. The tools implement the main relevant capabilities defined in the methodologies. Finally the ARISTOTLE platform integrates everything according to modern enterprise service oriented approaches.

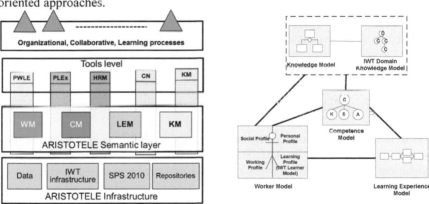

Figure 3: The ARISTOTELE Conceptual view (left) and the ARISTOTELE Models (right)

2.3.1 The Models

The ARISTOTELE Models are the primary inputs for the methodologies and play a central role because enable the "semantic" and intelligent features of the ARISTOTELE platform. The following four models have been identified and defined:

- Knowledge Model (KM) – This model focuses on three aspects of knowledge:
  - ○ Training domain – Models domain knowledge that is relevant for the learning activities and it is based on the IWT Knowledge Model;

- o Enterprise activity and strategy domain – Models enterprise knowledge, e.g. strategies, processes, activities, documentation, and, in general, all output generated by workers in their activities, including social interactions with peers;
  - o Organizational knowledge – The set of lightweight ontologies that are exploited to provide a shared classification of the organization resources and entities.
- Competence Model (CM) – This model provides a representation of competences (expressed in terms of knowledge, skill and attitudes) and their relations to other models elements such as context, evidence and objectives.
- Worker Model (WM) – This model provides a representation of the worker including social, learning, working and personal goals.
- Learning Experience Model (LEM) – This model provides a representation of the learning experience needed to achieve a new competence or fill a competence gap.

In general, the ARISTOTELE models aim at representing various factors of human resources-related, socio-cultural, and practical nature typically called "soft issues". The rationale behind this is twofold. From one side, the need of modeling soft issues comes directly from the analysis of the inputs (e.g. project's scenarios and requirements, methodologies needs). From the other side, a proper representation and management of these issues appears to be more and more related to recent Open Innovation Paradigms that foresee a central role of intangible assets, social activities and collaboration among peers.

2.3.2   The Intelligent Web Teacher Platform

As evidenced both in the conceptual view and in the models themselves, IWT plays an important role in ARISTOTELE. IWT is a learning & Knowledge management system developed by MOMA SpA and available on the market.

IWT allows the definition and execution of personalized e-learning experience tailored on the basis on learners' cognitive status and learning preferences. IWT is based on three main models (Gaeta, et al., 2009): Knowledge Domain Model, Learner Model and Didactic Model, which substantially interact to define a personalized learning path (Capuano et al., 2007). Such models allow taking into account three different aspects: the knowledge regarding the domain that is relevant for the learning activities, the context where the learning process is executing, the learning preferences and the cognitive states of the learners who's the process is addressed.

The first one is based on the definition of ontologies used to model the knowledge of the didactic domain of interest. These ontologies are used to organize the learning objects that are building blocks exploited to create personalized e-learning experiences. The second one defines the rules by which a learner is profiled within the system. The third one defines the rules (basing on the most suitable learning strategy) by which the learning process happens and the knowledge is acquired by the learners.

2.3.3   The Methodologies

The ARISTOTELE have defined 5 methodologies, shortly describe below.

Knowledge Building methodologies: The methodology for Knowledge Building aims at supporting the self-organization, acquisition, processing and sharing of new information and knowledge among workers Outputs of methodologies for Knowledge Building will be useful to support some objectives in terms of knowledge management like tagging, conceptualization, categorization, semi-automatic organization of resources according to their content and annotation.

Methodology for decision support for Human Resource Management: this methodology provides a competence-based methodological support to some human resource management functions such as identification of competence requirements, competence gap, optimal team formation considering trust network and recruitment.

Methodology to foster PWLE (Personal Working and Learning Environment): This methodology defines how to approach and interact with the PWLE. The PWLE is a personal virtual environment aiming at supporting the workers during their daily activities targeted to knowledge worker. As such the PWLE is an environment that facilitates the process of sharing and acquiring knowledge and supports the organization of personal work. At this purpose, the PWLE is placed in between individual and organizational knowledge and supports a cyclic and continuous process that involves the knowledge transfer from organization to individual worker and vice-versa.

Methodology for Exploitation of Collaboration Networks: The methodology is aimed at providing to the overall ARISTOTELE framework an independent module that implements the functionalities of an advanced Recommender Systems (RS). The outputs generated by the methodology are related to definition of suggestions, in term of activities, knowledge improvement plans or other, which can increase the exploitation of social capital and affinities.

Methodologies for Learning Experiences generation: Methodologies for Learning Experience Generation are committed to define mechanisms and modalities for preparing, instantiating, deploying and executing learning experiences (i.e. sequence of events that a learner participates in to achieve one or more learning objectives).

2.3.4   The tools and the Integrated Platform

The tools match one to one with the methodologies. Essentially starting from the research work done at the methodologies level where algorithms and approaches are defined the tools have been instantiated. The ARISTOTELE platform defined leveraging on already existing enabling technologies (IWT and Microsoft SPS2010) integrate models and tools in order to be instantiated within real organization.

3   ARISTOTELE PROJECT IN SERVICE SCIENCE MANAGEMENT ENGINEERING + DESIGN (SSME+D) AND VIABLE SYSTEMS APPROACH (VSA) PERSPECTIVES

The integrated ARISTOTELE platform, as we conceived, is methodological coherent with two basic theoretical building blocks/perspectives: SSME+D and VSA.

First of all, SSME+D&VSA architectures and contents are linked and synergic; so we can consider their synthesis as a new trans-disciplinary approach – based on Computer Science, Operations Research, Systems, Engineering, Management, of Information Systems, Economic, Law, Industrial Engineering, Marketing, Operations management, Game Theory – to improve studies and researches in Service Systems community. In fact, SSME+D&VSA investigate on complex human systems to ensure a viable quality of services: *increasingly quality of life depends both on the quality of service we receive from these systems as customers, but also the types of job roles we find in these systems as working professionals* (Spohrer, 2010; Golinelli, 2010).

Given these premises, we can argue "why" and "how" ARISTOTELE – as project and as framework – represents an interesting experience in order to improve, on one hand skills, capabilities, competencies of actual and future working professionals; on the other hand, complex service systems in business and societal areas. In particular, it is relevant to underline that ARISTOTELE is logically linked to the high variability and complexity of global scenarios in which we take over the transaction from managerial strong specialization of professional profile – *I-Shaped professional* – to managerial multi-disciplinary variety knowledge, based on consonant subjective specialization of new professional profile – *T-Shaped professional* – (Spohrer et al., 2010; Spohrer et al., 2011; Macaulay et al., 2010).

Indeed, in the actual global economy, centered on Service Dominant Logic (Vargo and Lusch, 2004), two aspects are significant:

1. the fit skill requirements in meeting the demands of variable service economy (SSME+D);
2. the interdependencies and the challenges of socio-economic contexts in which organizations try to survive and to be viable (VSA).

Both, in an integrated SSME+D&VSA perspective, refer to the requirements of more flexibility in knowledge to be suitable to cooperate and collaborate in service systems as well as to improve subjective and organizational skills and processes for "living in complex changing scenarios". This highlights another compatibility between SSME+D&VSA methodology and ARISTOTELE framework: *value co-creation.*

As we conceive Service Systems:

1. a dynamic configuration of resources;
2. a set of value co-creation mechanism between suitable entities;
3. an application of competencies considering skills, knowledge and attitude any person(s) in job or stakeholder roles;
4. an adaptive internal organization responding to the dynamic external environment;

5. continuous learning and collaborative feedbacks to ensure mutual benefits or value co-creation outcomes;
6. a complex viable structure enable to respond to environmental change, where environmental change is mostly generated by other viable systems;

then, ARISTOTELE is a practical experience of a complex viable service system, in which the viability depends first and foremost on a government capability, for both internal self governance and external relationship governance, that creates value for the involved collaborative resources and stakeholders (customers, professionals, firms, society).

In complex environments, the search for viability and value co-creation means the capability of the government of a viable service system to make decisions on the basis of approximate knowledge, by means of shared competencies. In fact, organization government for ensuring viability to its service system, needs to put in value two main types of decisional processes: Decision-Making (DM) and Problem-Solving (P-S). DM provides guidance about which ends ("know what") to achieve, and P-S provides means ("know how") to those ends. Both types of decisions are the fundamental capabilities (cognitive assets) required to attain and maintain viability in each kind of organization (Piciocchi et al., 2009). While both are required, DM is the primary capability. Without knowing "how to make a decision" about the ends to pursue, all the effort applied on "how to solve a problem" may in fact be wasted effort.

As viable system entities change and learn, the resource allocation problem is the fundamental decision that repeatedly must be made to remain viable (March, 1991, 2008). Both DM and PS knowledge may be required, but DM is primary. To reduce the decision ambiguity isn't useful to focus on problem solving at all, but it is more necessary to focus on decision making. In other words, dealing with new levels of complexity requires new types of decision making in which we cannot necessarily use fixed models or stochastic methods to find a solution, but we need a pattern or schema suitable to a particular problem's complexity level.

So, considering the integrated SSME+D&VSA lens, our analysis on the ARISTOTELE project emphasize: (1) the connectivity between consonant and resonant entities, and (2) value co-creation mechanisms used by entities according to internal and external networks. This means, ARISTOTELE is a complex integrated service architecture, useful to enrich organization environment to improve effectiveness and efficiency and a complex system supporting social interactive teams, characterized by T-Shaped people involved and collaborative in the organizational and societal development.

4   CONCLUSION

In this paper we have provided a first attempt to look at a complex environment like one under development in the ARISTOTELE project from the SSME+D&VSA perspective. In fact, ARISTOTELE, as it conceived technically and functionally, represents a smart integrated platform responding to the actual organizational need to build and improve specialized collaborative team characterized by high quality of competencies. The results sound promising and will be further extended considering the outcome of the experimentation activities that will executed in the 2012 and 2013.

ACKNOWLEDGMENTS

The research reported in this paper is partially supported by the European Commission under the Collaborative Project ARISTOTELE "Personalized Learning & Collaborative Working Environments Fostering Social Creativity and Innovations Inside the Organizations" under Grant Agreement n.257886.

REFERENCES

Adler, P. S. and Heckscher, C. 2006. "Towards Collaborative Community". In. Heckscher, C. & Adler, P. S, eds. *The Firm as a Collaborative Community: Reconstructing Trust in the Knowledge Economy*. Oxford University Press.

Albano, G., Gaeta, M., Ritrovato, P. 2007. "IWT: an innovative solution for AGS e-Learning model". *International Journal of Knowledge and Learning*. Inderscience Publisher, 2/3(3): 209-224.

Capuano, N., Gaeta, M., Marengo, A., Miranda, S., Orciuoli, F., Ritrovato, P. 2007. *LIA: an Intelligent Advisor for e-Learning. Interactive Learning Environments*. Taylor & Francis, 3(17): 221-239.

Gaeta, M., Orciuoli, F., Ritrovato, P. 2009. "Advanced ontology management system for personalized e-Learning". *Journal of Knowledge-Based Systems*, 4(22): 292-30.

Golinelli, G.M. 2010. *Viable Systems Approach (VSA). Governing Business Dynamics*. Cedam, Kluwer.

Hargadon, A. and Sutton. R. I. 2000. "Building an innovation factory". *Harvard Business Review*, 78: 157.

Macaulay L., Moxham C., Jones B. and Miles I., 2010. "Innovation and Skills". In Maglio, P., Kieliszewski, C., Spohrer, J. eds. *Handbook of Service Science*. Springer.

Maglio, P., Kieliszewski, C., Spohrer, J. 2010. eds. *Handbook of Service Science*. Springer.

March J.G. 2008. *Explorations in Organizations*. Stanford Business Books.

Schultze, U. and Leidner, D. E. 2002."Studying Knowledge Management in Information Systems Research". *Management Information Systems*. 26: 213-242.

Spohrer, J., Golinelli, G.M., Piciocchi, P., Bassano, C. 2010. "An integrated SS-VSA analysis of changing jobs". *Service Science*, Vol. 2, No. (1/2): 1-20.

Spohrer, J., Piciocchi, P., Bassano, C. 2011. "Three frameworks for service research: exploring multilevel governance in nested, networked systems". *Proceedings of the Naples Forum on Service. Service Dominant Logic, Service Science and Network Theory: integrating three perspectives for a new service agenda,* Capri, Italy, June 14-17.

Vargo, S. L. and Lusch, R. F. 2004. "Evolving to a new dominant logic for marketing". *Journal of Marketing*, Vol. 68: 1-17.

Wenger, E. C. and Snyder, W. M. 2000. "Communities of practice: The organizational frontier". *Harvard Business Review*, 78: 139-146.

CHAPTER 36

Human Resources for Governing Business Dynamics. The Viable Systems Approach

G.M. Golinelli[1], C. Bassano [2]

[1]"La Sapienza" University of Rome (ITALY)
gaetano.golinelli@uniroma1.it
[2]"Parthenope" University of Naples (ITALY)
clara.bassano@uniparthenope.it

ABSTRACT

The aim of the paper is to examine further and define "how" and "why" the Viable Systems Approach (VSA) may contribute to a rational and conscious need of governance for service systems (so called, government entities).

In his book "Viable Systems Approach. Governing Business Dynamics", Golinelli defines governance as the decision-making, strategically relevant, based mainly on the decision making processes, i.e. processes oriented to reduce the decision complexity and to ensure sufficient conditions of viability of the organizational system. Thus, if the business is qualified as viable system - that is able to survive in a context of supra systems - then its government (top decision maker or policy maker) is eligible to take the decision to reduce operational complexity and determine appropriate locations for the achievement of systemic goals. In this context, the methodological framework of the Viable Systems Approach, (VSA) is able to explain "how" business entity is able to live in its context and "why" a careful and enlightened governance is able to ensure the pursuit of corporate objectives ensuring adequate conditions for overall equi-finality. In fact, the action of the decision maker requires the maturation of experience and expertise to support the government - in his triple role identity: governance, management and administration - in the decisions which: 1) co-create systemic value; 2) characterize the sustainable "competitive advantage"; 3) define "results", i.e. to establish the conditions for systems effectiveness in the context.

Complex decisions of governance-management-administration allow adequate conditions of elasticity and flexibility, and, therefore, provide a diachronic viability to the business system as a result of co-evolutionary dynamic equilibrium.

The homeostatic balance of the system, in fact, is guaranteed by the verification of a dual effect: that of consonance (towards the external environment) on the one hand, and that on competitiveness (perception of the market with regard to competitors) on the other had.

Keywords: Government/Top Decision Maker, Consonance, Competitiveness, Governance, Viable Systems Approach.

1 INTRODUCTION

In an increased awareness of the role of social responsibility, we understand how, over the last two decades, thanks to the evolutionary pressure coming from competition, the way firms are governed has been marked by an emphasis on developing human resources, knowing that growth in competence and professionalism is a driving force in the search for lasting and defendable competitive advantage. These are reached thanks to the behaviour of the personnel, guided by the values they live and share.

Outside the business system, the need for greater cooperation with a multitude of actors (suppliers, research groups, competitors, clients, etc), symptom of an increasing degree of openness, as well as the need to acquire, keep and increase consent concerning strategic and operational initiatives, has led to a careful consideration of the needs, not only the economic ones, of different stakeholders the system has relationships with, even when their contracting ability is vastly inferior to that of the individuals that manage the firm itself.

Given this premise, the aim of the paper is to examine further and define "how" and "why" the Viable Systems Approach (VSA) may contribute to a rational and conscious need of governance for service systems (so called, government entities).

To this end, the VSA, and the qualification of the firm as a viable system improve understanding of how business systems evolved from the end of the 19th century on (Golinelli, 2000, 2010). This evolution has concerned, on the one hand, the relations and interactions between the two macro-components of *government* and *operative structure* that qualify the firm as a viable system. The former drives the emergence of the system from the structure, and manages its evolutionary dynamics; the latter appears as a complex of qualitative and quantitative features by which the entrepreneur can innovate and change his firm; government thus. Moreover, the systems approach relates to value creation theory. Indeed, improving the chances of survival implies value creation. If the government's action is to create value, the system's chances of survival increase over time.

But the question is: how to pursuit Survival in order to be Viable? Survival asks government to be able to read and grade the expectations of external systems entities differently.

To this end, we first examine some conceptual categories to better understand the VSA framework. In particular, we illustrate the *structure* and *system* dichotomy as well as the *governance decisions* and *management routines* distinction which are the bases for a general model of the business's evolutionary dynamics as a result of rational and conscious governance. Second, if we consider the firm as a viable system, then we have to focus on the firm's government who represents the logical component of the decision-making process. Third, through the notions of consonance and resonance, the systems approach provides reliable foundations for studying inter-systems relations, as well as for clarifying the relations between the firm and its external context. Therefore, by this means, it is crucial the role of individuals for the governance of business dynamics. In fact, this action stems from a broad set of skills based on which the government modifies its structure dynamics in an evolutionary sense. Finally, we highlight that the equilibrium of the system is thus ensured from a twofold perspective, that of consonance (towards the external context) on the one hand, and that of competitiveness (market perception regarding competitors) on the other hand.

From this, it is clear that one of the main conditions in the study of business dynamics, of vital importance for what we shall say later on, is resonance between subsystems and between the firm-system and the suprasystems, which in a traditional view, are generally defined as the external environment, in order to guarantee the firm-system the greatest possibilities of survival.

2   VIABLE SYSTEMS APPROACH (VSA) CONCEPTUAL FRAMEWORK

In coherence with the aim of this study, the methodology we are proposing and the qualification of the firm as a viable system improve understanding of how organizations evolve highlighting the relevant role played by individuals in business dynamics and the importance of relationship between individuals and business as social and cognitive system.

In his book "Viable Systems Approach. Governing Business Dynamics", Golinelli defines governance as the decision-making, strategically relevant, based mainly on the decision making processes, i.e. processes oriented to reduce the decision complexity and to ensure sufficient conditions of viability of the organizational system. Thus, if the business is qualified a viable system - that is able to survive in a context of supra systems - then its government (top decision maker or policy maker) is eligible to take the decision to reduce operational complexity and determine appropriate locations for the achievement of systemic goals.

In this context, the methodological framework of the Viable Systems Approach, is able to explain "how" business entity is able to live in its context and "why" a careful and enlightened governance is able to ensure the pursuit of corporate objectives ensuring adequate conditions for overall equi-finality.

This study aims to examine the firm from a systems point of view, focusing in particular on its government. For this reason, and also to fully understand the significance of the systems approach for the governance of the firm, it is necessary to illustrate some of the conceptual and methodological premises which represent, albeit in summary and not exhaustively, the scientific context in which systems thinking developed, setting out its principles and showing its epistemological importance (Emery, 1956; Ashby, 1958; von Bertalanffy, 1968; Parsons, 1971).

A first key theoretical insight in Golinelli's VSA is the dichotomy between *structure* (or structural dimension) and *systems* (or systems level of analysis). Without purpose there is only structure in an environment, but with purpose the dynamics of the firm become more or less viable, and open to the shaping power of governance decisions and management routines. A second and potentially controversial point comes from the distinction between *governance decisions* and *management routines*.

Even in an era dominated by more and more flexible, decentralized and network-based business models, VSA argues persuasively for the invariable need for governance (so called, government entities). Multi-stakeholder environments of highly developed and highly competitive viable systems require government. Competition is an incentive for accelerated learning of highly developed viable systems. This assumption should not be viewed as excessively restrictive or be read to be an obsolete way of thinking; it just says that even in the most decentralized and apparently "democratic" businesses – a superior government (governance decisions) will likely have fundamental advantages (for getting outside the box). So, the primacy of government does not conflict with other systems approaches that highlight processes of self organization (Maturana and Varela, 1980).

Golinelli strives especially hard to clarify the variety of decision-making processes in an organization. While, in Beer's model (1972), decisions are all included in the "management" category and there is a basic distinction between decisions and mere executive operations; Golinelli defines "governance" only as the strategic decisions taken by government and not routine processes. To greatly over simplify, government strategically decides what game(s) to play, and management operates to decide what will achieve the purpose and win the game. So, highly developed viable systems need government to keep them on a positive evolutionary path, and ensure competition is occurring within a strategically important game or viable path forward. A third feature of VSA, is the focus on a firm's relationships with its environment, more than simply on its inner organization and management. The approach highlights the richness of real-world interdependencies and the challenges of real socio-economic environments in which firms try to survive and remain viable. Environment in VSA is perceived as an ecology of more or less developed viable systems known as "suprasystems." Firm government has to constantly monitor the environmental composition and trends; it must compete while sustaining harmony with a viable fraction of relevant suprasystems.

We think that VSA provides a new lens to view service systems as a type of highly developed viable system in an ecology of other viable systems (Spohrer et al, 2010; Maglio et al, 2010). Structure/systems dichotomy as well as governance/management distinction are especially interesting, in this regard.

3 GOVERNMENT'S ROLE AND ACTION

If we consider the firm as a viable system, then the firm's government represents the logical component of the decision-making process. The firm's government defines the strategic guidelines and establishes the appropriate mechanisms for the integration and co-ordination of the firm's dynamic components, in order to preserve the system's unity and integrity (Fazzi, 1984).

Effective government action depends on a wide range of variables, which shows how complex the government of a firm can be. Government's freedom of action - allowed by ownership - (Veblen, 1923), the influences coming from other suprasystems, government composition, quality and skills, etc., are interconnected elements, all of which may contribute to an effective government action.

Summing up, an effective decision-making process – and therefore the ability to attain the highest possible level of overall resonance – is highly dependent on: 1. the nature of the relationship established between the firm's government and its ownership; 2. government structure and internal dynamics; 3. suprasystem expectations, pressure and corrective influence; 4. the relationship the firm's government establishes with the operative structure.

Figure 1 highlights the above influences and the special role played by the firm's ownership in determining the effectiveness of government activity .

Government's main task is attaining a high level of overall resonance, which we are going to define in the next session as the result of suprasystems and subsystems resonance.

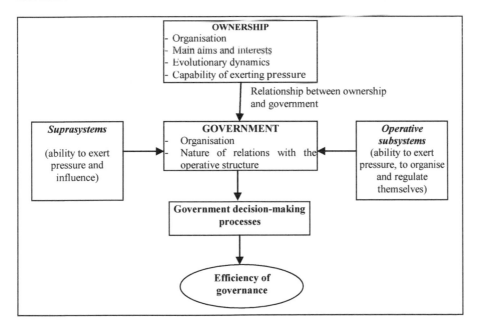

Figure 1 – Determining elements for the effectiveness of governance. (From Golinelli G.M. 2010. Viable Systems Approach (VSA). Governing Business Dynamics. Cedam, Kluwer.)

4 COMPETITIVENESS AND CONSONANCE FOR GOVERNING BUSINESS SYSTEMS

According to the basic assumptions of the VSA, governance is fundamentally, the adoption of decisions and actions to enable the system to pursue its goals. That purpose, with respect to business organizations, can be interpreted in many ways: in our approach, the firm hopes to survive in a general environment and in a more specific context, qualified by a growing complexity changing over time. The pursuit of this purpose requires to the top decision maker (government) to implement, adapt and renew continuously the structural envelope, in particular knowledge of the company, in harmony with the dynamics and environmental context. Here, we point out that, in search of the increased probability of survival of the system, governance is moved and guided by two fundamental drivers or goals, among them co-essential and complementary: the competitiveness and consonance.

Where the first reveals the tension of the continuous improvement in performance that will qualify system dynamics; the second emphasizes the aspiration for an harmonious relationship with sub and supra systems. The real achievement of the objectives of competitiveness and consonance requires the implementation of appropriate governance processes. Observation and contextualization of environmental trends on the one hand, and structural work on the equipment business, on the other hand, appear parallel and interrelated moments: the question of relationship between the firm and its environment is thus reduced to a unitary perspective, and co-evolutionary dynamic equilibrium.

To improve understanding viable systems governance, in this section we look at the government of a firm, its decision-making process (i.e. the activity aiming to achieve an acceptable level of overall resonance), and the conditions for its effectiveness. By this means, we consider that a firm's ability to survive through the creation of competitive advantages and the generation of value depends on a dual set of conditions:

– establishing consonance and – if the firm's government sees fit – resonance with the suprasystems;

– establishing consonance and – if possible – resonance among the firm's subsystems.

In the first case, a firm can only survive over time if it correctly perceives and satisfies the hopes and expectations of its suprasystems of the L+1 order. According to the second condition, the firm must also meet the hopes and expectations of its various subsystems of the L–i order (i = 1, 2, ..., n). This ensures that all the firm's subsystems components operate and evolve in an orderly fashion, consistent with the firm's objectives. The two sets of conditions are closely related: on the one hand, failure to meet the suprasystems' expectations jeopardizes the firm's very survival, thus making it impossible to fulfill the subsystems' expectations; on the other hand, consonance and/or resonance with the suprasystems depend on the balanced evolution of the firm's subsystems.

Therefore, an effective system is one able to balance the expectations of the suprasystems and the subsystems, thus achieving high degrees of resonance both outside and inside the firm. This is a precondition of the firm's viability and ability to survive. As a result, overall resonance – deriving from suprasystems and subsystems resonance – stems from the appropriate configuration and evolution of the relationships established between the firm and the suprasystems on the one hand, and among the firm's various subsystems, on the other. Moreover, from a structural view point, high levels of consonance with the suprasystems and among the subsystems are a necessary but not sufficient condition for overall resonance.

These observations are illustrated in Figure 2 below.

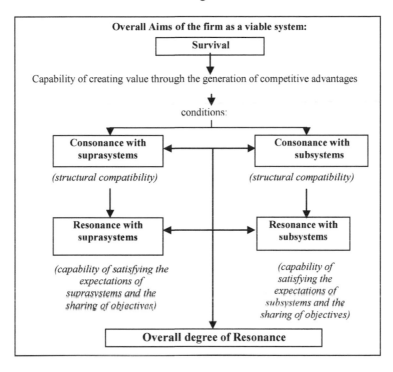

Figure 2 – Overall resonance as a result of suprasystems and subsystems consonance and resonance. (From Golinelli G.M. 2010. Viable Systems Approach (VSA). Governing Business Dynamics. Cedam, Kluwer.)

At a deeper level of analysis, a high degree of overall resonance (expression of systems effectiveness), critically depends on (see figure 3):

– an effective government process, that is the ability to identify those suprasystems whose expectations should be given priority, given the specific firm and environmental conditions, and those whose expectations are either incompatible with, or secondary to, the firm's purpose. Indeed, an effective government process is crucial in identifying the most suitable avenues for achieving high levels of resonance;

– an effective operation, whereby the firm's components are able to translate the strategic guidelines established by the firm's government into efficient actions.

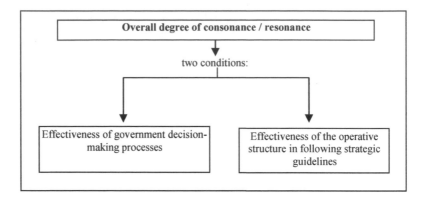

Figure 3 – The conditions for a high level of overall resonance. (From Golinelli G.M. 2010. Viable Systems Approach (VSA). Governing Business Dynamics. Cedam, Kluwer.)

Therefore, overall resonance is an outcome of both decision-making processes (by the firm's government) and dynamic processes (by the operative structure and its components), aiming to assure the highest possible consistency between the external and internal variables generally affecting the firm's evolution.

As we said, government (top decision maker) presides over his role in directing and guiding first identifying the drivers of competitiveness in its action and in consonance.

The consonance, understood as a line of action to the government of the system, regards the implementation/maintenance of conditions of harmony, correspondence, alignment and dialogue with the frame of reference: it expresses the basic need of the system to correspond to values, cultures and needs of the surrounding society and be recognized and considered in the various entities that populate it (the action taken with regard to human resources that drive the business structure). By this means, we emphasize the centrality of individuals in organizations and the importance of the link between the individual and the enterprise system as a cognitive and social (Luhmann, 1984). The action of the government, with a view to enhance the competitiveness and the consonance of the business system compared to the magnitude of the context must be thus that first influence on the set of attitudes and predispositions (such as rationality, creativity, loyalty, etc..) and on the necessary built knowledge of the individual working in the organization and then stimulate the collaboration between people and the sharing of knowledge, aimed at generating new knowledge and skills.

In this regard, the VSA allows us to draw at least two useful indications of government action. First, the individual knowledge can be read as a variety of cognitive endowment resulting from the combination of three basic components (Barile, 2011):

- *information units*, which represent the basis of information owned;
- *synthesis interpretation scheme*, resulting from the composition of sub-sets of elementary units of information and representing the methods of approach to the problems;

– the *categorical values*, which are the result of the elementary units of information connected to each other and with reference to synthesis interpretation scheme. The categorical values are the expression of the logic of composition of information and patterns in common as they refer to a same issue, to the same cognitive aim.

Basically, the individual with the skills and the provision of knowledge that are proper, acquires information, processes it through diagrams and synthesizes them into categories, by implementing the initial knowledge base. Secondly, the systems approach implies a sort of cycle or progression, or "resources-capabilities-competencies" in the process of defining/redefining of the structure, where the government is now pregnant, the resources are the subject of combination and recombination and are transformed into structural capabilities and systems competencies which determine the dynamics of the business system.

The government, in the process of defining/redefining of the structure, must first develop the skills and knowledge of individual equipment (resources) through the combination of them, so as to implement the individual skills (decision-making skills, interpersonal, analytical, operational etc..), and then encourage the dissemination and sharing at the level of individual organizational units and the overall company. This enhancement process still involves the formation/evolution of individual skills, through learning and experience derived from training on the job, allow the development of skills over time more and more consolidated, specialized and distinctive. The individual skills thus become, in turn, resources for the company and, through the dynamic capabilities of the government, generating its distinctive competencies (Pfeffer and Salancik, 1978).

The close link between the business system and the individuals who animate the whole can be re-interpreted paying attention to cognitive processes, i.e. the formation of knowledge. Information, skills, expertise, experience, learning, etc.. are, in fact, all aspects, phenomena or processes that are explained in the generation and maintenance of knowledge over time. Hence, the act of government (governance) can be an essential reference and synthesis of the total knowledge used, valued and regenerated by the system in its dynamic evolution.

In a vision of governance that goes beyond the short term, the consonance is revealed as a complement to competitiveness for the system to implement and maintain competitive advantage, a harbinger of value creation.

5 CONCLUSIONS

Awareness of the two drivers (consonance and competitiveness) of governance is a force, a stimulus, a voltage to the best knowledge of the magnitude of the systems context and identification of structural changes aimed at giving the system the necessary skills to be competitive and to link up effectively with this connection. In contrast, short-sighted governance, without taking into consideration suprasystems' expectations, pressures, trends and guidelines, it is surely destined to record the strong increase in risk, the true key to graduate corporate decisions in time and to focus on competitiveness and the consonance allowing the government to achieve the above mentioned competitiveness.

To sum up, governance in line with the irreversible environment trends, tense to the search of consonance with the context entities and competitiveness understood primarily as a tension to the continuous improvement in their performance, and constantly on the exploitation/exploration of the set of capabilities built into the specific business structure is the basis of value co-creation and implementation of the probability of survival of the business system (Cyert and March, 1963).

It is so explained the idea of how a company that fulfills its function and generate wealth and economic well-being for all those with whom it interacts more or less directly (full-risk investors, creditors, customers, suppliers, partners, employees etc..), maintains and implements over time its centrality in the institutional context and social life.

REFERENCES

Ashby, H.R. 1958. "General Systems Theory as a New Discipline". In. *General Systems*. (Yearbook of the Society for the Advancement of General Systems Theory), vol.3: 1-6.

Barile, S. 2011. "A viable system conceived as a universal decision maker". In. *Contribution to theoretical and practical advances in management. A viable systems approach*, eds. Barile, Bassano, Calabrese, Confetto, Di Nauta, Piciocchi, Polese, Saviano, Siano, Siglioccolo, Vollero, International Printing Srl, Avellino.

Beer, S. 1972. *Brain of the firm. the managerial cybernetics of organization*. The Penguin Press.

Capra, F. 1996. *The Web of Life: A New Scientific Understanding of Living Systems*. Anchor/Doubleday Books, New York.

Cyert R.M., March J.G. 1963. *A behavioral theory of the firm*. Prentice Hall.

Emery, F.E. 1956. *General System*. (Yearbook of the Society for the Advancement of General System Theory).

Fazzi, R. 1984. *Il governo d'impresa*. Vol. II, Giuffrè.

Golinelli, G.M. 2000. *L'approccio sistemico al governo dell'impresa*. Cedam.

Golinelli, G.M. 2010. *Viable Systems Approach (VSA). Governing Business Dynamics*. Cedam, Kluwer.

Luhmann, N. 1984. *Soziale systeme*. Suhrkamp.

Maglio, P., Kieliszewski, C., Spohrer, J. 2010. eds. *Handbook of Service Science*. Springer.

Maturana, H.R., Varela, F.J. 1980. *Autopoiesis and Cognition: The Realization of the Living*. Reidel, Boston.

Parsons, T. 1971. *The System of Modern Societies*. Prentice-Hall, Englewood Cliffs.

Pfeffer, J., Salancik, G.R. 1978. *The external control of organizations. A resource dependence perspective*. Harper & Row.

Spohrer, J., Golinelli, G.M., Piciocchi, P., Bassano, C. 2010. "An integrated SS-VSA analysis of changing jobs". *Service Science*, Vol. 2, No. (1/2): 1-20.

Veblen,T. 1923. *Absentee ownership and business enterprise in recent times*. The New American Library Publ.

von Bertalanffy, L. 1968. *General System Theory: Foundations, Development, Applications*. George Braziller, New York.

Williamson, O.E. 1964. *The economics of discretionary behavior: managerial objectives in a theory of the firm*. Prentice Hall.

CHAPTER 37

Modeling Cooperative Behaviors in Innovation Networks: An Empirical Analysis

Laura Prota*, Maria Rosaria D'Esposito**, Domenico De Stefano***, Giuseppe Giordano**, Maria Prosperina Vitale**

*Macquarie University (Australia), **University of Salerno (Italy), ***University of Trieste (Italy)
Email address: laura_p@fastmail.fm

ABSTRACT

The aim of this paper is to combine evolutionary economics and social network analysis to empirically evaluate the impact of network based policies. The evaluation of these policies is undertaken in the framework of behavioral additionality. In the case study we concentrate on a techological district located in southern Italy to investigate whether the institution of the district has altered the mechanisms by which research collaborations are established amog actors. Results from structural holes analysis and blockmodeling show that indeed a mechanisms of global bridging expanded research collaborations beyond the previous local trajectory.

Keywords: behavioral additionality, policy evaluation, social network analysis, technological districts

1   INTRODUCTION

Networks of collaborations linking firms, universities and local governments have since long been considered the loci of innovation (Powel, Koput, and Doerr, 1996). To favor the development of such networks of knowledge, an increasing number of Technological Districts (TDs) has been established worldwide through policy interventions. The present paper aims at evaluating the impact of these networks based policies (NBPs) in southern Italy.

The evaluation of NBP policies is undertaken in the framework of behavioral additionality (BA) (Georghiou and Roessner, 2000; Capuano et al., 2011). The concept of BA was formulated to overcome the shortcomings of linear methods such as input and output additionality. Input additionality is used to evaluate whether the allocation of extra resources (funds) stimulates participants to undertake more projects or to enlarge project scopes. Similarly, output additionality is used to assess the effects of public incentives on project outputs, such as the number of patents and research articles (Gok and Edler, 2011). In contrast with input and output additionality which assume a linear relation between policy interventions and outcomes, BA aims at capturing the persistence of changes in the system behavior after policy interventions. In this respect, BA introduces a structuralist element in the analysis. The question is therefore how the innovation process changes and whether the changes can be considered persistent and sustainable over time. In so doing, BA opens up the so called black-box of intra-firm cooperation (Gok and Edler, 2011).

The study adopts an evolutionary framework to conceptualize NBPs as shocks to the system of R&D collaborations and to formulate hypotheses on the development trajectories of the system. In particular, the evolutionary perspective by Gluckler (2007) is combined with social capital and structural holes theories (Walker, Kogut, and Shan, 1997) to identify what are the underlying social processes of link formation and sever. Persistent changes in the behavior of the system are defined as changes able to deviate the system from its previous development trajectory. Network statistics derived from Social Network Analysis are used to describe the extent of the deviation and to evaluate its effects (see for instance Del Monte et al., 2011).

The paper is organized as follows: in section 2 an evolutionary perspective to the analysis of BA is presented; in section 3 the case study is illustrated along with the characteristics of TDs in Italy; section 4 reports the main results of the network analyses; section 5 presents some preliminary concluding remarks.

2   BEHAVIORAL ADDITIONALITY: AN EVOLUTIONARY PERSPECTIVE

The mechanisms of link formation and sever have been at the center of a widening and multidisciplinary literature since the 1990s. Broadly speaking, the debate on network evolution can be divided in two main streams of research.

On the one hand, the seminal paper by Jackson and Wolinsky (1996) modeled network evolution as the result of rational individual choices based on costs and benefits (for a review of the economic literature related to this subject see Jackson, 2005). A critique to this approach can be found in Hummon (2000) and Doreian (2006). In particular, Doreian advances three critiques: *i)* rational actors employ bounded rationality; *ii)* the sequencing of events is critically important for the resulting network outcome; *iii)* costs and benefits vary among actors depending on what position they occupy in the network and at what time their choice is due. Doreian's critique is rooted in a longer tradition of network analysis in sociology which conceives network evolution as a cumulative causation process, strongly dependent from prior conditions. In such conceptualization, sequencing determins an uneven propensity to form and sever new ties. Key to this literature is the concept of social capital and structural holes accredited to Coleman (1990) and Burt (1992). According to these theories, actor utility varies in relation to the position and the structure of the network. Network ties can become instrumental to access resources otherwise inaccessible (social capital) or being associated with benefits derived by the opportunity of bridging structural holes.

Building on these theories, a second stream of research has developed in sociology and business looking at the social processes influencing network evolution in the empirical cases. A major field of investigation has been research collaborations among firms in high-tech sectors. Studying technology-intensive industries for instance, Walker, Kogut, and Shan (1997) hypothesize two main forces influencing ties formation: on the one hand, social capital pushes toward the reproduction of existing ties creating areas of dense, cohesive collaboration; on the other hand, the intensification of collaborations within clusters automatically creates structural holes, opening up the opportunity for enterpreneurs to exploit brokerage opportunities.

In a more recent study on unequal growth rates and geographical agglomeration of innovation, Gluckler (2007) explains network evolution as the result of three evolutionary forces: *selection, retention* and *deviation. Selection* is not intended as the ability of an organization to survive in a competitive market, but it refers to the mechanism by which an organization chooses its collaborators, forming in this way new ties. *Retention* (path-dependence) is theorized as the result of three sub-mechanisms of system self-reinforcement: preferential attachment, embedding and multi-connectivity. Each of these mechanisms is conservative insofar as it specifies a way by which new ties become progressively more likely within a given trajectory. In the case of preferential attachment, for instance, new ties will more likely be extablished with already central actors within the network. Similarly, in the hypothesis of embedding new ties will form among actors already bounded by strong, repetitive interactions built on trust and indirect referrals (Gulati, 1999). Finally, multi-connectivity is identified as the third indicator of path-dependence. The more organizations seek diversity of resources and relations the more they will be likely to form cohesive subgroups (Powell, Koput, and Doerr, 1996; Walker, Kogut, and Shan, 1997). *Deviation* from the existing trajectory occurs when there is a change in the selection mechanism. For the purpose of this analysis, the focus will be on two mechanisms of variations proposed by Gluckler (2007): global and local bridging. The former hypothesizes that densely connected local network will connect with global clusters in order to seek complementary resources. This mechanism is crucial to avoid technological lock-in and it is based on the idea of bridging structural holes. Conversely, the latter mechanism of variation, local bridging, stresses the way by which innovation springs from an original use of existing local resources. Different, complementary local clusters will establish ties among themselves producing in this way an innovative re-combination of resources.

3   CASE STUDY

The paper focuses on territorial innovation networks, where a variety of actors (firms, institutions and research centers) are involved in research activities. Institutional policies have been implemented in many industrial economies to set up Technological Districts. In Italy, TDs were instituted in 2003 in the framework outlined by two national laws: L.317/91 and L.140/99. The aim of the laws was that of producing incentives for research organizations to behave cooperatively instead of strategically, generating in this way positive externalities through the network effect. In other words, the law aimed at favoring the formation of research collaborations among firms and research centers co-located in a region and specialized in complementary fields. Moreover, the connection of these local technological clusters with global partners was encouraged through funding and public support. A question arises whether public fundings have succeeded in creating persistent, structural changes in the way cooperation is carried out and whether such changes are sustainable in the long run.

In the following, the focus is on a TD located in southern Italy. It is of particular interest as it brings about the issue of growth rates convergence between the northerner and the southern regions of Italy. Among the others, the development objective of the policy was that of strengthening the technological capabilities of the south, which has remained behind in its economic development process for a number of historical and political reasons. The examined TD has achieved important results in its field of specialization obtaining prominence both nationally and globally.

3.1   RESEARCH HYPOTHESES

The main assumption of the study is that new links are formed and severed according to a logic inspired by underlying social and economic processes. There are therefore some organic patterns in the formation of a network that can be broadly summarized as the opposition of the two forces: on the one hand, there are retention mechanisms through which the network self-reinforces its existing structure; on the other hand, there are mechanisms able to divert the system from its trajectory. A crucial factor for change is the existence of structural holes within the network as it provides bridging opportunities. Actors sitting on the hedges of structural holes can exploit the opportunities inherent in their position either to bridge parts local clusters with global networks or to recombine local resources to produce innovation. Bridging structural holes is therefore an entrepreneurial action that rewards bridges with temporary benefits in a Schumpeterian logic. At the same time, as soon as the gap is filled, the overall structure of the network change. Subsequently, retention mechanisms will reinforce the deviation, by selecting ties compatible with the new regime. It is this particular type of structural change in the overall network structure that can be consider as a deviation from its previous trajectory (Walker, Kogut, and Shan, 1997; Gluckler, 2007).

Given two state points of the network, if a variation in the process of link formation and sever has occurred between time 1 and time 2, the following hypotheses are stated:

H1. Strong differences among actors with respect to their structural holes measures will signal the possibility for a few actors to exploit the advantages of their structural position. These behaviors will characterize the first phase of research collaborations (time 1), initiating in this way a process of variation;

H2. At time 2, the occurred variation will be reinforced by retention processes (such as preferential attachment and multi-connectivity);

H3. Finally, these changes will have a structural effect on the network as whole.

3.2   DATA AND METHODS

The development of the collaboration network is analyzed by comparing an initial phase characterized by projects activated after the institution of the TD since 2005 (time 1) with a second wave of projects started in 2008 (time 2). The study uses secondary data to reconstruct the network of co-participation in projects among associated members and partners. The data on research projects published on the TD website have been integrated with information directly obtained from TD members. The co-participation in research project encompasses 13 public grants funded by the Italian Ministry of Education, Universities and Research (MIUR) and 2 European funded projects. At time 1, the TD had 20 associated members (private firms, universities and public research institutes) and 5 partners. In order to have finer grain results, the universities departments have been considered separately, obtaining in this way 25 associated members. At time 2, the TD had 18 associated members and 20 partners. As before, two associated members (universities and public research departments) have been exploded to represent the seven departments within each research organization, resulting in a total of 30 associated members.

Collaboration data are extracted from the set of p research projects and from the set of n members arranged in two affiliation matrices. Let A_{t1} and A_{t2} (nxp) be the affiliation matrices in which the generic element a_{ij} is equal to 1, if the $i\text{-}th$ member participates at the $j\text{-}th$ research project and 0 otherwise. From A_{t1} and A_{t2}, two adjacency matrices of size (nxn) are derived representing undirected adjacency matrices, G_{t1} and G_{t2}. The two matrices report the number of research projects each pair of members have co-participated in. It has to be noted that in matrix G_{t1} values range from 0 to 12; matrix G_{t2} instead has values ranging from 0 to 1. In the following, network indices, both at global and actor level, and structural holes analysis are computed starting from a binary version of both adjacency matrices, setting all entries greater than zero to "1".

The techniques proposed in the framework of Social Network Analysis (Wasserman and Faust, 1994) are used to describe the network characteristics at time 1 and time 2 on the basis of patterns of ties among social entities. Specifically, the structural holes measures developed by Burt (1992) are used to test hypothesis

H1 and H2. The software UCINET is used to calculate the potential structural advantage associated with individual positions within the network at time 1 and to evaluate to what extent such potentials have been exploited at time 2. Furthermore, blockmodeling analysis (Doreian, Batagelj, and Ferligoj, 2005; Ziberna, 2007) is used to evaluate the changes occurred in the collaboration network at macro-level, as stated in research hypothesis *H3*.

4.   NETWORK RESULTS

The network visualizations show the collaboration patterns elicited by the data in the two periods considered (see Figure 1, Figure 2). First of all, it has to be noted that the 12 research projects at time 1 are all financially supported by national grants. Conversely, the network at time 2 encompasses 2 research projects funded by the European community and 1 supported by national grant.

At time 1, the collaboration network (Figure 1) is characterized by an area of denser activity (on the right) and a group of less connected actors (on the left). Freeman degree centrality (1979) highlights that a central position in the network is played by two research centers, R_center2_1 (scoring 22), R_centrer1_3 (scoring 18), and a group of firms, Firm2, Firm3 and Firm4 (scoring 11). The network density, equals to 0.21, reveals a quite good cohesion in the net. Only five partners are however involved in these first phase of research collaborations.

At time 2, the network shows a different configuration. Several partners enter the network, both national and international ones, thanks to the two EU research projects in which the district is considered as a partner. Indeed, the TD (with degree equals to 28) is the actor that connects groups of members involved in the three projects. Several associated members result as isolated with a reduction of overall network cohesion (the density is equal to 0.13).

By comparing the collaboration networks in the two periods, it is possible to notice a change in the TD governance. A time 1, the network is characterized by a closure as there are strong cooperative behaviors between few big firms and two research centers; at time 2 the network opens up to include external actors in the framework of projects that go beyond national funding schemes.

4.1   STRUCTURAL HOLES ANALYSIS

To evaluate the different potential roles the network offers to TD members, an analysis of structural holes (Burt, 1992) is undertaken through the definition of two measures: *effective size* measures the number of non-redundant ties in each ego-network; and *efficiency* norms the effective size of ego's network by its actual size (ranging from a minimum of 0,when all ties are redundant, to a maximum of 1 in presence of no redundant ties).

Average effective size and efficiency in the network at time 1 are 2.21 and 0.33, respectively. At time 1, two actors (R_center2_1 and R_Center1_3) have above average scores on both measures. R_center2_1's ego network effective size is 15.9 (non-redundant ties) and its efficiency value (0.72), indicating that about 70 percent

375

of contacts are non-redundant. Similarly, actor R_Center1_3's ego network has an effective size of 11 and an efficiency of 0.63. Hence, the two research centers occupy structurally different positions within the networks and have the potential to bridge structural holes. This result seems to confirm the research hypothesis *H1* and *H2*, insofar as the network at time 1 appears vulnerable to changes.

At time 2, the overall structure of the collaboration network has changed (see Figure 2). A new actor emerges with a dominant position in the network: the Technological District (TD) itself, involved in international collaborations as partner. The TD's ego network presents larger effective size and efficiency (scoring 18 and 0.66) with respect to average values (scoring 1.35 and 0.13, respectively).

Before testing the impact of these changes on the overall structure of the network, it is important to derive some basic conclusions about structural holes results. The two actors R_Center2_1 and R_Center1_3 at time 1 had a potential opportunity to bridge disconnected parts of the networks; at time 2, TD members have all the identical potential opportunities to bridge and the network appears more stable. In contrast with this homogeneity among members' structural holes measures, the TD at time 2 emerges as the most entrepreneurial actor linking members and international partners.

Furthermore, at time 1 the network is mainly characterized by collaborations linking the members of the TD itself. The involvement of the TD at time 2 as partner in EU projects is somehow an indirect formal acknowledgment of the TD expertise attained in the field of specialization at time 1. Finally, the network at time 2 is characterized by a larger participation of non-local partners in the collaboration network. The capacity to attract global partners is an important indicator of the fact the TD has managed to function as a global bridge linking the local technological capabilities with the wider international community (Glucklei, 2007).

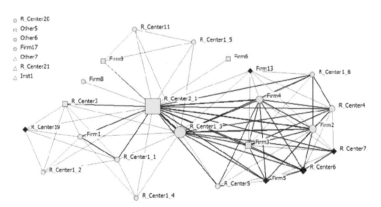

Figure 1 Collaboration network at time 1 among actors (associated members and partners) according to the participation in the 12 research projects started in 2005.
Node size = Number of projects per actor; Node symbol: different associated members (circle, square and up triangle) and partners (diamond); Node color: associated members (grey), partners (black). Edge size: Number of projects shared by pairs of actors.

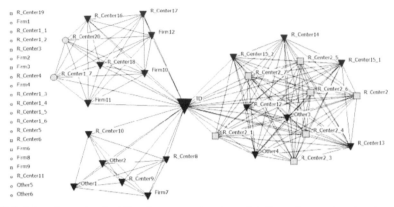

Figure 2 Collaboration network at time 2 among actors (associated members and partners) according to the participation in the 3 research projects started in 2008.

Node size = Number of projects per actor; Node symbol: different associated members (circle, square and up triangle) and partners (diamond); Node color: associated members (grey), partners (black). Edge size: Number of projects shared by pairs of actors.

4.2   STRUCTURAL DIFFERENCES AT NETWORK LEVEL

The analysis of structural holes carried out in the previous section aimed at capturing the differences characterizing the individual positions of actors within the network. A question arises whether these changes have had any impact on the overall structure of the network. An *a posteriori,* indirect blockmodeling analysis for valued adjacency matrices is then used. Clustering (through Euclidian distance for obtaining dissimilarity matrix and Ward as agglomerative method) and blockmodeling optimization of the partitions derived in the clustering approach are performed using the package blockmodeling of R software. According to the regular equivalence concept, three block types are permitted: "null", "regular" and "complete". Using the valued procedure described in Ziberna (2007) a threshold value of 5 is set for G_{t1} (an actor is linked to another actor if they co-participate at least in 5 projects) and 1 for G_{t2}.

According to the block types described in Doreian, Batagelj, and Ferligoj (2005) and the block composition, these preliminary results show that at time 1 the network structure can be described almost as a center-periphery structure with core positions internally cohesive and connected with other groups. Conversely, at time 2, the network structure is characterized by cohesive subgroups (i.e. ranked clusters) on the main diagonal with no links among groups. While in the cores identified at time 1, both research centers and firms were present in each block, at time 2 cohesive subgroups are predominantly composed by research centers alone. This might be interpreted as the result of TD specialization. These results confirm the changes in the overall network structure, as stated in research hypothesis *H3*.

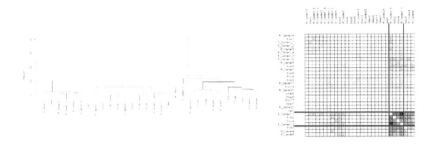

Figure 3 Dendrogram and permuted adjacency matrix **G**$_{vt1}$ obtained according to the 3 groups derived by clustering approach and optimized by blockmodeling approach.

Figure 4 Dendrogram and permuted adjacency matrix **G**$_{vt2}$ obtained according to the 5 groups derived by clustering approach and optimized by blockmodeling approach.

5.   DISCUSSION AND CONCLUSION

The study has combined evolutionary theories in economics and social network analysis perspective to empirically measure the behavioral additionality of network based policies. The case study concentrated on the particular case of a TD located in southern Italy. To describe the collaborations into projects among associated members and partners of the TD in the two periods, first the structural holes analysis allowed to identify the actors that had the potential to bridge unconnected areas of the networks, influencing in this way the overall development of the system. Then, the comparison of the collaboration networks showed that two actors have bridged the structural holes at time 1; while the TD itself assumed a dominant position at time 2 being partner in two EU projects. The blockmodeling analysis demonstrated that a structural change has occurred: from an almost center-periphery structure to a ranked clusters model with cohesive groups on the diagonal and no ties linking different groups.

A question remains as to what extent these changes are positive for the research development of the TD. From one point of view, it can be argued that the TD has acquired during time a more international role. On the other side, a closer investigation of the individual organization within the network has to be carried out to evaluate which organizations were left behind and which have most profited by the changes. While the study has a number of limitations, it contributes important new insights on the way behavioral additionality can be measured and on how the structural impact of public policies can be evaluated.

REFERENCES

Burt, R.S. 1992. *Structural Holes*. Cambridge: Harvard University Press.
Capuano, C., D. De Stefano, A. Del Monte, and M.P. Vitale. 2011. A strategy to analyze network additionality for territorial innovation: the case of Italian Technological District. *CLADAG proceedings*.
Coleman, J.S. 1990. *Foundation of Social Theory*. Cambridge: Harvard University Press.
Del Monte, A., M.R. D'Esposito, G. Giordano, and M.P. Vitale. 2011. Analysis of Collaborative Patterns in Innovative Networks. In. *New Perspectives in Statistical Modeling and Data Analysis*, eds. S. Ingrassia, R. Rocci, and M. Vichi, 77-84. Heidelberg: Springer-Verlag.
Doreian, P., V. Batagelj, and A. Ferligoj. 2005. *Generalized Blockmodeling*. Cambridge: Cambridge University Press.
Doreian, P. 2006. Actor network utilities and network evolution. *Social Networks* 28: 137-164.
Freeman, L.C. 1979. Centrality in social networks: Conceptual clarification. *Social Networks* 1: 215-239.
Georghiou, L. and D. Roessner. 2000. Evaluating technology programs: tools and methods. *Research Policy* 29: 657-678.
Gluckler, J. 2007. Economic geography and the evolution of networks. *Journal of Economic Geography* 7: 619-634.
Gok, A. and J. Edler. 2011. *The use of behavioural additionality in innovationpolicy-making*. Working Paper, Manchester: University of Manchester.
Gulati, R. 1999. Network location and learning: The influence of network resources and firm capabilities on alliance formation. *Strategic Management Journal* 20: 397-420.
Hummon, N.P. 2000. Utility and dynamic social networks. *Social Networks* 22: 221-249.
Jackson, O.M. 2005. A survey of models of network formation stability and efficiency. In. *Group Formation in Economics: Networks, Clubs, and Coalitions*, eds. G. Demange and M. Wooders, 11-88. Cambridge: Cambridge University Press.
Jackson, M.O. and A. Wolinsky. 1996. A strategic model of social and economic networks. *Journal of Economic Theory* 71: 44-74.
Powel, W.W., K.W. Koput, and L.S. Doerr. 1996. Interorganizational collaboration and the locus of innovation: networks of learning in biotechnology. *Administrative Science Quarterly* 41: 116-145.
Walker, G., B. Kogut, and W. Shan. 1997. Social capital, structural holes and the formation of an industry network. *Organization Science* 8: 109-125.
Wasserman, S. and K. Faust. 1994. *Social Network Analysis: Methods and Applications*. Cambridge: Cambridge University Press.
Ziberna, A. 2007. Generalized blockmodeling of valued networks. *Social Networks* 29: 105-126.

CHAPTER 38

Introduction of Computer Supported Quality Control Circle in a Japanese Cuisine Restaurant

Ryoko Ueoka*, Takeshi Shinmura* **, Ryuhei Tenmoku*, Takashi Okuma*,

Takeshi Kurata*

*National Institute of Advanced Industrial Science and Technology
Tsukuba Japan
**Ganko Food Co. Ltd.
Osaka, Japan

ABSTRACT

The activity of quality control circle (QCC) has been conducted in restaurant business to improve service productivity. However, it was difficult to show the evidence of improvement objectively. For the purpose of developing the new QCC method specified in service industry, we propose Computer Supported Quality Control Circle (CSQCC). In this study, as a first step of CSQCC we carried out a pilot study in a Japanese cuisine restaurant and evaluated potential capabilities of CSQCC with which we can not only measure behaviors of workers over a month but also show statistical data combining workers' trajectories with point-of-sale (POS) accounting data in a 3D manner. As members of the QCC shared waitresses' behaviors, they could effectively find some patterns of traffic in working area correlating to productivity, which shows potential of CSQCC as a new method of service observation.

Keywords: Quality Control Circle (QCC), Service Engineering, Visualization, Human Behavior Sensing

1   INTRODUCTION

QCC is defined as a small group of employees who come together to discuss with the management issues related to either quality control or improvement in production methods. These employees usually work in the same areas, and voluntarily meet on a regular basis to identify, analyze and solve their problems (Watanabe, 1991, EZINE, 2008). QCC was first introduced in manufacture such as automobile industry (Monden, 2000). In this industry, observational techniques such as measuring behavior of workers at a fixed place were fairly effective in identifying problem and adopting techniques such as the *kaizen* method for improving work efficiency (Imai, 1986). However in service industry, collecting objective data is not easy due to the cost and technology for dealing with relatively-large indoor fields. Therefore most of reasonable data being used for conventional QCC is questionnaire to employees or customers, amount of daily sales and so on. So for the purpose of developing the new QCC method to observe both productivity and subjective values specified in service industry, we propose Computer Supported Quality Control Circle (CSQCC) (Figure 1).

In this study, as a first step of CSQCC we carried out a pilot study in a Japanese cuisine restaurant and evaluated the potential of CSQCC for observing service productivity. This was done by using measurement devices, including Pedestrian Dead Reckoning sensors (PDR), RFID (Radio Frequency IDentification) tags (Ishikawa, Kourogi, Okuma and Kurata, 2009, Kourogi, Ishikawa and Kurata, 2010) , to measure the behaviors of workers for a month. Also by using 3D visualization tool showing 3D visualization and statistical data which combines workers' trajectories with point-of-sale (POS) accounting data during a phase of pilot discussion of QCC, members of the circle shared waitresses' traffics which helped to find some patterns of traffic in working area which relate to productivity. It shows CSQCC helps to identify and analyze problems of interfering productivity.

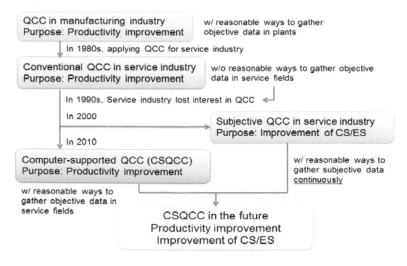

Figure 1 History of QCC in the Service Industry in Japan and CSQCC.

2   RELATED RESEARCHES

A range of devices including wearable sensors and mobile phones have been used to measure human behavior, and a variety of analytical methods have been proposed including organizational analysis and the generation and analysis of interpersonal topological network diagrams. For example, the SocioMeter designed at MIT Media Lab (Choudhury and Pentland ,2003) and Hitachi's Business Microscope (Hitachi Cooperation, 2011) employ IR transmitters as sensors to measure face-to-face communication between office workers, and this data is then used to analyze personal interaction in organizations and topological behavior between groups. The SocioMeter analyzes connectivity between people and how long they converse using sensor data, while the Business Microscope provides visualization of organizations as an index of office work value creation through face-to-face communication. When measuring face-to-face communication with sensors, the point of contact with the other person is the critical point that is sensed, and fine-grained positional accuracy is much less important. In reality, office work is generally categorized by area in the office—meeting room, showroom, section room, and so on—and more detailed information about specific work within an area is typically measured using PC logs or some similar mechanism. In the service industry that is the focus of this study, there is relatively little work involving face-to-face communication with other employees. Rather, the work primarily involves work in different areas and communication with customers. Thus, differences in types of food depending on place, differences in customers depending on service location, and other considerations regarding position and paths in areas are significant factors for work in the service industry. While detailed movement data within an area is required, the accounting data such as POS can be used to replace communication history with customers. Because the types of sites we are interested in differ from those addressed in previous studies, this means that we are looking for different types of data.

Reality mining permits us to visualize social structures around human connectivity (Eagle and Pentland, 2009, Eagle, Pentland and Lazer, 2009). The purpose of this research is to build generative models that can be used to predict what a single user will do next as well as model behavior of large organizations using log data collected from mobile phones. Here they have analyzed the interpersonal behavior of individuals around connectivity with others in society, and attempted to model the social activity of individuals based on their mobile phone usage history. Since modeling service providers and customers in the service industry field differs in terms of social scale and members, the type of sensors used, the type of data collected, and the model assumptions used in this study all differ.

A number of studies measuring human behavior in the service industry have been conducted including an analysis of changes in the selection of shopping centers based on the type of customer at a shopping mall (Dennis, Marsland, Cockett and Hlupic, 2009) and a shopping promotion study that seeks to induce purchasing using an experiment environment that simulates store shelves (Sae-Ueng, Pinyapong, Ogino and Kato, 2007). However, the focus of these studies of purchasing behavior is on the customer, whereas our study focuses on the behavior of the service providers.

Naya *et al*. proposed using sensors and a behavior recognition algorithm to detect the behavior of nurses as they go about their work at a health-care facility (Naya, Ohmura, Yakayangi and Noma, 2006). The authors employ the similer method we use in our study of attaching sensors to the clothing of the subjects to measure their behavior, but Naya *et al*. are primarily interested in the structuring of nursing work. The purpose of our study is fundamentally different. Rather than structuring the work of individual workers, in our study we integrate the measured behavior data with POS accounting data with the goal of discovering work objectively linked to service value and elevating service quality.

3   HUMAN BEHAVIOR SENSING AND VISUALIZATION TOOL

For this study, we continually monitored the behavior of about 20 workers in a two-story Japanese cuisine restaurant over a one-month period. Staff can be divided into three main job categories: kitchen staff, waitresses, and assistant waitresses who convey food and dishes from the kitchen to the pantry, prepare drinks, and do other miscellaneous tasks. On any given day there are at least 30 employees working in the restaurant, but we had limited experimental device so we selected 20 workers to participate in the study.

Measurements were recorded using PDR sensors attached at the waists of the subjects that measured their walking paths. And we compiled electronically recorded POS data including what was ordered in the restaurant, the number of customers and orders, when the orders occurred, where in the restaurant the orders occurs, the times that customers paid their bills, the names of waitresses associated with each order, etc.

Figure 2 illustrates how the sensors were attached to the subjects. Because the study was conducted over a one-month period at the restaurant, the sensor system was implemented in such a way that it would not interfere with work and could be easily removed by the subjects themselves. The waitresses wear kimonos when they are on duty, and the sensors were incorporated into accessories that are attached to the sash (*obi*) that goes around her waist.

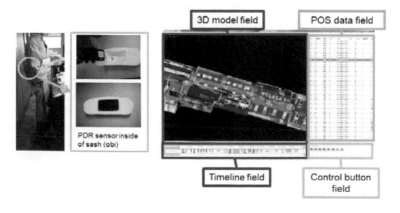

Figure 2 Human behavior sensing image (left) and developed 3D visualization tool (right)

Also we developed a visualization tool that permits synchronized playback of all the measured behavioral data (estimated walking path results for each employee, POS and other work-related data). One can see in Figure 2 that the visualization tool consists of the following components:

- 3D model field: Displays a 3D model of the experimental environment, including walking paths and indicators of the subjects' positions as well as customer icons showing the location and the number of customers in the time period. (Figure 3 (a),(b)).
- Timeline field: Shows where the subject is in the environment and voice activity on the time line.
- POS data field: Displays POS log data in a spreadsheet.
- Control button field: Displays buttons to control playback, fast forward, and rewind of data shown in each field.

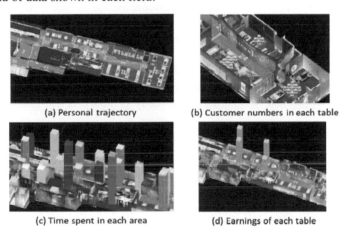

(a) Personal trajectory   (b) Customer numbers in each table

(c) Time spent in each area   (d) Earnings of each table

Figure 3 Example of visualized analysis

In addition to synchronized playback, the tool has a number of other important capabilities: you can input queries then search for places in the measurement data corresponding to the queries, and the tool calculates and displays various statistical values such as number of items ordered per customer, the proportion of time employees spend in different areas of the restaurant during working hours, etc. Regarding statistics of measured data, there are person, time, and area axes as multi-dimensional data, so by inputting queries, we are able to compare measured data for each person, each unit of time, and each area. Some of the statistical values that can be compared include amount of movement, number of times waitresses pass through an area, number of orders, sales, and face-time with customers. Comparing these statistics for each area, we can generate graphic representations superimposed on a 3D model as illustrated in Figures 3(c) and (d).

4 PILOT STUDY OF CSQCC

4.1 Discussion Phase

In order to assess how CSQCC works in actual service fields, we conducted a pilot study to evaluate the ability of the restaurant staff to identify problems, suggest improvements, and actually implement the proposals at the restaurant through CSQCC. In this pilot study, we first used the 3D visualization tool to show the behavior measurement data in the first week such as the movements of the waitresses, the amount of time they spent in the dining area, and so on. Then, once the staff had a good objective understanding of the situation based on the visualization results, they sorted out the problem areas and proposed improvements. In the fourth week of the behavior measurement study, the staff actually implemented their suggestions for improvement. And we compared staff behavior before and after the improvements were implemented to see if there were any discernible changes.

Details of the QC circle activity are summarized as follows:

- Date: February 1, 2011
- Participants: Three executive team managers, and three local staff members.
- Data used for visualization: Seven waitresses for four or more days during the first week of the behavior measurement study. (January 12-18, 2011).

During the QC circle activities, the participants discussed problems and issues on productivity and quality control while seeing data replayed in real time, and also while seeing graphs and tables showing one week's data that had been statistically processed in advance using the visualization tool.

As an example of traffic patterns of peak time, Figure 4 shows the movements of waitresses over one hour period from noon to 1:00 PM on two floors of the Japanese restaurant (floors B1 and B2). Note that peak time is a particularly busy time during the lunch hour rush, and we were able to verify the extent that waitresses stayed in service area based on actual data. By playing back the traffic data, we could see that the waitresses frequently shuttled back and forth to the back office on floor B1 even though this was the busy lunch hour. (Figure 5) Feedback from the manager revealed that when a waitress in the service area takes a call with her cell phone from a customer who wants to make a reservation, she has to make a trip to the backyard to check the reservation ledger. This finding led to a future agenda to introduce computerized reservation ledges for improving efficiency of the work not directly linked to serving customers.

These findings motivated participants to discuss whether the waitresses actually focus on customers in the service area.

As one can see in Figure 6 (left), the first week measurement results revealed that the average ratio of time stayed in service area per hour is 40%, which was not enough proportion as a waitress. It was thought that this is because the waitresses also work as assistant waitresses conveying the food before it reaches the dining area, and as one can see in Figure 6 (right), they move around other areas doing work other than serving customers in the service area.

While no major changes were made in terms of shifts or system modifications as a result of the QC circle pilot study in a short time, an improvement plan was nevertheless implemented in the fourth week of the study with the goal of increasing time staying in service area by having the waitresses consciously increase the time spent for interacting with customers. The plan was implemented on floor B1 where many walk-in customers were coming without reservations during the time frame from 2:00 to 4:00 PM when there are relatively few customers. Because the restaurant is not busy during those hours, this made it easier for the waitresses to take care of customers and increase face-time with them by encouraging them to have a low-cost dessert or coffee after their meal, or by chatting with them. As one can see in Figure 6 (left), the first week measurement results revealed that the average ratio of time stayed in service area per hour fell from 40% throughout the day on average to 37% around the off-peak hours of the targeted time frame. Although time stayed in service area is reduced simply because there are fewer customers during these hours, it was decided based on the QC circle activities to improve services by increasing the time stayed in service area per customer during this time frame.

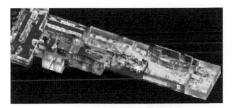
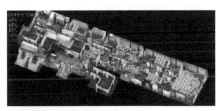

Figure 4 Example of peak time traffic of waitresses on two floors

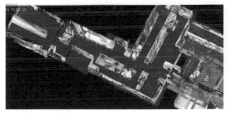

Figure 5 Example of traffic to and from the backyard during peak time on Jan 16 and 17,2011

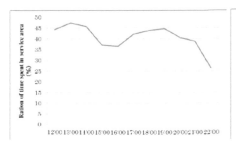
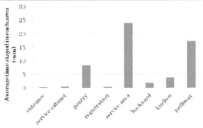

Figure 6 Ratio of time stayed in service area of each time frame (left) and time stayed in each area (right)

4.2   Comparison of Before and After CSQCC

As a pilot study for evaluating the effects of this service improvement, we first compared time spent in service area during the first week (before improvement) and the fourth week (after improvement) of the experiment.

Figure 7 shows the box plot of the ratios of time spent in service area from 14:00 to 16:00 during the first week and the fourth week. Being compared to the median time ratio of 22% in the first week, the ratio increased by 1% to 23% in the 4th week. No big change has been found but variance of time decreased in the 4th week, which may say that service quality became stable in the fourth week.

Figure 8 shows that the total time spent in the service area per customer had increased from 2.4 minutes in the first week to 2.6 minutes in the fourth week. This too reflects a gradual increase of improvement in the latter week.

In a follow-up with the managers, we confirmed that these results were actually reflected in changes at the restaurant since during the fourth week, a manager tried to tell to the waitresses to spend more time in the dining area and to communicate with the customers more.

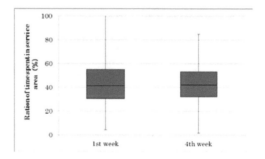

Figure 7 Boxplot of the ratio of time stayed in service area in 1st week and 4th week

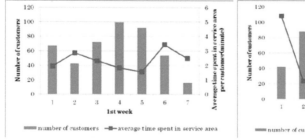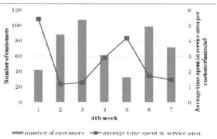

Figure 8 Number of customers and average time stayed in service area per customer (min)

5 CONCLUSIONS

As a first step of CSQCC we carried out a pilot study in the Japanese cuisine restaurant and evaluated the potential of CSQCC. By using measurement devices to measure the behaviors of workers for a month, we proved our proposing system was able to measure employees' behavior in service industry over a long period of time.

Also by using 3D visualization tool, members of the circle shared waitresses' traffics which helped to find notable patterns of traffic in working area which relate to productivity during a phase of pilot discussion of QCC.

It shows CSQCC has the potential capabilities to help to identify and analyze problems of interfering productivity. In the end, based on the discussions in the circle, they tried to increase waitresses' staying time in service area as a pilot study of quality control. And we evaluated before and after the activity out of workers' measuring data objectively. It succeeds to show potential of CSQCC as a new method of service observation.

Usually, it is said to take a few weeks to understand the current situations before identifying the problem with traditional QCC. In the future study we would like to compare the effectiveness of traditional QCC and CSQCC by seeing how long it takes to identify problems and issues with each methodology.

Also for further understanding of correlation between workers' behavior and productivity, we are currently working on developing the algorithm to estimate the types of service operation from sensor data in order to break down service operation as a context (Tenmoku, Ueoka, Makita, Shinmura, Takehara, Tamura, Hayamizu and Kurata, 2011). By adapting this method, it is expected to understand the relationship quantitatively between each service operation and productivity.

Moreover, by combining behavioral data and feedbacks from customers, we can link workers' behavior and customers' satisfaction, which will make it possible to interactively analyze the relation between productivity and subjective values for improving service quality in future.

ACKNOWLEDGMENTS

This research was entrusted by the Ministry of Economy, Trade and Industry (METI). We thank all the staffs in Ganko Food Restaurant at Ginza 4-chome for cooperating with our experiment. We also thank Mr. Ogawa and Mr. Tsunemori of executive managers in Ganko Food Restaurant Tokyo branch office for cooperation of our research.

REFERENCES

Choudhury, T. and Pentland, A. 2003. Sensing and Modeling Human Networks using the Sociometer. *Proceedings of ISWC* :216-222.

388

Dennis, C.,Marsland, D., Cockett, T.,Hlupic, V. 2003. Market segmentation and customer knowledge for shopping centers. *Proceedings of the 25th International Conference on Information Technology Interfaces*: 417-424.

Eagle Nathan, Pentland S. Alex 2006. Reality mining: sensing complex social systems. *Journal of Personal and Ubiquitous Computing*,10,4,Springer-Verlag London, UK :255-268.

Eagle Nathan, Pentland S. Alex, Lazer David 2009. Inferring friendship network structure by using mobile phone data. *Proceedings of the National Academy of Sciences of the United States of America*,106,36: 15274-15278.

EZINE articles "Quality Control Circle- A Quality Get Together" Accessed October 1, 2011, http://ezinearticles.com/?Quality-Control-Circle---A-Quality-GetTogether&id=1330739

Hitachi Corporation "Business Microscope" Accessed October 1, 2011, http://www.hitachi-hitec.com/global/business microscope/solution/microscope.html/.

Imai Masaaki. 1986. Kaizen: The key to Japan's Competitive Success. New York U.S.A: Random House.

Ishikawa T., Kourogi M., Okuma T., and Kurata T. 2009. Economic and Synergistic Pedestrian Tracking System for Indoor Environments. In *Proceedings of Int. Conf. on Soft Computing and Pattern Recognition* :522-527.

Kourogi M., Ishikawa T., and Kurata T. 2010. A Method of Pedestrian Dead Reckoning Using Action Recognition. *Proceedings of IEEE/ION PLANS 2010 Position Location and Navigation Symposium* : 85-89.

Monden Yasuhiro. 2000. Target Costing and Kaizen Costing in Japanese Automobile Companies. *Japanese cost management*, Imperial College Press, London U.K.:99-117.

Naya Futoshi, Ohmura Ren, Yakayanagi Fusako, Noma Haruo, Kogure Kiyoshi. 2006. Worker's Routine Activity Recognition using Body Movements and Location Information. *Proceedings of ISWC* :105-108.

SAE-UENG Somkiat, PINYAPONG Sineenard, OGINO Akihiro, KATO Toshikazu. 2007. Consumer-friendly Shopping Assistance by Personal Behavior Log Analysis on Ubiquitous Shop Space. *Proceedings of the 2nd IEEE Asia-Pacific Service Computing Conference* :496-503.

Watanabe Susumu. 1991. The Japanese quality control circle: Why it works. *Int'l Labour Review*,130:1 .

Tenmoku R., Ueoka R., Makita K., Shinmura T., Takehara M., Tamura S., Hayamizu S., and Kurata T., "Service-Operation Estimation in a Japanese Restaurant Using Multi-Sensor and POS Data", International Journal of Virtual Reality, In Proc. APMS 2011 conference, Parallel 3-4: 1, 2011.

CHAPTER 39

Leading Change by Playing: Design and Implementation of a Management Game for Aviation Ground Services

Janine Kramer[1], Frieder Mayer[2], Michael Gassner[2], Rainer Nägele[1]

[1]Fraunhofer Institute for Industrial Engineering
Stuttgart, Germany
janine.kramer@iao.fraunhofer.de/rainer.naegele@iao.fraunhofer.de

[2]Cost Aviation by Stuttgart Airport
Stuttgart, Germany
M.Gassner@stuttgart-airport.com

ABSTRACT

There is big potential for improvement and optimization when looking at aviation ground services. Like airports that have once been deregulated by EU law facing now high cost pressure of the travel industry, struggle with making their aviation ground services competitive and efficient. But how shall change processes be implemented so that employees support those changes and understand the necessity and effectiveness? Within this paper we describe our developed method that simulates aviation ground services with flows of passengers, luggage, and service tasks with the help of LEGO bricks in the shape of a management game. Within the game aviation ground processes can be analyzed for delays, bottlenecks and conflict potential and continuously improved.

Keywords: aviation ground services, management game, change management, process analysis, process improvement, employee integration

1 CHALLENGES AND POTENTIAL FOR IMPROVEMENT FOR AVIATION GROUND SERVICES

1.1 Industry Trends

During the last decade the conditions for airports and ground services providers at airports changed a lot. Key elements which lead to this development have been the low cost movement within the airline industry and the deregulation of the ground service provider market by European Union (Council Directive 96/67/EC).

In an environment of growing passenger numbers and aircraft movements established ground handling companies face new challenges: both new competitors and airline customers' new spirit generate tremendous cost pressure. At the same time airlines claim high quality standards as they are a requirement for efficient operation and customer satisfaction.

A lot of airports decided to found union-free subsidiaries to save at least employment cost. Other ground handling providers went for cost saving and process re-engineering projects to find new ways of production to match the goals.

1.2 Challenges for Human Ressource Planning

Airport processes are driven by the flight plan, which cannot be considered as a stable schedule but as an ongoing planning process of fluctuating events and demands.

At airports which do not operate at their capacity limit, aircraft movements pile up for peaks at certain times due to night-curfews, customer needs, airline strategies, etc. whilst the airport operation nearly comes to standstill at others. This induces a deeply volatile demand for human resource, which cannot be covered by a standard roster system effectively. At the same time the industry comprises seasonality regarding weekdays and months as well as high sensitivity regarding world's economic, political or epidemic events, which requires additional flexibility.

1.3 Challenges for Ground Handling Operations

But even if the demand for HR is accurately planned the operation remains highly sensitive for weather conditions, technical failures, passenger delays etc. And none of the ground processes can be postponed – the aircraft has to be serviced, when it is on ground, doesn't matter if delayed, before or on time. This leads to only roughly predictable workload and high demand for flexibility. All processes must be able to be controlled in real time conditions - this includes communication needs between several working groups/companies involved in ground handling of one aircraft and their dispatching centers.

As local infrastructure differs a lot there is no industry standard regarding the processes organization and necessary communication. But there are customizable

IT-systems available for resource planning and process dispatching which seem to solve the problems. However often the true potential of these systems is not recognized – finally introduction of ITC is a good option to add value to the company in order to cope with the industry's development.

1.4 Initial Status and Objectives of the Management Game

Fraunhofer IAO and Stuttgart Airport's consulting branch Cost Aviation established a close cooperation for consulting projects in the aviation context dealing with issues like improvement of flexibility, introduction of ITC systems, process re-engineering, company organization, developing KPI systems and customer orientation.

Within these consulting projects the management game approach shall generate awareness and positive attitude towards the introduction of new procedures and technology either before project kickoff or as a tool for implementation. (Please note, that in the following paragraphs the words management game, business game, and simulation game are used as synonyms.) Goal of the management game is to simulate basic aviation ground handling services according to a predefined flight schedule with iterative stages of expansion, where employees involved in these very tasks in real life are players within the game. The business game will therefore deliver an experimental learning environment for those employees and pave the way for implementation of new procedures and/or technology.

2 THE MANAGEMENT GAME METHOD

2.1 Management Game Theory and Value Added

The history of modern management games dates back to 1955, when the RAND Corporation developed a simulation game for the U.S. Air Force (Faria, 1998). Since then the number and variety of business games grew rapidly, due to the fact that the suitability and potential of the method as training and learning-method has been discovered. Simulation games are tools for simulation of situations that need planning, in order to better understand, experience and assess them (Blötz, 2002). They simulate courses of action and decision processes within organizations and deliver a model for fast, riskless, and cheap experimentation (Schweizer and Rally, 2006; Lu, 2008). Simulation games are used to convey knowledge, methodological and social competencies. They simulate a real world situation for decision-making or alternative evaluation and allow the player to experience cooperation and teamwork (Deshpande and Huang, 2008). The participant is part of a system and recognizes his dependencies, so that customer-supplier relationships between different divisions are becoming visible (Schweizer and Rally, 2006).

Our developed simulation game is a haptic simulation game. They are based on objective models to touch and are often called board games (Schweizer and Rally, 2006); in our case the haptic element is represented by LEGO bricks. Furthermore, it is made up of different phases/scenarios (iterations, cycles) which can be found in most business games. Those phases include changes within the simulation setting, such as new tasks or methods and result from the development and learning successes of the players. In our case the consecutive phases represent process innovations, unlike other games, where phases mostly represent time progress.

Simulation games are an adequate alternative to traditional methods for teaching theoretical content. Some of the advantages of simulation games are: they are a method of organized experiential learning that incorporates an element of fun in the learning process, they possess the ability to alter attitudinal positions when designed in accordance with theory (Williams, 1980), they open up dynamic participation that lessens resistance to accept innovative ideas and concepts (Petranek, 1994), they give immediate feedback to the participant making the interaction with the game a learning process rather than an evaluation process (Crown, 2001), they provide participants an opportunity to face the consequences of the results of the decision making process applied and not just be an observer (Torres and Macedo, 2000), and they show greater retention over time than the traditional classroom instruction (Randel et al., 1992).

Simulation games are particularly suitable in adult education, as they are not as much used to frontal instruction as students. In our business game, where participants are practitioners, they need and benefit from a very close relation of the simulation to real situations within their work environment. Simulation game participants do not only improve their professional competence, but they also train abilities like methodological competencies, social competence, self-learning and self-reflection competence (Windhoff, 2001)

2.2   Learning Targets and Target Groups

With the help of management games, practical learning shall take place by accessing existing expertise and simultaneously implementing new learning experiences (Schweizer and Rally, 2006). Investigations have shown that 75 percent of the imparted content of business games is memorized enduringly, while only 39 percent are achieved by traditional learning methods (Schweizer and Rally, 2006).

According to Blötz, 2002 there are four different learning targets of management games: objective actions such as planning, decision making, debriefing, analysis, and construction; social actions such as debating, presentation, and conflict-solving communication; objective experience such as experience of interdependencies and relationships between different matters, and experience of the behavior of the system over some development periods; and social experience such as experience of personal and foreign behavior in objective and social situations. The achievement of these learning targets depends on some factors. First of all, the management game method has to be accepted by the participants, as they have to bring in personal commitment. Furthermore the effectiveness of the game will depend largely on the quality of the simulation in representing the behavior of the real world (Lu, 2008). One should also take into account that decisions are reached by a group process (Lu, 2008), thus the team work and interaction of the group influences the learning process. Additionally, it is essential that participants receive a thorough introduction at the beginning, so that they understand the game background, rules and data; otherwise no significant conceptual or strategic management learning can take place (Knotts and Keys, 1997).

The target groups of management games are usually ether students (within an educational environment) or practitioners (within a working environment). In our case the target group will be employees in ground aviation services at middle management level with some executive functions and managerial responsibility. The business game is especially suitable for new hires that come mostly from particular functional areas and lack the overview of all aviation ground services and their interdependencies.

3 MANAGEMENT GAME CONCEPT

3.1 Process Simulation and Layout

The business game simulates basic aviation ground processes and jobs associated with the landing and departure of aircrafts after a predefined flight schedule. Those services are: check-in, security controls, gate, luggage distribution, and carrier, bus service, loading team, ground power units, push back, fuelling, and planning and event processing. Those tasks are carried out by one or two players within the game according to a specific set of rules and timetables. Passengers, luggage and devices or resources (e.g. fuel) are simulated with LEGO bricks. For instance one passenger entering the system is composed of up to eight bricks: a 2x6 brick in the color of the booking class, a 2x6 brick in the color of the airline, a 2x2 brick in the color of the flight, a 2x2 brick standing for the carrying of hand luggage or laptop and/or fluids, a 2x4 brick standing for a piece of baggage that has also got two brick resembling the airline and flight on top. With all these information the passenger is assignable to a flight event and processed through the system.

The playing field is divided into several functional areas shown by different tables: the check-in counter, security, gate, baggage distribution, laptop control, equipment hall, fuel depot, and aircraft positions 1 till 4. All passengers and flight events are processed at those different stages according to FIFO principle.

One game turn lasts 30 minutes, where passengers enter the system, flights come in, aircrafts are serviced, and airplanes take off again. The underlying flight schedule includes fluctuations in passenger arrival and flight arrival as they occur in real life. Tests have shown that an optimal number of players are 12, distributed in the following roles: two check-in players, two players at security check, one player at the gate, two bus drivers, one baggage driver, two baggage loaders that also operate the power ground unit, and one push back player. Each play round starts with an introduction, is followed by the actual 30 minutes of playing (simulation), and concluded by a debriefing. During the simulation a video camera shoots the action, which can be useful in the debriefing, so that participants can have a look at what happened at the other tables, when they were busy fulfilling their tasks.

394

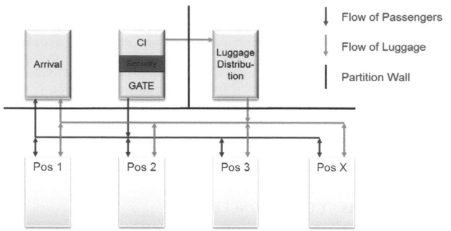

Figure 1: Game Layout

3.2 Game Phases and Course of Action

The business game has been designed as a basic setup on which several game scenarios are based upon. The development of the basic setup took into consideration that a representative selection of aviation ground services shall be existent by turning attention to the fact that the complete process chain of arriving and departing passengers/aircraft can be reproduced and that one sample of all kind of process characteristics are represented (for example fuelling where the physical filling level of the truck must be considered). The selection of processes had to be satisfying to drive four to five scenarios and its variations of the production processes. At the basis scenario, participants become acquainted with the business game and simulate all necessary ground aviation processes (except for event planning) via a fixed flight schedule by using a low level of technology. Quality of aircraft service is intentionally low in the beginning, so that participants clearly realize the need for process improvements.

Within the five game scenarios participants shall implement process improvements and realize interdependences and their effects, when procedures are changed. Each scenario procedure is designed as a plan-do-check-act (PDCA) cycle, while planning will be carried out at the beginning (introduction stage), doing happens during the simulation, checking is carried out at the debriefing, and action is done in the subsequent scenario. The group work between simulations will also be used to deliver some theoretical background and to introduce new scenario features (e.g. the use of handheld devices, or advanced rules and tasks). The contents of the five game scenarios are:

Scenario 0: Initial situation; can be little worse than actual status quo at the respective airport. The scenario shall generate acceptance. Players should be able to identify themselves with the processes, and realize mistakes and/or problems.

Scenario 1: Introduction and implementation of a planning process, as well as first methods for the transition to flexible shift lengths and multi-qualification.

Bottom line realization: effectiveness and quality of production is obviously bad, but there are no objective data available.

Scenario 2: Implementation of new order rules which are delivered by paper system, so production becomes measurable by evaluating the backflow afterwards. Bottom line realization: data analysis is helpful for planning, but the paper system is a time killer.

Scenario 3: Introduction of an IT system for automated planning and steering of resources. Bottom line realization: IT has to deliver a contribution to saving resources; this can be done by an effective reporting system which leads to improved management decisions.

Scenario 4: Introduction of customer information for a customer analysis in order to improve the satisfaction of customer needs (in this context customers are airlines) and/or to improve resource allocation with cost savings. Behavior patterns must be identified and procedures adapted accordingly. But individual treatment causes customer related operating expenses as well. Therefore reporting and accounting must be enhanced. With the help of customer segmented profit and loss statements the identification of contracts which have to be adjusted to reach the breakeven point is possible. Bottom line realization: customer orientation – different customers have different needs – therefore they must be treated individually.

The improvements of the different scenarios shall guide participants through a learning-by-doing process that highlights various methods for process innovations that deliver a better outcome.

3.3   Requirements and Roles of ICT

In order to develop a management game that depicts real aviation ground services and processes, an integration of ICT functions into the haptic gaming world is a key factor, as ICT solutions are common in this branch promising huge optimization potentials. This part of the management game should sensitize the participants on how to generate added value by using ICT solutions. For an optimized experimental learning the role of ICT in the game underlies a continuous evolution from scenario to scenario, taking up past and on-going developments in the aviation branch.

In the early game phases the role of ICT is reduced on publishing flight schedule data via screen or video projection in real-time to all participants, illustrating the branch standard in terminals or ground handling dispatch centers. Characterized by one-way communication without feedback channel, the added value is that all actors base their individual actions on the same level of information.

But as personnel costs are mainly determined in the planning stadium this is only a small improvement. In later phases ICT supports a detailed simulation in the planning phase, for an optimized resource management based on the demand to minimize costs, supplemented by a real-time process dispatching with mobile devices, tablet computers, mobile phones, or laptops as used in real operations. Due to the two-way-communication, tasks can be assigned to individual participants with a real-time feedback of timestamps, allowing now a process evaluation and monitoring afterwards. In an interactive learning-by-doing environment the participants can experience the effectiveness and potentials of a feedback loop between planning and dispatching to realize sustainable improvements.

Supporting the evaluation phase and making the benefits of change projects visible ICT is furthermore the base for the PDCA cycle as the core of all gaming phases. All production processes and provided services can be surveyed by a meaningful set of KPI's, indicating if the target of being a high quality and profitable ground handling service provider can be reached.

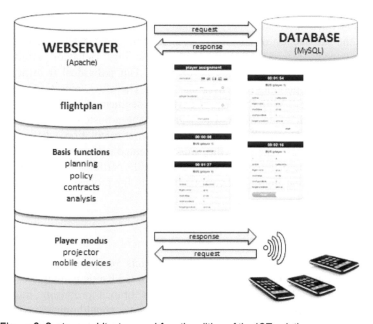

Figure 2: System architecture and functionalities of the ICT solution

ICT shall intentionally not dominate but support the haptic business game. The focus during the development was therefore an intuitive understanding and usage for a real hands-on-experience and a successful knowledge transfer. The developed solution is based on a web server with a MySQL database hosting a platform independent PHP web solution. Consisting of a laptop and an access point for the mobile devices the ICT gaming package is completely portable.

3.4   Implementation Considerations

The business game has been designed for the use within middle management levels, as a certain level of professional knowledge and ICT competencies is necessary. If it is to be used within operational levels, adaptations have to be made. Furthermore, assigning players to roles that they are not carrying out in the real world enhances the learning experience of the players, as they have to slip into another role.

Since the use of ICT is significant and essential within the game it is absolutely necessary to assure that all technology is working properly. Also the room where the management game takes place has to be big enough, so that players generally don't handicap each other when they operate between game stations (tables).

Another important factor that came up during game tests is the technique players use to disconnect LEGO bricks. It is far faster to bend it to the side and remove it, than trying to pull it off straight, which can curse enormous delays at the gate for instance. Also strongly recognizable in the tests was the learning curve of the players in every round, as they get more and more familiar with the layout, the haptic elements, and game rules, when scenario zero was played several times in a row. This effect will be slowed down during a real simulation game by the fact that every new round has new features as mentioned in chapter 3.2 that have to be understood and integrated.

4   CONCLUSIONS

4.1   Results and Management Implications

First of all, the developed management game can give all participants an overview of basic aviation ground services and how they are linked together. It integrates middle management employees when it comes to identification of potential for increase of productivity and implementation of change processes. Participants learn and try out different production methods and process innovations, as well as technology improvements that can result into higher productivity and increased service quality within the aviation ground handling processes. During a step by step phase approach, different scenarios focus various aspects like: advanced ICT, transition to flexible working hours, and customer orientation.

The most benefit of the business game can be achieved, when it is used to prepare employees for the introduction of change processes as a first step or door opener in a consulting project. The game is able to generate awareness for the need for consulting and gain acceptance from relevant people, since concepts are introduced as needed, and improvements to the system can be seen, measured, and documented by all participants (Baburdeen et al., 2010). Within the consulting project it is very important to clearly link the activities following the management game to the methods and outcomes of the simulation game, so that the participants see the importance of their gained knowledge within the game, since those very methods and innovations are then transferred to the real world.

4.2   Prospects

The management game is still in the development stage, as the further design of the scenarios is now being defined and tested. Within these scenarios, the basic setup and ICT part gained the highest progress so far. Playing with the use of mobile devices is already possible. One basic step also remains regarding the development of a continuous and complete storyline for the 2.5 days lasting management game training. This includes the distribution of the game phases, reviews, general feedback, breaks and team building aspects. Group work phases and whiteboard presentations have to be designed to complete the PDCA cycle and to impart additional knowledge which cannot be included in the management game. In addition, play equipment and instructions should be improved regarding a professional and appealing design

Further work has also to be done regarding the design of a flight schedule for different group sizes, as it is the most important trigger of player workload. That way the game can be adapted to different organizational needs. These have to be tested for the individual phases. Additionally, the whole simulation game, with all rounds and their introduction, simulation and debriefing phases, has to be tested and validated as well.

REFERENCES

Baburdeen, F. and Marksberry, P. et al. 2010. Teaching Lean Manufacturing with Simulations and Games: A Survey and Future Directions. *Simulation and Gaming* 41: 465-486.

Blötz, U. (eds.) 2002. Planspiele in der beruflichen Bildung, Vol. 2. Bielefeld: Bertelsmann Verlag GmbH & Co. KG.

Crown, S. W. 2001. Improving visualization skills of engineering graphics students using simple JavaSpript web based games. *Journal of Engineering Education* 90: 347-355.

Desphande, A.A. and Huang, A.H. 2008. Simulation Games in Engineering Education: A State-of-the-Art Review. In. *Computer Applications in Engineering Education* Vol. 19 No. 3: 399-410.

Faria, A. J. 1998. Business Simulation Games: Current Usage Levels – An Update. In. *Simulation and Gaming* Vol. 29 No. 3: 295-308.

Knotts, U. S. and Keys, J. B. 1997. Teaching Strategic Management with a Business Game. In. *Simulation and Gaming* 28: 377-384.

Lu, L. 2008. Overview of business games. Paper presented at International Conference on Wireless Communications, Networking and Mobile Computing, WiCOM 2008, Dalian, China. Art. No. 4681098.

Petranek, C. 1994. Amaturation in experiential learning: Principles of simulation and gaming. *Simulation and Gaming* Int. J. 25: 513-522.

Randel, J.M. and Morris, B.A et al. 1992. The effectiveness of games for educational purposes: A review of recent research. *Simulation and Gaming* 23: 261-276.

Schweizer, W. and Rally, P. 2006. LIFE! – Haptische Planspiele im Industrieeinsatz. In. *Technologiemanagement in der Praxis*, eds. D. Spath. Stuttgart: Fraunhofer IRB Verlag, 91 – 97.

Torres, M. and Macedo, J. 2000. Learning sustainable development with a new simulation game. *Simulation and Gaming* 31: 119-126.

Williams, R.H. 1980. Attitude change and simulation games: The ability of a simulation game to change attitudes when structured in accordance with either cognitive dissonance or incentive models of attitude change. *Simulation and Gaming* 11: 177-196.

Windhoff, G. 2001. Planspiele für die verteilte Produktion. Aachen: Mainz 2001.

CHAPTER 40

Incorporating Kano's Model and Markov Chain into Kansei Engineering in Services

Markus Hartono[#], Tan Kay Chuan*, John Brian Peacock*

*Department of Industrial Engineering, National University of Singapore
Singapore, Singapore
[#]Department of Industrial Engineering, University of Surabaya
Surabaya, Indonesia
markushartono@nus.edu.sg; markus@ubaya.ac.id

ABSTRACT

Nowadays, customers concern themselves more on fulfilling their emotional needs/Kansei instead of focusing only on functionality and usability. Products and services need to be attractive, delightful and appealing to consumers' emotions. In dealing with this, Kansei Engineering (KE) has been applied extensively. KE is useful in several regards. The first is its ability to translate customer emotions into concrete product/service design parameters. The second is its capacity to optimize properties that are not directly detectable or visible. The third is its flexibility to grasp and accommodate 21st century's trends including hedonism, pleasure and individuality.

This study focused on attractive attributes of service quality as the drivers of customer delight and loyalty. Kano's model is used to exhibit the relationship between service attribute performance and emotional response. Customer preferences change over time. This study developed a means to respond to these changing needs. Markov chain can be applied towards this end.

This study provides an integrative framework. It has two objectives. The first is to conduct a survey of luxury hotel services. Singaporean and Indonesian tourists served as the subjects. The second it is to enable service designers to prioritize their customer service improvement programs. A comprehensive interview and survey involving 181 Indonesian and 170 Singaporean tourists who stayed at luxury 4- and 5-star hotels was carried out. Luxury hotels were chosen since they focus much on delighting customers.

A finding of this study shows the following three service attributes to be important: i) "the outdoor environment is visually clean", ii) "the employees are never too busy to respond to your requests" and iii) "the employees are consistently courteous with you". Subjects rated the attribute "the employees are never too busy to respond to your requests" as the most important. A house of quality (HOQ) was used to illustrate this. This study determined that the proposed improvements in response to this attribute are related to personnel management, general affair management, employee training, complaint responses and information services.

This study offers several contributions. First, the results can be used as a prioritization tool in service quality improvement efforts where resources are limited (e.g., limited budget and time). Second, guideline for practitioners can be constructed to determine service attributes that are significantly sensitive to customer delight. Third, with the use of Markov chain, practitioners can be provided with information to understand the dynamics of customer needs over time and to prepare appropriate response strategies early.

Keywords: The Kano model, emotions, Kansei Engineering, Markov chain, services, house of quality

1   INTRODUCTION

In today's competitive business, it is imperative for companies to provide innovative and differentiated products and services.Competitive pricing, performance and features have become relevant factors in deciding which products to buy (Schifferstein and Hekkert, 2008). Products and services, therefore, need to offer features which can make them distinguishable and attractive to customers.

The sale of many products tends to decrease over time as compared to their first launch. Efforts such as quick model changes and even price reduction have been taken into account to reduce turnover. However, these are no longer sufficient solutions (Shaw and Ivens, 2002; Schütte et al., 2004). This condition indicates a lack of understanding of customer needs.

Designing for customer is increasingly important for the success of new services and products. Companies must listen closely and carefully to the voices of their customers, especially their latent needs. These are the unspoken emotional needs that customers seek in products and services. According to Nagamachi and Lokman (2011), good quality products are not enough to win market competition. However, products and services that consider peoples' feelings and emotions will sell (Hartono and Tan, 2011a; 2011b).

The focus of customers has shifted from objects to product and service experience. This refers to the switch from functionalism to product semantics and emotions (Krippendorff, 1995). In dealing with customer emotions in product design and development, Kansei Engineering (KE) has been proposed since the 1970s. KE is conceived as the only tool that can quantify customer emotions and translate them into product properties (Nagamachi, 1995). Its applications cover both product design and service quality improvement (Nagamachi and Lokman, 2011). Recent research (see Hartono and Tan, 2011a; 2011b) has extended the application of KE into international-class services and cross-cultural studies.

In today's competitive business environment, to satisfy customers is not sufficient. More importantly, companies need to focus on how to delight their customers beyond expectation (Hartono and Tan, 2011a). According to Yang (2011), there would be a significant benefit for a company that always fulfills the unexpected and memorable experiences of customers.

Basically, customer satisfaction is positively and linearly related to improved product or service features. However, in several cases, the relationship between product or service quality and customer impression is no longer linear (Kano et al., 1984; Chen and Chuang, 2008). As a consequence, radical improvement on products or services might not be sufficient. In dealing with this, the Kano model has the potential to eke out the latent customer needs, the satisfaction of which may lead to customer delight. Hartono and Tan (2011a; 2011b) have revealed the potential of the Kano model in KE applied to services. Their study presented the integrative framework of the Kano model and KE. By focusing on the Kano attractive service quality, their work offered competitive benefits in terms of prioritized improvement and efficiency. The Kano attractive category is found to have the greatest influence on customer Kansei and being an important tool in KE (Hartono and Tan, 2011a; 2011b).

Traditional approaches to product and service designs focus on present customer needs. However, in order to compete effectively in the long-term, a company should take into account as well future customer needs. The future voice of the customer (VOC) may enable companies to provide better services and products so that they can delight customers effectively (Wu and Shieh, 2006).Studies on future VOC have been done by several scholars, such as developing fuzzy trend analysis (Shen et al., 2000) and applying grey theory to analyze VOC combined with HOQ (Wu et al., 2005). Their studies have been extended by Wu and Shieh (2006) through incorporating Markov chain into the HOQ.

Apart from the Kano model, this study utilizes Markov chain incorporated into KE to analyze the dynamics of customer preference from a probability viewpoint. By using Markov chain, the importance of customer requirements can be further adjusted based on new information from a survey (Wu and Shieh, 2006).

This study has two objectives. The first is to conduct a survey of luxury hotel services involving Singaporean and Indonesian tourists. The second is to present a framework for service designers in prioritizing their improvement programs. This paper is organized as follows. Following the introduction, a brief review of Kansei Engineering, the Kano model and Markov chain, is presented. It is then followed by the framework development and case study. Thereafter is a discussion section. Conclusions and recommendations wrap up the paper.

2 LITERATURE REVIEW

2.1 Kansei Engineering in services

The trend of the 21st century is in hedonism, pleasure and individuality (Nagamachi, 1991, 1995, 2002b). Such notions stimulate customers to change their concern on emotional experience rather than functionality and usability (Helander, 2003). To accommodate customer emotional needs and experience, Kansei Engineering (KE) has been intensively used (Nagamachi, 1995).

Basically, the KE methodology is useful in several regards (see Hartono et al., 2012): i) KE is able to translate customer emotions into concrete design parameters through engineering aspects (Nagamachi, 2002a; 2002b), ii) KE tries to minimize subjectivity by building a mathematical model between emotional responses through all the human senses and their respective external stimuli, iii) KE has demonstrated a relationship model between cognitive and affective experiences, and iv) an internet KE system can work as a catalyst for innovative ideas during product design process (Ishihara et al., 2005).

In complex circumstances, customersexperience both mixed physical and non-physicalobjects. For instance in a restaurant, customeremotions may be influenced not only by the cleanlinessof meals and other physical/tangible stuffs, but alsoby the friendliness of the staff, accuracy of bills andprompt service. KE has to be capable of conductingexamination of both physical products and services in a singlestudy (Schütte et al., 2004; Nagamachi and Lokman, 2011).

2.2 Kansei Engineering and the Kano model

The Kano model categorizes customer attributes into the following three different types (see Kano et al., 1984): must-be (M), one-dimensional (O) and attractive (A). Basically, a must-be (M) is related to something taken for granted by customers. The absence of it will cause significant dissatisfaction while the presence of it will not give any significant impression. A one-dimensional (O)attribute serves the linearity relationship betweencustomer satisfaction and performance of the attribute.The better the performance, the higher the level ofcustomer satisfaction is. While the attractive/delighting (A) attribute is beyond customer expectation. A little fulfillment on it brings a great deal of impression. Free wireless ultra-speed internet access in a hotel may serve as attractive attribute.

In identifying which Kano category a particular service attribute falls under, the Kano questionnaire is used (Kano et al. 1984). A subject is faced with two Kano situations. The first is where the quality of the service attribute is present or functional. The second situation is where the quality of the service attribute is absent or dysfunctional. By integrating these two responses, the service attribute criterion can be identified as attractive (A), must-be (M) or one-dimensional (O).

The integration of the Kano model with KE have been initially proposed by Llinares and Page (2011) and Hartono and Tan (2011a; 2011b). According to Hartono and Tan (2011a; 2011b), the Kano categorization incorporated into KE offers competitive benefits in terms of prioritized improvement and efficiency. The Kano attractive attribute was found to have the greatest influence on customer Kansei. In addition, it was the first step of prioritization of service quality improvement with regard to limited resources (e.g., budget, time, workforce and other technical reasons). A practical contribution is presented by providing a guideline for service providers in investigating which service attributes are significantly sensitive to customer delights and given a priority for improvement.

2.3 The Markov chain model for future customer preference

Markov chain is useful for monitoring the evolution of systems over repeated trials where the state of a system in any particular period is uncertain. The Markov process assumes that a system starts in an initial state, but the initial state will change over time. A special model of Markov process, Markov chain, is then used to study the short- and long-run behaviors of certain stochastic systems (Taha, 1997).

Markov chain is a model with a regular transition probability matrix. It is introduced as a model of customers' performances and preferences in the long-term period or steady-state such that a better strategy can be made based upon the most updated customers' surveys. In order to promote the long-term behavior of a system like customers' preferences to certain services, a limiting distribution and a stationary distribution are used. Let define n step transition probability P_{ij}^n to be the

process in state i(e.g., customers' preference in i) will be in state j(e.g., customers' preference in j) after n additional transitions. Let P denotes the matrix of one-step transition probability P_{ij}which shows the transition probabilities from state i to j (Wu und Shich, 2006):

$$P = \begin{bmatrix} p_{00} & p_{01} & p_{02} & p_{03} & \cdots \\ p_{10} & p_{11} & p_{12} & p_{13} & \cdots \\ p_{20} & p_{21} & p_{22} & p_{23} & \cdots \\ p_{30} & p_{31} & p_{32} & p_{33} & \cdots \\ \vdots & \vdots & \vdots & \vdots & \end{bmatrix} \qquad (9\text{-}1)$$

$and \sum_j P_{ij} = 1$. In the long-term behavior, it is defined as follows:

$$\pi_j = \sum_{i=0}^{\infty} \pi_i P_{ij}^n, \; j \geq 0 \Leftrightarrow \pi = \pi P \qquad (9\text{-}2)$$

$$\sum_{j=0}^{\infty} \pi_j = 1 \qquad (9\text{-}3)$$

where π is called a stationary (steady state) distribution for P.

3   FRAMEWORK DEVELOPMENT AND CASE STUDY

Essentially, there are two types of service attributes, i.e., physical and non-physical attributes. Whatconstitutes the physical part may be referred to as'servicescape' (Bitner 1992) or the 'tangible'dimension of the SERVQUAL model(Parasuraman et al. 1988). Lin (2004) classified the tangible attributeinto several parts, such as visual cues (e.g., colour,lighting and layout), auditory cues (e.g., musical andnon-musical sounds) and olfactory cues (e.g., scents,ambient odours). Regarding non-physical attributes,this includes the application of specializedcompetencies (knowledge and skills) throughprocesses, activities and interaction (Lovelock 1991,Vargo and Lusch 2004).

This study uses 39 service attributes and 16 Kansei words used by Hartono and Tan (2011a; 2011b) and extends their study by incorporating Markov chain. Empirical data were collected through a field study involving 181 Singaporean and 170 Indonesian tourists who stayed in luxury hotels. A face-to-face questionnairewas used as the primary means of data collection. The questions were pre-tested byseveral experienced participants and a Kansei expert to increase the face validity of theresearch instrument.

This study starts with Kansei evaluation. It begins with the analysis of its structure. The exploratory factor analysis (EFA) was performed. It aims toidentify the number of underlying factor structure without imposing any preconceivedstructure on the outcome (Child, 1990). The findings show that there were two mainlatent variables (factors) formed in 'Kansei' construct:i) servicescape-based Kansei (i.e., emotions caused by physical surroundings,such as happiness, cleanliness and elegance) and ii)interaction-based Kansei (i.e., emotions influenced by the interaction between the customerand the employee, such as welcomeness, satisfaction and friendliness).

The respondents were then asked to rate the importance, expectation and perception of the 39 service quality items using a five-point Likert scale. Thereafter, the Kano questionnaire was used to rate the service attribute performance (please refer to Hartono and Tan [2011a; 2011b] for the details of Kano categorization).

To ensure the constructs' validity and reliability, the proposed properties of the constructs were tested using confirmatory factor analysis (CFA). AMOS^{TM} version 16 was used to verify the factor structure of a set of observed measures. Overall, the finding demonstrated adequate validity and reliability (see Hartono and Tan [2011a] for details).

To begin the analysis for future customer preference, an integrative framework is provided (see Figure 1). The participants were then asked to evaluate the importance of each service attribute and how each service attribute would change in the near future.Initial probability was determined by the past choice of customer preference, while the transition/conditional probability was extracted from both the past and present customer preferences. In addition, the initial expected weight for each service attribute was computed. For instance, for SQA$_1$ (i.e., service attribute 1 in Indonesian group), the transition probabilities were:

$$P \begin{bmatrix} 0.13 & 0.74 & 0.13 \\ 0.35 & 0.53 & 0.12 \\ 0.5 & 0.25 & 0.25 \end{bmatrix},$$

where, for instance, $P_{21} = 0.35$ represents the 1-step transition probability of shifting from 'medium' to 'high' levels. It was assumed that the probabilities were stationary over time. Afterwards, the future condition/steady-state probability and expected weight of each service attribute were computed using Gauss-Jordan method and solved by Microsoft® Excel solver. For instance, the steady-state probabilities for SQA_1 at Indonesian group were $V_1 = 0.32$, $V_2 = 0.57$ and $V_3 = 0.11$. Hence, its expected weight became $5(0.32) + 3(0.57) + 1(0.11) = 3.41$.

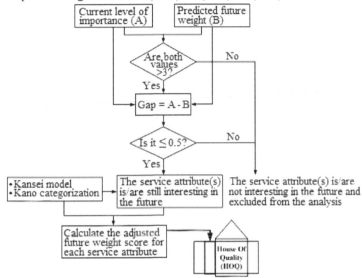

Figure 1 Integrative framework of future customer preference study

In order to analyze whether particular service attributes were still important in the future, their predicted steady-state weights were compared with their current importance values (see Table 1). It is called 'weight gap' and assumed that 0.5 is the threshold value. A service attribute is considered interesting/important if the weight gap is below or equal to 0.5. In addition, a medium level of importance (with a score of 3) was used for a complementary threshold. The findings show that the most preferred service attribute in the future was SQA_6 "the atmosphere of restaurant is inviting appetite" (in Indonesian) and SQD_4 "the hotel provides a safe environment" (in Singaporean).

Table 1 Comparison between current and future weight values

Item	Indonesia			Singapore		
	Current weight	Future weight	w-gap*	Current weight	Future weight	w-gap*
SQA1	4.00	3.41	0.59	4.03	3.16	0.87
SQA2	3.86	2.64	1.22	3.95	3.16	0.79
SQA3	3.98	3.28	0.69	4.16	3.32	0.84
SQA4	4.21	3.70	0.51	4.04	3.32	0.72
SQA5	4.08	3.57	0.50	3.97	3.41	0.56
SQE6	4.13	3.83	0.29	3.99	3.76	0.22

Note: *w-gap = current weight − future weight

Through an inclusion of only the interesting service attributes, we analyzed their importance incorporating the Kano categorization, Kansei model and house of quality (HOQ).This was then followed by the related technical responses for improvement strategy as adopted from a study by Chang and Chen (2011) on luxury hotel services. There are two steps in adjusting the importance of the 'whats'. First, we identify and choose service attributes which have significant relationships with particular Kansei. Second, we determine the scores for each service attribute (it is called 'adjusted future weight score') by incorporating expected importance weights, the Kano score and the number of Kansei words involved.According to Tan and Pawitra (2001), values of '4', '2' and '1' are assigned to the Kano (A), (O) and (M) qualities, respectively. This paper shows the HOQ for Indonesian group (see Figure 2).

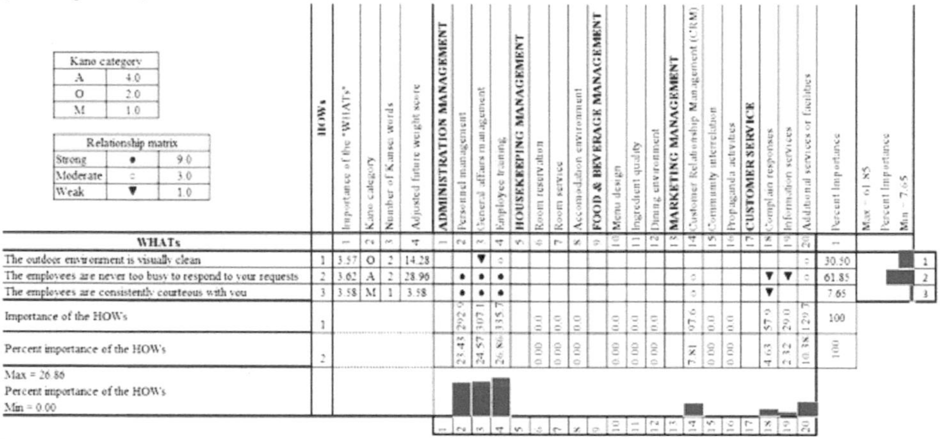

WHATs	Importance of the 'WHATs'	Kano category	Number of Kansei words	Adjusted future weight score	Percent Importance
The outdoor environment is visually clean	1	3.57 O	2	14.28	30.50
The employees are never too busy to respond to your requests	2	3.62 A	2	28.96	61.85
The employees are consistently courteous with you	3	3.58 M	1	3.58	7.65

Figure 2 HOQ in Indonesian group

4   DISCUSSION

This KE study introduces the Kano model with a focus on attractive qualities (i.e., delighters). This category is of interest to fulfill customer Kansei. According to Yang (2011), it provides several benefits such as to drive customer loyalty, differentiation and total customer delight. From a business perspective, Collins and Porras (2004) argue that it is essential to invest in proactive and generative markets which are reflected through customer delight. The other two common Kano categories (i.e., one-dimensional and must-be categories) are considered as primary features to be satisfied on a regular basis. Since these qualities are less sensitive and less satisfaction-driven (Yang, 2011) they are less unlikely to be associated with stronger emotions.

The more the Kansei are significantly influenced, the more important the service attributes are. In order to engage customer loyalty and satisfaction in the future, Markov chain model and HOQ were adopted. The research findings show that, for example, there were three service attributes important in the future for Indonesians, i.e., "the outdoor environment is visually clean", "the employees are never too busy to respond to your requests" and "the employees are consistently courteous with you". Using a modified HOQ,the service item "the employees are never too busy to respond to your requests" should be given the greatest priority for improvement. Improvement here relates to personnel management, general affair management, employee training, complaint responses and information services.

5   CONCLUSIONS AND RECOMMENDATIONS

The trend of the 21stcentury is hedonism and pleasure. Delighting customers is an essential key to achieving total emotional satisfaction and customer retention in services. KE has shown its ability to deal with customer emotional needs in products and services. The application of KE to hospitality services is recognized as Kansei quality management (KQM) (Nagamachi and Lokman, 2011).

This study offers some potential contributions. First, the results can be used as a prioritization tool in service quality improvement where resources are limited (e.g., budget, time, workforce and other technical reasons). Second, this study provides a well grounded theoretical contribution to the academic literature on Kansei ergonomics, service science, quality management and cultures. Third, by using the Markov chain model, it provides valuable information for managersand decision makers to understand the dynamics of customer needs as time goes by, so that appropriate strategies can be prepared at the early stage.

Future studies could test the framework applicability in different service settings (e.g., shopping centers, restaurants, airports and so on).

6.   ACKNOWLEDGEMENT

The research was carried out while the author, Markus Hartono, was a research scholar at NUS. This research work was supported by National University of Singapore, Singapore and University of Surabaya, Indonesia.

REFERENCES

Bitner, M.J. 1992. Servicescapes: the impact of physical surroundings on customers and employees. *Journal of Marketing* 56 (2): 57–71.
Chang, K.-C. and M.-C., Chen 2011. Applying the Kano model and QFD to explore customers' brand contacts in the hotel business: a study of a hot spring hotel. *Total Quality Management& Business Excellence* 22 (1): 1–27.
Chen, C.-C. and M.-C. Chuang 2008. Integrating the Kano model into a robust design approach to enhance customer satisfaction with product design. *International Journal of Production Economics* 114 (2): 667–681.
Child, D. 1990. *The Essentials of Factor Analysis*. London: Cassel Educational Limited.
Collins, J. and J. Porras 2004. *Built to Last: Successful Habits of Visionary Companies*. New York: Harper Collins Publishers.
Hartono, M. and K.C. Tan 2011a. How the Kano model contributes to Kansei engineering in services. *Ergonomics* 54 (11): 987-1004.

408

Hartono, M. and K.C. Tan 2011b. A proposed integrative framework of Kansei engineering and Kano model applied to services. *Proceedings of 2nd International Research Symposium on Service Management*: 484-492.

Hartono, M., K.C. Tan, S. Ishihara, and J.B. Peacock 2012. Incorporating Markov chain modeling and QFD into Kansei engineering. *To be published in International Journal of Human Factors and Ergonomics.*

Helander, M. G. 2003. Hedonomics-affective human factors design. *Ergonomics* 46 (13/14): 1269-1272.

Ishihara, S. et al. "Catalytic effect of Kansei Engineering system at collaborative design process". Paper presented at the 8th International Symposium on Human Factors in Organizational Design and Management, Maui, Hawaii, 2005.

Kano, K.H., et al. 1984. How to delight your customers. *Journal of Product and Brand Management* 5(2): 6–17.

Krippendorff, K. 1995. On the essential contexts of artifacts or on the proposition that "design is making sense (of things)". In. *The idea of design,* eds. Margolin, V. and R. Buchanan. Cambridge: MIT Press: 156-184.

Lin, I.Y. 2004. Evaluating a servicescape: the effect of cognition and emotion. *Hospitality Management* 23 (2): 163–178.

Llinares, C. and A.F. Page 2011. Kano's model in Kansei Engineering to evaluate subjective real estate consumer preferences. *International Journal of Industrial Ergonomics* 41 (3): 233-246.

Lovelock, C.H. 1991. *Services Marketing.* New Jersey: Prentice-Hall.

Matzler, K. and H.H. Hinterhuber 1998. How to make product development projects more successful by integrating Kano's model of customer satisfaction into quality function deployment. *Technovation* 18 (1): 25–38.

Nagamachi, M. 1991. An image technology expert system and its application to design consultation. *International Journal of Human-Computer Interaction* 3 (3): 267–279.

Nagamachi, M. 1995. Kansei Engineering: a new ergonomic consumer-oriented technology for product development. *International Journal of Industrial Ergonomics* 15 (1): 3–11.

Nagamachi, M. 2002a. Kansei engineering in consumer product design. *Ergonomics in Design – The Quarterly of Human Factors Applications* 10 (2): 5–9.

Nagamachi, M. 2002b. Kansei engineering as a powerful consumer-oriented technology for product development. *Applied Ergonomics* 33 (3): 289–294.

Nagamachi, M. and A.M. Lokman 2011. *Innovation of Kansei Engineering.* Boca Raton: CRC Press.

Parasuraman, A., L.L. Berry and V.A. Zeithaml 1988. SERVQUAL: a multiple-item scale for measuring consumer perceptions of service quality. *Journal of Retailing* 64 (1): 12–40.

Schifferstein, H.N.J. and P. Hekkert 2008. *Product Experience.* Oxford: Elsevier Ltd.

Schütte, S., J. Eklund, J.R.C. Axelsson, and M. Nagamachi 2004. Concepts, methods and tools in Kansei Engineering. *Theoretical Issues in Ergonomics Science* 5 (3): 214-232.

Shaw, C. and J. Ivens 2002. *Building Great Customer Experiences.* New York: Palgrave Macmillan.

Shen, X.X., K.C. Tan, and M. Xie 2000. An integrated approach to innovative product development using Kano's model and QFD. *European Journal of Innovation Management* 3 (2): 91–99.

Taha, H.A. 1997. *Operations Research: An Introduction.* New York: Prentice-Hall.

Tan, K.C. and T.A. Pawitra 2001. Integrating SERVQUAL and Kano's model into QFD for service excellent development. *Managing Service Quality* 11 (6): 418–430.

Vargo, S.L. and R.F. Lusch 2004. The four service marketing myths-remnants of goods based, manufacturing model. *Journal of Service Research* 6 (4): 324–335.

Wu, H.-H., A.Y.H. Liao and P.C. Wang 2005. Using grey theory in quality function deployment to analyze dynamic customer requirement. *International Journal of Advanced Manufacturing Technology* 25 (11-12), 1241-1247.

Wu, H.-H. and J.-I. Shieh 2006. Using a Markov chain model in quality function deployment to analyze customer requirements. *International Journal of Advanced ManufacturingTechnology* 30 (1–2): 141–146.

Yang, C.-C. 2011. Identification of customer delight for quality attributes and its applications. *Total Quality Management & Business Excellence* 22 (1): 83–98.

CHAPTER 41

A Comparative Analysis of Lean Techniques

Elizabeth A. Cudney, Ph.D.

Engineering Management and Systems Engineering Department
Missouri University of Science and Technology, Rolla, MO 65409, USA
cudney@mst.edu

ABSTRACT

The long-term health of any organization depends on their commitment to continuous improvement. The principles, practices, and techniques embodied within continuous improvement form a comprehensive organizational philosophy that strives to effectively fulfill customers' needs. Within continuous improvement efforts, learning is driven by the use of scientific methods (i.e., tools to monitor and analyze work processes such as control charts and Pareto charts). Within continuous improvement projects, various techniques (e.g., flowcharts, brainstorming, cause-and-effect diagrams, etc.) are used to help teams effectively utilize their collective knowledge to develop shared understandings and solve problems. Today, organizations often use Lean philosophies to improve operational performance and quality, which has been well documented across a variety of industries. This paper addresses the effectiveness of specific Lean techniques and tools based on the experiences and insights obtained through a survey administered to practitioners working in a variety of industries. The objective of this research is to assess and understand the performance of continuous improvement philosophies based on the specific techniques used, as well as the methods used to implement these techniques, and to evaluate the reasons they have or have not been effective. The results of this survey were analyzed using a variety of statistical techniques.

Keywords: Lean, Continuous Improvement, Manufacturing, Service

1. INTRODUCTION

Today, irrespective of the business domain, corporations must focus on speed, efficiency, and customer value to be globally competitive. Lean principles have enabled corporations to achieve significant economic benefits while improving quality, costs, and cycle time. Lean has its origins in the teaching and writings of Total Quality Management (TQM) and Just-In-Time (JIT). As discussed by Shah and Ward (2007), lean principles originated from Ford's practices, and beginning in 1937, Taiichi Ohno of Toyota studied Ford's original thinking in detail and invented the concepts and tools of the Toyota Production System (TPS). The origin and general evolution of the lean philosophy and some of the basic techniques first brought into use by the Toyota Corporation are discussed by Papadopoulou and Ozbayrak (2004). The so-called Toyota Production System (TPS), which includes the JIT philosophy and associated tools, is the original lean implementation.

To make the best of lean principles, the organization must look for new opportunities. Lean espouses the idea of "delighting the customer through a continuous stream of value adding activities (Womack and Jones, 1996)." Specifically, it is an extension of the idea of *"world class"* as defined by Schonberger (1986): "… adhering to the highest standards of business performance as measured by the customer." Lean as a philosophy therefore is not about just doing better than competitors (Hartwell and Roth, 2006); it is about going beyond and being the best in every process and product.

The lean approach is focused on the identification and elimination of waste in production, product development and service industries. Although lean principles were originally developed by Toyota for automobile manufacturing, they are increasingly being applied to businesses with many routine processes in support functions. Lean focuses on eliminating waste and improving flow using various proven methods. Lean is applied to improve the flow of information and material. Waste stems mainly from unnecessary delays, tasks, costs, and errors. The seven wastes of lean include overproduction, transportation, inventory, processing, waiting, motion, and defects. These wastes can also be applied to support functions such as procurement, engineering, invoicing, inventory control, order entry, scheduling, accounting, and sales.

A well-designed lean system allows for an immediate and effective response to fluctuating customer demands and requirements. Lean manufacturing tools that are most commonly used to eliminate waste and achieve flow are: value stream mapping, standard work, 5-S housekeeping, single minute exchange of dies (SMED), and visual management.

Lean implementation within an organization has shown considerable financial results. However, expanding the lean implementation to the supply base has shown to be more difficult and often haphazard. Part of the challenge is due to the fact that applying lean tools requires tailoring to different production contexts (Abdulmalek and Rajgopal, 2007) and also affects the social as well as technical systems of the organization. Lean implementation must be approached with both a context-specific viewpoint and a systems viewpoint that covers the extended supply chain. Based on the current status of lean implementation in industry, this research attempts to address several objectives in the realm of lean supply chains.

412

2. LITERATURE REVIEW AND BACKGROUND

Research on the implementation of lean thinking in support functions indicates that the lean philosophy is not restricted to the manufacturing industry alone and can be applied to any business domain. It is the awareness of the external and internal developments, which enables a company to incorporate changes and survive.

Implementing the lean philosophy is a continuing and long-term goal (Kindler et al., 2007). Kotter (2003) discusses eight steps to organizational transformation. The Army Enterprise Solutions Competency Center adopted the eight success factors that must be considered prior to initiating a change program. However, the change methodology suitable to one company may not be suitable to another (Roth, 2006). Some organizations learn as they implement, as was the case of Rockwell Collins. It moved through distinct phases from methods applied within departments to crosscutting activities using lean tools to better identify improvement opportunities to operating along value streams as a way of organizing and managing. This resulted in a deeper knowledge of lean methods (Roth, 2006).

Brown et al. (2006) noted that fear of losing jobs was a result of the Kaizen events. Therefore, lean was initially viewed as another downsizing method. Such misconceptions must be handled with care. Case studies at various organizations have revealed multiple enablers that have been used for the smooth implementation of lean. Lean implementation may not reach its intended purpose if there is a) small-term commitment, b) wrong focus, c) improper planning, d) lack of employee involvement, or e) inappropriate training methods and knowledge transfer. Sometimes, lean is not an ongoing engagement of workforce knowledge and creativity (Barrett and Fraile, 2005). If lean is only partially implemented, a study predicts that performance will languish. By starting new programs and changing direction without fully implementing lean, benefits will diminish from its change program investments (Roth, 2006). Also, in tight economic situations, a common corporate response is to restrict or cancel most scheduled training and development activities.

Lean can be implemented in the process industry just as it can be implemented in the manufacturing industry. Abdulmalek et al. (2006) argue that this has been very slow. They suggest that the process industry contains discrete portions at some point and that the specific challenge is to identify lean techniques from discrete manufacturing that can be adapted easily to the process sector. A classification of the process industry based on (a) product characteristics, (b) process characteristics, and (c) the point of discretization helps to understand whether lean can be applied to a specific industry (Abdulmalek et al., 2006). Based on the current literature, this research was developed to quantitatively measure lean technique implementation.

3. METHODS

A survey was created to collect data relevant to lean techniques implemented, success rates for industrial organizations who implemented lean, and also to gain insight on suppliers' role in lean techniques in the supply chain. The survey was distributed online via Zoomerang.com to informed individuals of lean philosophies and techniques. Members of the Lean Division of the Institute of Industrial Engineers (IIE) and also select individuals from industry, who are active in the field of lean through conferences and other outlets of lean information sharing, were asked to participant in the survey. Student members of the Lean Division of IIE who responded to the survey were not included in survey analysis to ensure responses were individuals who were experienced with lean techniques in an industrial setting. To prevent redundancy in responses through these two avenues of respondents, the survey was restricted so that each respondent could respond to the survey only once. Incomplete survey responses were also not included in this analysis. This analysis includes responses from 75 respondents from both pools the survey was distributed to. Using a standard sample size calculation (z=0.075; s=1.15, e=0.20), approximately 69 respondents are needed for a valid sample size. Since a total of 75 respondents participated in this survey, a valid sample was obtained for this study.

The distribution of industries making up the data set used in this study is shown in Figure 1. The manufacturing industry represented the majority (37%) of responses, followed by the Consulting industry (14%). The manufacturing industry is perhaps one of the largest implementers of lean philosophies and techniques (Hartwell and Roth, 2006).

To ensure the credibility, and familiarity with lean philosophies, of the individual survey respondent representing the industries in Figure 1, the respondents were asked to provide their individual functional work areas. The majority response was "engineering" (25%), followed by "consulting" (22%) and "operations/production" (15%), as shown in Figure 2. These three functional areas typically are quite familiar with lean philosophies and make up approximately two-thirds of the respondent pool. The respondents were also asked to provide their occupational title within the organization. Thirty-two percent of the respondents indicated they were a manager or assistant manager while fourteen percent indicated they were a CEO, president, owner, general manager, or partner, as shown in Figure 3. This question was asked in order to ensure that the level of involvement within the organization for each respondent would ensure they would be actively aware of any lean philosophies or techniques being employed in the organization.

Figure 1: Distribution of Respondent Industries

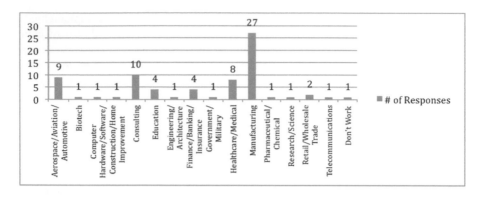

Figure 2: Individual Respondent Work Areas

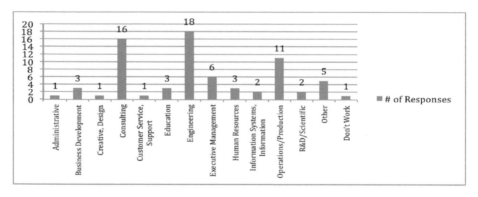

Figure 3: Individual Occupational Title

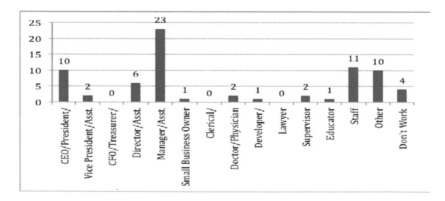

Figure 4 represents the length of time each respondent has spent working with lean philosophies during their career. The majority of the respondent population (62%) has spent more than 3 years working with lean philosophies. In contrast, a small portion (5%) of the survey respondents had no direct experience working with lean but were familiar with lean topics. These participants were retained in the results to provide a representative sample in quantifying perceptions on lean. In addition, the years of experience were utilized rather than the participants work title since highly experience individuals may be relatively new to lean but have an executive level title. This solidifies the credibility of the respondent population in responding to the remaining questionnaire about successes and failures with lean philosophies and how lean has affected their relationships with suppliers.

Figure 4: Respondents' Length of Time Working with Lean

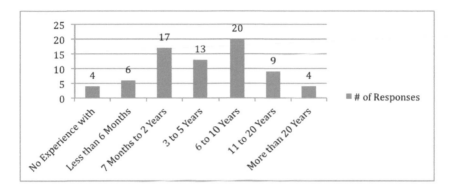

In order to further understand the organization's extent at implementing lean, the participants were asked what lean techniques their organizations had successfully implemented. This research question investigated which lean methods experienced practitioners have implemented successfully, whether lean techniques have different success rates in production versus service environments, and what factors contribute to the successful implementation of these techniques. Figure 5 shows the percent of respondents that agreed/strongly agreed that their organization had successfully implemented a particular lean technique. Of the methods successfully implemented in the organizations represented by survey respondents, the most common, in descending order, were 5S/visual factory, value stream mapping, standardized work, kanban, and poke-yoke. More than 70% of the practitioners surveyed said that their organizations successfully implemented value stream mapping, 5S/visual factory, and standardized work. In addition, more than 50% indicated that poke-yoke and kanban were used successfully within their organizations. The techniques that the least number of practitioners said were successfully implemented within their organizations included heijunka (33%) and SMED (38%).

For comparison purposes, the survey results were re-examined for production and service operations separately. As also shown in Figure 5, production operations had double the success rate compared with service environments in terms of the successful implementation of several lean techniques, including value stream mapping, 5S/visual factory, kanban, and heijunka. Within service operations, more than 40% of the practitioners surveyed indicated that their organizations successfully implemented 5S/visual factory and standardized work. In addition, more than 30% said that value stream mapping and poke-yoke were used successfully within their organizations. The techniques that the least number of service oriented practitioners indicated were successfully implemented within their organizations included heijunka (19%), SMED (19%), and kanban (26%).

Figure 5: Successful Lean Implementation by Industry Type

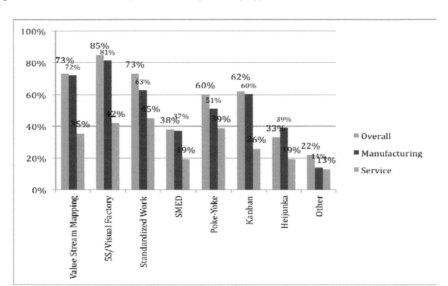

After establishing basic demographics regarding the respondents' area of employment, experience with lean philosophies, and size of organization, further exploration into the organization's involvement with lean philosophies and their suppliers was pursued.

In order to increase the participation in lean techniques by suppliers, organizations may provide training for them to lean their processes and ultimately provide more value and better pricing to the organization (Roth, 2006). Figure 6 reports the various types of lean technique training organizations provided to their suppliers. "No Training Provided" was the majority response, followed by "Value Stream Mapping". In terms of organizations providing training to their suppliers, the percentage is relatively equal for manufacturing and service industries. Value Stream Mapping (VSM) is a process for identifying all activities required to produce a product or product family from conception to launch, order to delivery, and raw materials to the customer. This is usually represented pictorially in a value stream map (Cudney, 2009). The purpose is to document the entire process (of a supply chain, company, or a department) on a single sheet of paper to encourage dialogue and understand the process better. VSM is often considered the foundation for implementation any lean technique. Therefore, it was expected to be the major tool provided to suppliers.

Outside of the organization providing lean training directly to its supply base, some organizations indicated that they had provided consultant services to their suppliers by a third party. Although the majority of the organizations indicated that they had not provided this service to its suppliers, some provided consultants in the areas of Value Stream Mapping and 5S/Visual Factory areas.

Figure 6: Lean Training Provided to Suppliers

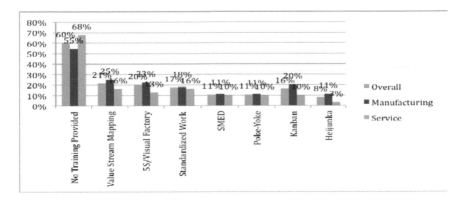

After training was provided to suppliers either by the organization itself, or via a third party consultant, the organization would like to realize some value or benefit experienced by providing the training. Figure 7 outlines the responses organizations had to the benefit of lean technique training. The majority of responses indicated "Significant Time Saving Benefits" as the largest benefit received, followed by "Significant Financial Benefits." The benefits experienced with approximately equal value between manufacturing and service were: significant financial benefits, insignificant or no financial benefits, significant time savings, and insignificant or no quality increase. However, manufacturing industries has over four times the response for insignificant or no time saving and significant quality increase. The respondents were also invited to expound on other benefits they perceived having received as a result of providing lean training. The common theme with these responses revolved around the benefit received being hard to quantify or that it was unknown to date.

The aforementioned data outlines a weak implementation and understanding of lean techniques both by the organization itself and their suppliers. However, some organizations did use methods to encourage their suppliers to start the use of lean techniques to benefit both parties. Some of the organizations used a method by which they dictate prices to suppliers in order to keep the organizations business with that supplier. This would force the supplier to adopt methods to reduce cost to keep business. Sharing savings is also another useful method used. This method says that if the organization helps the supplier implement lean techniques, they will share the savings gained from the leaner processes. Bonus process implementation is beneficial to some organizations in that if the organization helps the supplier implement lean techniques on certain production lines, those lines remain dedicated to that organization.

Figure 7: Benefits Experienced from Suppliers Leaning Processes

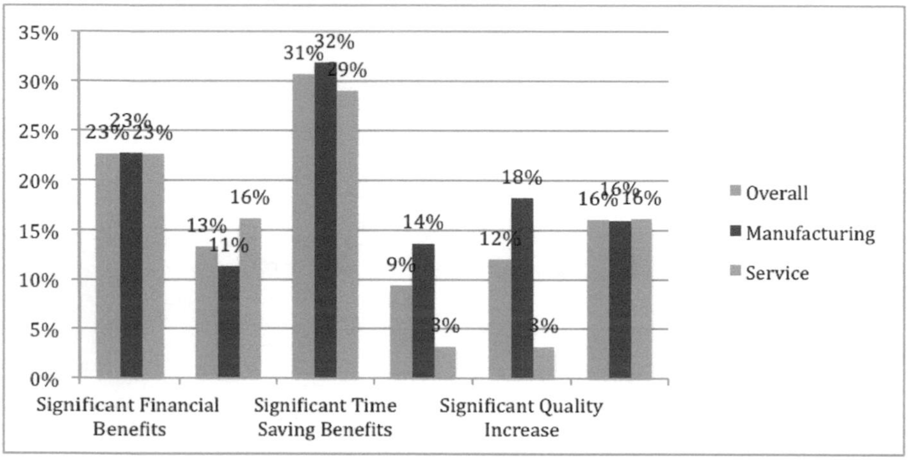

4. CONCLUSIONS AND FUTURE RESEARCH

Industries such as computer technology, construction, design and engineering, government and military, and finance are all areas in which the survey showed little or no lean involvement. Some of these industries (e.g. government) have a reputation for waste. It is here that lean advancement is needed to discover how lean techniques can be beneficial. With industries outsourcing products and services, it is a necessity for companies to extend their lean practices to their supply chain. As highlighted in the survey, the majority response was that organizations are not extending their lean training to their suppliers. The benefits gained from suppliers leaning their processes shown significant time and financial benefits. As expected this supports the need for organizations to expand their lean training offerings to their supplier base.

A more extensive survey with a larger respondent pool should be performed to provide more statistically significant evidence for the qualitative conclusions reached here. Additional information about lean experiences in industries with little lean involvement will be crucial to understanding how these industries can increase efficiency and reduce waste as other industries have done using lean philosophy. The area of lean implementation offers many avenues for further research. Development of a standard system for creating a lean company remains a high priority for continued research. The unique situation of every entity demands that this system will involve extensive research, including the examination of extensive case studies from pioneering businesses in many fields. Global markets and supply chains also affect the way a lean corporation does business. Cultural and language barriers introduce particular challenges for inclusively implementing lean techniques. These hurdles will complicate a lean roadmap and require the integration of lean skills and global management skills to facilitate the global understanding lean philosophy and encourage its acceptance.

REFERENCES

Abdulmalek, F.A., and Rajgopal, J., (2007) "Analyzing the Benefits of Lean Manufacturing and Value Stream Mapping via Simulation: A Process Sector Case Study," International Journal of Production Economics, 107, pp. 223-236.

Abdulmalek, F., Rajgopal, J., and Needy, K., (2006) "A Classification Scheme for the Process Industry to Guide The Implementation of Lean," Engineering Management Journal, Vol. 18(1), pp. 15-25.

Barrett, B., and Fraile, L., (2005) "Lean at the C-5 galaxy Depot: Essential Elements of Success," Lean Enterprise Change Research Case Studies.

Brown, C., Collins, T., McCombs, L., (2006) "Transformation from Batch to Lean Manufacturing: The Performance Issues," Engineering Management Journal, 18 (1), pp. 3-14.

Cudney, E., (2009) Using Hoshin Kanri to Improve the Value Stream, Productivity Press.

Hartwell, J., and Roth, G., (2006) "Case Study Rockwell Collins and IBEW Locals 1362 and 1634 Investing in Knowledge Skills and Future Capability in an Uncertain Business Environment," Lean Enterprise Change Research Case Studies.

Kindler, N., Krishnakanthan, K., and Tinaikar, R., (2007) "Applying Lean to Application Development and Maintenance," The McKinsey Quarterly, Number 3, pp. 99-101.

Kotter, J., (2003) "The Power of Feelings," Leader to Leader, Number 27, pp. 25-32.

Papadoupoulou, T.C. and Özbayrak, M., (2004) "Leanness: experiences from the journey to date", Journal of Manufacturing Technology Management, 16(7/8), pp 784-807.

Roth, G., (2006) "Rockwell Collins Lean Enterprise Change Case Study," Lean Enterprise Change Research Case Studies.

Schonberger, R., (1986) World Class Manufacturing: The Lessons of Simplicity Applied, The Free Press, New York, New York, pp. 38-46.

Shah, R., and P. T. Ward, (2007) "Defining and developing measures of Lean production," Journal of Operations Management, 25(4), pp. 785-805.

Womack, J., and Jones, D., Lean Thinking: Banish Waste and Create Wealth in Your Corporation, Simon & Schuster, New York, New York, 1996.

Womack, J.P., Jones, D.T., and Roos, D. (1990) "The Machine that Changed the World", Rawson Associates, Maxwell Macmillan, New York, NY.

CHAPTER 42

An Analysis of the Impact of Lean on Safety

Elizabeth A. Cudney, Ph.D., Susan L. Murray, Ph.D., and Pankaj M. Pai

Engineering Management and Systems Engineering Department
Missouri University of Science and Technology, Rolla, MO 65409, USA
cudney@mst.edu

ABSTRACT

The lean philosophy has proven potential to help businesses improve productivity and reduce its losses. Lean can give businesses a cutting edge in this age of global competition. The fundamental principle of lean is to identify wastes in the system and reduce or eliminate them. There is a concern that during lean implementations the focus on productivity may result in health and safety issues being ignored or worse, changes driven by lean may introduce new hazards. The relationship between lean and safety is not clearly understood. Lean and safety should be compatible. Both strive to improve processes. Both are against safety hazards and accidents; safety by definition and lean because the money spent on compensation claims is a waste.

When there is a passionate effort to lean processes, there is a danger that lean facilitators might overlook health and safety issues or even introduce new hazards. For example, it is possible that during an attempt to minimize cycle times that a redesigning of a process or workplace could result in protective machine guards being removed. This would compromise safety and potentially lead to an accident. Accidents bring with them the indirect costs including compensation claims as well as forced shut down of machines and processes. These are counter to the fundamental principle of lean, minimizing wastes in addition to everyone's general disdain of accidents. Lean and safety should not be viewed as having conflicting goals but should be addressed simultaneously. The integration of lean and safety can help companies achieve a competitive edge that is critical while providing a safe workplace. Despite the synergistic nature of lean and safety, researchers have found conflict or at least neglect to consider safety in lean implementation. The process changes associated with lean have an effect on safety whether related or not.

An online survey was conducted to gauge the effects of lean initiatives on safety and understand the level of integration of the two. Results have been provided in the lean areas for value stream mapping (VSM), one piece flow, material handling, and single minute exchange of dies (SMED). As lean and safety have the common goal of reducing wastes, there are natural opportunities where they integrate into each other.

Keywords: Lean, Continuous Improvement, Safety, Ergonomics

1. INTRODUCTION

Lean is a collection of ideas, tools, techniques, and initiatives that companies can apply to increase productivity and throughput. The central premise of lean is to identify wastes (non-value adding activities) in a process and to eliminate or reduce them. In 1926, Henry Ford said, "The longer an article is in the process of manufacture and the more it is moved about, the greater its ultimate cost" (Manos et al, 2006). This belief lead Ford to develop the idea of flow production and to be the first to integrate an entire production process (Lean Enterprise Institute, 2007). As discussed by Shah and Ward (2007), lean principles originated from Ford's practices, and beginning in 1937, Taiichi Ohno of Toyota studied Ford's original thinking in detail and invented the concepts and tools of the Toyota Production System (TPS). The main focus of TPS is cost reduction through waste elimination, which encourages using a just-in-time (JIT) production system to produce only what is needed, when it is needed, in the quantity that is needed (Ohno, 1988).

Today's companies are under tremendous pressure to perform at the lowest cost, highest quality, and fastest pace; therefore, lean has emerged as a popular management philosophy for companies to attain that competitive edge (Womack and Jones, 1996). Later, Womack et al. (1990), coined the term "lean production," which is a direct descendent of TPS, and described a lean system and its underlying concepts. Womack and Jones (1996) later explored lean principles further and extended these principles to the enterprise level.

Lean is generally associated with methods for eliminating waste and improving flow. The transfer of lean principles from Japan to the U.S., however, occurred in a piecemeal and informal fashion over the course of many years. As a result, until recently there was not a precise way of defining lean. According to Shah and Ward (2007), "Lean production is an integrated socio-technical system whose main objective is to eliminate waste by concurrently reducing or minimizing supplier, customer, and internal variability." In short, lean focuses on eliminating waste and improving the flow of work using proven methods, and lean tools or techniques are most often applied to improve the flow of information and/or materials. To do this, lean suggests the use of tools such as value stream mapping (VSM), kanban, kaizen, pull systems, 5S, one-piece flow, poka yoke, jidoka, and others (Womack, et al. 1990).

Lean focuses on eliminating waste and improving flow using various proven methods to improve the flow of information and material. Waste stems mainly from unnecessary delays, tasks, costs, and errors. The seven wastes of lean include overproduction, transportation, inventory, processing, waiting, motion, and defects. These wastes can also be applied to support functions such as procurement, engineering, invoicing, inventory control, order entry, scheduling, accounting, and sales. Unfortunately, these wastes are inherent in every process and lean principles provide the underlying philosophy, tools, and techniques to reduce them. These methods range from a simple housekeeping process such as 5S to more complex kanban systems, which help to control inventory levels.

The primary focus of lean is on the customer, to address value-added and non-value-added tasks. Value-added tasks are the only operations for which the customer is ready to pay. The idea in creating flow in lean is to deliver products and services just-in-time, in the right amounts, and at the right quality levels at the right place. This necessitates that products and services are produced and delivered only when a pull is exerted by the customer through a signal in the form of a purchase. A well-designed lean system allows for an immediate and effective response to fluctuating customer demands and requirements (Cudney, 2009). Lean implementation within an organization has shown considerable financial results (Cudney and Elrod, 2011).

Strong proof of the power of lean systems is evident through the continued success of Toyota over the past two decades. Over time lean principles have migrated throughout the entire enterprise and are now proving to be beneficial in service applications as well. As a result, the tools and principles of lean thinking have spread beyond manufacturing to service organizations such as retail, healthcare, and even government (Lean Enterprise Institute, 2007).

However, in a passionate effort to lean processes, there is a chance that lean facilitators might overlook health and safety issues. For example, it is possible that during a lean redesign of a process or workplace with the goal of minimizing cycle times that protective guards or other safety devices might get ignored and removed. This may compromise safety and can lead to accidents and, in extreme cases, even death. Accidents bring with them the indirect costs of compensation claims as well as a forced shut down of a machine or process. These are against the fundamental principle of lean to minimize wastes. Hence, lean and safety should not be viewed as having conflicting goals but should be addressed simultaneously. The integration of lean and safety can help companies achieve that competitive edge.

2. LITERATURE REVIEW AND BACKGROUND

Extensive literature is available on lean as well as safety. However, there is little literature that integrates the two areas and addresses the relationship between lean and safety. In recent years some authors have investigated the relationship between the two and have published their work.

Kovach et al. (2011) examined the implementation and effectiveness of continuous improvement methods that practitioners in the U.S. use to improve productivity and solve problems. The research found that the majority of practitioners surveyed indicated that 5S/visual factory, VSM, standardized work, kanban, and poke-yoke were among the lean techniques their organizations had successfully implemented. It is likely that 5S/visual factory was often reported as being successfully implemented due to its relatively simple methods and low cost. Since 5S in its simplest form involves keeping workplaces well organized and clean, it involves few changes to the manufacturing process; thus, it requires minimal resources (i.e., time, material, etc.) for implementation (Hirano, 1996). Also, VSM is one of the most fundamental lean practices, and although this technique can be very involved, it is often a first step in estimating how much waste exists in a process so that projected savings from a lean program can be estimated (Cudney and Kestle, 2010). In addition, since standardized work is already in place to some extent in many work environments, it may not require significant changes. Therefore, it may be relatively easy to implement this technique in many settings by simply documenting existing processes. In addition, Kovach et al. (2011) found that few practitioners reported that their organizations had successfully implemented techniques such as SMED and heijunka, which are more complex lean methods that require greater organizational change to implement (Cudney, 2009).

An attempt to classify 22 critical lean implementation elements into four categories was published by Shah and Ward [2], putting lean tools under the headings of 'total preventative maintenance', 'total quality management', 'human resources management', and 'just-in-time'. This approach attempts to associate the type of waste with the category of tool to be applied to correct it. An approach with similar goals was published by Pavnaskar et al. [3] which organized lean manufacturing tools and metrics according to their abstraction, their location of application in the organization, as well as type of wasted resource addressed and what it does to that waste (identify, eliminate, etc.) with the end goal of associating lean tools with specific manufacturing problems. Worley and Doolen (2006) investigated roles of management support in lean implementation. Management support and communication are identified as important variables in a lean manufacturing implementation and to continuing lean planning and deployment.

Wilson (2005) states that the integration of ergonomics into the lean process should begin in the planning stages. Ergonomic metrics should be used in the value stream mapping risk assessment process. Main et al. (2008) discuss how the misapplication of lean techniques can hamper safety issues. They state that, "lean focuses on minimizing waste in the system while safety focuses on minimizing risk. Optimizing for one or the other can lead to suboptimal solution for the overall system". Manuele (2007) states that accidents cause a loss to the organization; hence, they are fundamentally a waste. Therefore, preventing accidents should be an integral part of lean implementation. Manuele categorizes unsafe activities as muda or non-value added waste. Walder and Karlin (2007) stated, "Successful implementation of lean thinking and ergonomics includes the redesign of work, standardizing work, and reduction or elimination of muscular-skeletal disorders MSD factors". The utilization of material handling assist devices is a very useful application for the same.

In the ANSI B11 Technical Report 7: Designing for Safety and Lean Manufacturing (TR7) released by Association for Manufacturing Technology (AMT), the B11 secretariat is an extensive and complete work in the field of integrating safety and lean. It was developed and released in May 2007 based on materials provided by The Boeing Company, Deere & Company, General Motors, Liberty Mutual Group, Design Safety Engineering Inc., and Tenneco Inc. This report concurrently addresses lean and safety concerns of machinery. It gives a brief overview of lean and challenges to concurrently address it with safety. It introduces a risk assessment framework on how to address lean and safety concerns. It also provides design guidelines to meet lean objectives without compromising safety. Though this report was initially developed for the machine tool industry, it can be used by many other industries. However, the existing research does not attempt to identify relationships between safety and learn or to be a decision-making tool to ensure safety measures are accounted for during lean implementation.

3. METHODS

A survey was developed to investigate lean practitioners' awareness of the relationship between lean and safety. People associated with lean implementation in industry were surveyed to understand how lean initiatives affect safety. A wide array of professionals participated in the survey. The participants' age ranged from 18 to 55 years and their educational background ranged from high school to Ph.D. In addition, the survey was administered internationally with 71% of the participants were from USA, 12% from Europe, and 12% from India. The companies for which the participants worked ranged from a less than 49 employees to more than 5,000 employees. As part of the survey, the participants were asked which lean techniques their organization had successfully implemented. The lean techniques that were successfully implemented are shown in Figure 1.

Figure 1: Participants' Experiences with Lean Implementation

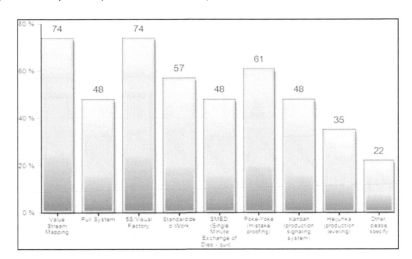

One-piece-flow is an important concept of lean. It is the key to optimizing material and work in progress (WIP), which is considered as waste, as it does not add value to the process. To realize one-piece-flow, the work cell must be redesigned. However, while redesigning a work cell, human capabilities and limitations should also be taken into account. A work cell that is not anthropometrically designed will lead to wasted motion for the employees, which lead to wasted time. The placing of tools, gauges, jigs, and fixtures at strategic locations in the work area will reduce excessive as well as hazardous motion. The appropriate illumination and provision of displays and control panels leads to less strenuous operations for the employees. Therefore, there will be fewer accidents and compensation claims in a well designed work cell. Also, the flow of material will be uninterrupted which satisfies the central premise of lean.

The survey results indicate that there is a trend in lean implementation to provide better tools, material handling equipment, and equipment to facilitate fast pickup or set down. 70% of the responders said that during the work cell design process they had investigated the location of tools, gauges, jigs, and fixtures. In addition, 57% investigated opportunities for material handling devices and 61% investigated easy and fast pickup/set down equipment as shown in Figure 2. Whereas, a lower percentage of respondents said that they had investigated opportunities for displays and control panels, illumination, noise, vibration, and air contamination. Sharing of personnel, equipment, and utilities is also not being adequately considered during the designing of lean work cells.

426

Figure 2: Safety & Ergonomic Issues Considered during Work Cell Redesign

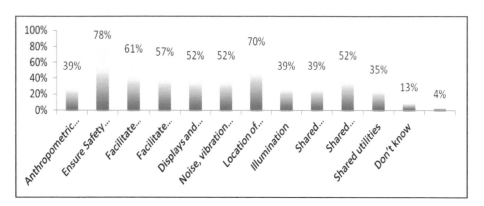

Efficient material movement is at the heart of the lean philosophy. Material handling equipment facilitates better movement of material on the factory floor, which leads to increased productivity. Many occupational risk factors can be reduced or eliminated by the appropriate use of material handling equipment. Different kinds of material handling equipment are used in industry depending on their applicability, adjustability, and usefulness. Based on the process and product material handling equipment such as scissor lifts, stackers, elevating platforms, container tilters, balancers for hand tools, small workstation cranes, vacuum hoists, self-leveling turntables, and anti-fatigue matting can be used. 67% of the survey respondents said that they were providing anti-fatigue matting in new layout design, which has the potential of reducing ergonomic injuries and operator fatigue. Scissor lifts, elevating platforms, small workstation cranes, and vacuum hoists are other commonly used material handling devices that can reduce physical demands on the worker. The survey showed some application of these material handling equipment devises during lean implementation but we were unable to draw overarching conclusions about their application.

The key to lean is the concept that companies must manufacture products at a faster pace and in smaller batches to remain competitive. This requires that changeover times be as small as possible. Although changeover is necessary, it is still a non-value added activity. Single minute exchange of die (SMED) refers to a single digit changeover time of less than ten minutes. While designing for SMEDs, care has to be taken that the process does not deviate from safety requirements. As in the case of work cell design and material handling, equipment anthropometric considerations should be taken into account to avoid overburden and unsafe working conditions. 48% of the respondents said that they had implemented SMED successfully in their organization. Of them, 62% of the respondents surveyed said that lockout devices such as switches and valves were incorporated while designing for SMED, 57% said that care was taken so that the switches would be placed at points that were easily accessible, whereas 43% said that they had considered placing the switches away from any hazardous area.

The ANSI technical report B11.TR7-2007 proposes a risk assessment process, which can be implemented in a wide variety of industrial or manufacturing settings. The fundamentals of the process are to identify hazards, assess risks, and reduce risk to an acceptable level and document the results. 94% of the respondents who said that their company follows some kind of risk assessment process said that their process is documented as shown in Figure 3. As for reviewing or updating the process, 21% said it is performed weekly, 43% said it is performed monthly, approximately 7% said it is performed quarterly, and 21% said it is performed yearly. Approximately 7% said it is never updated or reviewed.

Figure 3: Frequencies of Risk Assessment

	Respondents
Weekly	3
Monthly	6
Quarterly	1
Half yearly	1
Yearly	3
Never	4
No written plan	4

4. CONCLUSIONS AND FUTURE RESEARCH

Of the responders who took the survey, 88% said that they had observed a positive impact of their lean activities on the health and safety performance of their workers while 12% said that they experienced neither a positive or negative impact on the health and safety performance. None of the responders had observed a negative impact on the health and safety performance on their workers because of their lean activities. These values are based on self-assessment and may be biased, particularly if the respondents are not aware that they have introduced a safety or health hazard to their facility.

Respondents who said they experienced a positive impact on the health and safety performance showed the following statistics:
- 71% said they had health and safety personnel on their lean teams while 14% said they did not have any health and safety personnel on their lean teams.
- 75% said their health and safety personnel were given some kind of training on lean methods and concepts and 10% said no such training was given.
- 70% said their lean facilitators were given training in safety principles and 20% said their lean facilitators were not given any training in safety principles.
- 73% said that during the value stream mapping process waste related to safety issues was considered and 27% said waste related to safety was not considered during the value stream mapping process.

Respondents who said they experienced no impact on the health and safety performance showed the following statistics:

- 100% said they did not have any health and safety personnel on their lean teams.
- 67% said their health and safety personnel were not given any kind of training on lean methods and concepts.
- 33% said their lean facilitators were given training in safety principles and 67% said their lean facilitators were not given any training in safety principles.
- 100% said waste related to safety was not considered during the value stream mapping process.

These trends lead to the conclusions that accommodating health and safety officials on lean teams is beneficial for a positive impact on the health safety performance of the workers during the implementation of lean principles. Also, the lean team members should be given training in safety and the health and safety officials should be trained in lean philosophies.

Lean, having been developed on the fundamental idea of removing waste, inherently brings some degree of safety concerns in newly designed processes. However, it is still not clearly known among lean practitioners the quantified implication of their lean initiatives on safety. Lean initiatives such as the introduction of material handling equipment and work cell design for one-piece-flow necessitate the integration with ergonomic considerations for the design to be safe. However, knowledge among lean facilitators is frequently lacking with respect to safety requirements and considerations. Including health and safety officials on the lean teams and providing training of lean facilitators and health and safety officials on safety concepts and lean philosophies, respectively, will be beneficial to achieve the goal of integrating lean and safety.

REFERENCES

ANSI Technical Report for Machines – Designing for Safety and Lean Manufacturing, B11.TR7 – 2007

Cudney, E., 2009, Using Hoshin Kanri to Improve the Value Stream, Productivity Press, New York, NY.

Cudney, E., and Elrod, C., 2011, "A Comparative Analysis of Integrating Lean Concepts into Supply Chain Management in Manufacturing and Service Industries", International Journal of Lean Six Sigma, Vol. 2 (1), pp. 5-22.

Cudney, E., and Kestle, R., 2010, Implementing Lean Six Sigma Throughout the Supply Chain: The Comprehensive and Transparent Case Study, Productivity Press, New York, NY.

Hirano, H. (1996) 5S for operators: 5 pillars of the visual workplace. New York: Productivity Press.

Kovach, J., Cudney, E., and Elrod, C., (2011) "The Use of Continuous Improvement Techniques: A Survey-based Study of Current Practices", International Journal of Engineering Science and Technology, Vol. 3 (7), pp. 89-100.

Lean Enterprise Institute (2007) 'A brief history of lean', Available online at: http://www.lean.org/WhatsLean/History.cfm (accessed 25 February 2012).

Main, Bruce, Taubitz, Michael and Wood, Willard, January 2008, "You Cannot Get Lean Without Safety: Understanding the Common Goals", Professional Safety, Vol. 53, No. 1, pp. 38-42

Manos, A., Sattler, M. and Alukal, G. (2006) 'Make healthcare lean', *Quality Progress*, Vol. 39, No. 7, pp. 24-30.

Manuele, Fred A., August 2007, "Lean Concepts: Opportunities for Safety Professionals", Professional Safety, Vol. 52, No. 8, pp. 28-34

Ohno, T., 1988, Toyota production system: Beyond large-scale production, Productivity Press, New York, NY.

Shah, R. and Ward, P. T., 2007, "Defining and developing measures of lean production", Journal of Operations Management, Vol. 25, No. 4, pp. 785-805.

Walder, Jon and Karlin, Jennifer, November 2007 "Integrated lean Thinking & Ergonomics: Utilizing Material Handling Assist Device Solution for a Productive Workplace", http://www.mhia.org/search/jennifer+karlin.

Wilson, Robert, April 2005, "Guarding the LINE", Industrial Engineer, Vol. 37, No. 4 pp. 46-49

Womack, J., and Jones, D., (1996) Lean Thinking: Banish Waste and Create Wealth in Your Corporation, Simon & Schuster, New York, NY.

Womack, J.P., Jones, D.T., and Roos, D. (1990) "The Machine that Changed the World", Rawson Associates, Maxwell Macmillan, New York, NY.

Worley, J.M., and Doolen, T.L., (2006) "The role of communication and management support in a lean manufacturing implementation", Management Decision, Vol. 44, No. 2, pg. 228-245.

Section VII

Value Co-Creation and Customers

CHAPTER **43**

Value Orchestration Platform: Model and Strategies

Kyoichi Kijima,* Timo Rintamki,** and Lasse Mitronen**

*Tokyo Institute of Technology, Japan; **University of Tampere, Finland

ABSTRACT

New service businesses such as Amazon, iTunes, and Rakuten have a common characteristic: they play a role of platform that orchestrates and facilitates value co-creation by customers and providers. In the value co-creation process customers and providers interact each other and co-create new value, while the value orchestration platform invites customers and providers to "get on board" and facilitates the process and leaves the control entirely to providers and sometimes to customers as well.

The purpose of the present paper is to model value co-creation process and value orchestration platform from service science perspective and, then, to argue value orchestration management strategies in terms of them. To achieve the purpose, we first examine several major perspectives from which service has been argued so far and clarify our research position in service science. Then, we propose a process model of value co-creation consisting of four phases, i.e., co-experience, co-definition, co-elevation and co-development. Finally, by developing value orchestration platform model and relating it to the process model, we argue three management strategies for orchestrating value co-creation, i.e., SIPS, curation and empowerment, by referring to recent typical businesses.

Key Words: Service System Modeling, Value Co-Creation Process, Value Orchestration Platform, Curation

1. INTRODUCTION

With the establishment of sophisticated logistics and the rise of information technology, new service businesses, whether real or virtual, have become increasingly important. One common characteristic of these service businesses is that they have two layers (Figure 1).

In the value co-creation process, customers and providers interact with each other and co-create new values. The lower layer invites customers and providers to "get on the board." It facilitates and orchestrates new value co-creation by customers and providers, but leaves the control of the process entirely in the hands of providers and sometimes of customers as well. We call this layer the value orchestration platform.

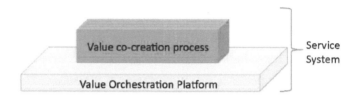

Fig. 1 Value Co-Creation and Value Orchestration Platform

Amazon, eBay, iTunes, and Rakuten primarily serve as value orchestration platforms; indeed, they connect tens of thousands of providers (sellers) to millions of customers (buyers). For example, the strength of the Apple App store lies in its function as a value orchestration platform. The store is interested in inviting as many users and appropriate developers as possible onto it, but it takes no physical or full legal "possession" of the software it distributes. The essential idea of a value orchestration platform dates back several decades, and its traditional and well-know examples include credit cards and shopping malls (Refer to Table 1).

Table 1. Some Examples of Value Orchestration Platform

Service	Value Orchestration Platform	Customers	Providers
Credit Card	VISA	Customers	Restaurants, Hotels
Shopping	Shopping Mall, e-commerce	Buyers	Tenants, Sellers
iPhone App	Apple App Store	Users	App Developers
Portal/search site	Google	Web surfers	Advertisers

By observing these concrete examples, we may confirm the legitimacy of our definition of value orchestration platform: a business model that entirely emphasizes value orchestration and allows customers and providers to interact with each other and co-create new values, often using information and communication technology (ICT).

This paper proposes a model of the value co-creation process and value orchestration platform from service science perspective, and then discusses value orchestration management strategies by referring to current typical businesses.

2. VALUE CO-CREATION IN SERVICE SCIENCE PERSPECTIVE

While the concepts of, interests in, and approaches to service are quite diversified among the disciplines, service science is an emerging area of study that draws on decades of pioneering work in the research area of service marketing, service operations, service management, service engineering, service economics, and service computing (Spohrer and Maglio, 2008), (Cambridge, 2008), (Barile and Spohrer, 2010).

Service science tries to define service as a phenomenon observable in the world in terms of a **service system** with value co-creation interactions among entities (Maglio and Spohrer, 2008) by taking a bird' s eye view of various perspectives in which service system entities can be people, businesses, non-profits, government agencies, and even cities.

As a specialization of systems sciences (Spohrer, 2009), it tries to shed light on a scientific approach to understanding social value and identifying propositions that can be formulated and theories that can be empirically tested (Barile and Spohrer, 2010). Indeed, service science is a study of service value co-creation phenomena among service system entities (Ng and Maull, 2011).

The goal of service science is to promote innovation in service and increase service productivity. Innovation is a key to productive service and is born from the intersection of different types of knowledge. To this end, promoting an interdisciplinary approach is crucial to the field. Measuring value co-creation is complex and involves many rational and experiential dimensions.

Service itself has been characterized by several perspectives in service science. In the realist research tradition and according to the good-dominant logic, service is characterized as an outcome that can be measured by attributes and variables in a functional domain.

According to service-dominant (S-D) logic, interpretative consumer research, and consumer cultural theory (CCT), service is defined as an **experience** and a phenomenon. It claims, "Value is always uniquely and phenomenologically determined by the beneficiary" (Vargo and Maglio, 2008).

The Nordic School Approach and New Service Development (NSD) understand service as a **process** (Rintamaki et al., 2007). Value co-creation is an active, creative, and social process based on collaboration between the provider and customer that is initiated by the provider to generate value for customers. However, it is a form of collaborative creativity of customers and providers aimed at enhancing the organization's knowledge acquisition processes by involving the customer in the creation of meaning and value, although it is initiated by providers. Such collaborative value co-creation often requires greater effort on the part of both the customer and provider than does a traditional market interaction. The people on both sides must think about what they want to get out of a cooperative relationship.

Whatever value is co-created, we may be able to identify hierarchical structure in it. For example, Rintamaki et al. (2007) identifies four layers in value in service, i.e. economic value (e.g. focus on price), functional value (e.g. focus on solutions), emotional value (e.g. focus on experience) and symbolic value (e.g. focus on meanings). The four dimensions of value reflect the spectrum from utilitarian to hedonic value, which is a widely accepted categorization (Smith and Colgate, 2007).

The hierarchical structure has at least three implications. From the customer perspective, economic and functional value represent often key criteria used when assessing service offerings, and meeting these criteria represents a prerequisite for buying. The role of emotional and symbolic criteria is more subjective and often represents a substantial value added to the core service offering. In many cases, emotional and symbolic value may be the true drivers for customer' s decision making, although economic and functional value are used for argumentation.

3. FOUR-CO-PHASE MODEL OF THE VALUE CO-CREATION PROCESS

To open up the concept of value co-creation, we identify value co-creation interaction as an active, creative, and social process based on collaboration between the provider and customer that is initiated by the provider to generate value for customers. It is a form of collaborative creativity of customers and providers that is used to enhance the organization's knowledge-acquisition processes by involving the customer in the creation of meaning and value, although it is initiated by the provider.

Such collaborative value co-creation often requires greater efforts on the part of both customer and provider than does a traditional market interaction. People on both sides must think about what they want to get out of a cooperative relationship. Customers need to trust the provider to not misuse the information they provide or unfairly exploit the relationship. Providers need to actively manage customer expectations about how the relationship will evolve. Providers must provide capabilities for co-creation and also receive the tools and training necessary to efficiently co-create.

However, it may be too simple to assume that both sides know about the other's preference, expectations, or capabilities when participating in the collaborative process. Rather, they may or may not need to learn about each other to share internal models (mental models).

This consideration leads us to the idea of service as a dynamic interaction process in which customers and providers are mutually learning and collaborating by co-experience.

Now, we propose a new model called the "Four-Co-Phase Model of the Value Co-Creation Process" (Refer to Figure 2).

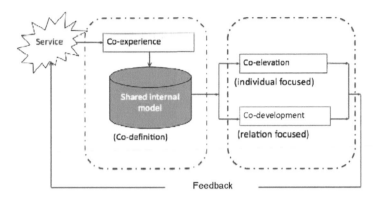

Fig. 2 Four-Co-Phase Model of Value Co-Creation

The model explicitly defines service as a value co-creation interaction between customers and providers and identifies four phases that occur in the process. The first two phases, co-experience and co-definition, are relatively short-range concepts for describing service appreciation, while the final two phases, co-elevation and co-development, refer to the long-range activities necessary for service innovation.

3.1 Co-Experience of Service

When participating in the collaborative value co-creation process, customers and providers may have little or no idea about the other's capabilities and expectations. Hence, rather than reducing the gap between the needs and seeds, by co-experience, the provider and customer share an internal model to co-define a mutual understanding about the service.

3.2 Co-Definition of the Shared Internal Model

By interacting with each other, the customer and provider may learn about the other's preference, capabilities, and expectations so that they may co-define and share a common internal model. Satisfaction for both sides is generated by the co-experience of the service and the co-definition of a shared internal model. For example, at a sushi bar, through conversation, the chef recognizes a customer's preferences, mental and physical condition, and appetite and the customer learns about the day's specialties and seasonal fish. If they are able to share a common internal model (i.e., understand the other's preference, capabilities, and expectations), then both are happy. This is a typical process of co-experience and co-definition.

3.3 Co-Elevation of Each Other

In general, a system is defined as a pair of entities and sets of relationships among entities. Hence, it is relevant to relate the value co-creation process to the entities of the service system as well as to the relationships among them. We call the former co-elevation, which focuses more on value co-creation led by entities in the system. Co-elevation is a zigzag-shaped spiral process of customer expectations and provider abilities. Higher expectations of service by intelligent and literate individuals lead to higher-quality service and greater social values (needs-pull). High-quality service, in turn, increases customer expectations (seeds-push).

3.4 Co-Development of Value

On the other hand, we call the later "co-development" because it pays attention to the co-innovation generated by simultaneous collaboration among various entities. Co-development of service innovation is usually carried out in the context of customers evaluating and assessing the value and providers learning from customer responses. Collaborative improvement of Linux software by anonymous engenderers and developers is a typical example of co-development.

4. VALUE ORCHESTRATION MANAGEMENT STRATEGIES

Now, we shift our attention to the value orchestration platform and management strategies (Figure 3). Platform orchestrator is primarily concerned with the methods to get appropriate customers and providers "on board" the platform and to vitalize interactions between customers and providers. Hence, strategies for the platform to attract and involve customers and providers to maximize profit are crucial. Indeed, one of the advantages of an online value orchestration business such as e-commerce is that they have no limitations on the number of customers who can participate.

While strategy for involvement of customers and providers focuses on how to attract customers and providers on the platform, value curation is essential for the platform to encourage customers and providers to co-elevate and co-develop.

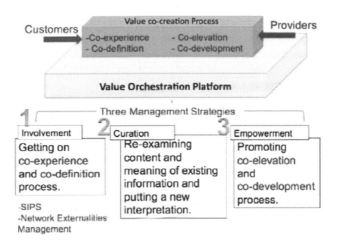

Fig. 3 Manage ment Strategies for Value Orchestration

Empowerment is another aspect of value orchestration, particularly for the co-elevation and co-development phases. Specifically, this refers to how a platform empowers customers and providers so that each side finds the other attractive and both are motivated to interact with each other.

These strategies are directly related to particular phases of the value co-creation process. For instance, involvement strategy deals with co-experience and co-definition phases in particular while empowerment targets co-elevation and co-development phases directly. However, the following matrix should be helpful as a checklist to get inspired appropriate actions for value orchestration. For example, by using it we could discuss what curation means for co-experience phase?

Table 2 Value Orchestration Strategy Matrix

		Value Co-creation process			
Value Orchestration Strategy		Co-experience	Co-definition	Co-elevation	Co-development
	Involvement				
	Curation				
	Empowerment				

4.1. SIPS: Involvement Strategies for attracting customers and providers

A cycle of Sympathize, Identify, Participate, and Share and Spread (SIPS) is useful for identifying how customers and providers become interested in a platform (Figure 4).

It generates interest among customers and the provider toward co-experience and co-definition phases.

SIPS proposes that the trigger for customers and providers to become interested in a service system is their having sympathy toward it. Presently, people are connected with each other through social media outlets such as Facebook and Twitter. They communicate through rather subjective comments about what they experience, and the comments that gain a certain level of sympathy for being useful and interesting spread quickly throughout these media platforms. As a result, the media triggers sympathy to, for example, a shopping mall and leads to its identification as an interesting place. In SIPS, participation does not necessarily mean purchase of some products or services. Rather, it emphasizes that the experience would lead to sharing and spreading through a common internal model.

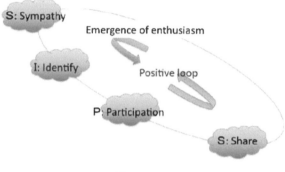

Fig. 4 SIPS

4.2. Value Curation Strategy

Curation can be defined as a highly proactive and selective approach of value orchestration that collects, selects, analyzes, edits, and reexamines the content and meaning of existing products, service, and information on customers and providers to provide a new interpretation of and a new meaning to them. Based on the newly developed interpretation and meaning, it facilitates a value co-creation process involving customers, providers, information, and technology.

To collect information, sufficient technology and methodology is necessary for scooping up appropriate information from an enormous amount of data on the Internet and databases. To provide a new interpretation of the information, it is necessary to combine human intelligence with technology to evaluate, understand, and process data; dig out information and value from that data; and visualize what the data indicates. Blending new content while filtering and managing other useful information is a productive and manageable solution for providing prospective customers with a steady stream of high-quality and relevant content. While pure creation may be demanding and pure automation does not engage, content curation can provide the best of both.

For instance, the East Japan Railway Company (JR East) is now well known for curating content. This large railway company employs 60,000 people and has lines covering the Tokyo and Tohoku areas. The company is now expanding to a new business by utilizing existing stations as shopping spaces. Under the slogan of "station renaissance," JR East has given a new meaning to its stations and established them as a "drop-in area" rather than a "transit point." It has also introduced unique shopping malls called "E-Cute," a coined word created out of a combination of "Eki" (station in Japanese) and "cute." The company has not simply divided the station space into rentable areas for appropriate tenants but edited the sales space on the basis of a unique concept: they curate the content of each shopping mall. The focus of JR East's value orchestration is identifying qualities in tenant candidates and editing those tenants for how much consistent value they will provide for each station.

4.3. Empowerment Strategies of Stakeholders

Customers are empowered by lifting up their aspiration level, while providers are empowered by referring to their capability of providing service.

For example, Rakuten seems keen on its balanced empowerment at the co-elevation phase. Rakuten's business model is simple: the company mainly earns money from merchants in the form of participation and consultation fees, but customers (buyers) join at almost no cost. Using a super database, they conduct cross-business analyses and predict user behavior. They employ sophisticated search methods and social media such as Twitter to deliver useful information especially for each customer, i.e., empowering the customer by personalizing service on individual basis. Due to this, a customer becomes more interested in Rakuten as he/she would know when it has anything he/she wants.

5. CONCLUSION

In this paper we introduced a comprehensive framework consisting of two new models of the value co-creation process and value orchestration platform, as inspired by recent service businesses, such as Amazon and Rakuten in e-commerce.

The process model of value co-creation opens up the concept of dynamic value co-creation and identifies four phases in it: co-experience, co-definition, co-elevation, and co-development. The value orchestration platform is a platform in which customers and providers are orchestrated and facilitated such that they can interact and co-create new values. Combining these two models, we discussed how there are three management strategies for orchestrating value co-creation: SIPS, curation, and empowerment. We referred to real cases to illustrate each of the three strategies.

The concept of a value orchestration platform is general enough to explain various service systems. Now, we are discussing its relevance to public and volunteer services. Particularly, since the earthquake, tsunami, and the nuclear power plant disaster that occurred on March 11, 2011 in Japan, we have recognized that the community service system in the damaged areas should become a value-orchestration platform involving politicians, bureaucrats, doctors and medical staff, local people, and volunteers to support the local people. However, the power structure and information flow among these groups and how it would affect such a platform is not as clear as it is in business cases.

REFERENCES

1. S Barile and J Spohrer. System Thinking for Service Research Advances. *Service Science*, 2010.
2. Cambridge. Succeeding through service innovation. *Cambridge White paper*, pages 1–33, June 2008.
3. P Maglio and J Spohrer. Fundamentals of service science. *Journal of the Academy of Marketing Science*, January 2008.
4. I Ng and R Maull. Embedding the New Discipline of Service Science. *The Science of Service Systems*, 2011.
5. T RintamÅNaki and H Kuusela. Emerald — Managing Service Quality — Identifying competitive customer value propositions in retailing. *Managing service quality*, 2007.
6. J Smith and Mark Colgate. Customer Value Creation: A Practical Framework. *The Journal of Marketing Theory and Practice*, 15(1):7–23, January 2007.
7. J Spohrer. Service Science and Systems Science. Proceedings of COE Final Symposium, Tokyo, 2009, March 2009.
8. J Spohrer and P Maglio. The emergence of service science: Toward systematic service innovations to accelerate co-creation of value. *Production and Operations Management*, 17(3): 238–246, January 2008.
9. S Vargo and P Maglio. On value and value co-creation: A service systems and service logic perspective. *European Management Journal*, 26:145–152, January 2008.

CHAPTER 44

Service Space Communication by Voice Tweets in Nursing

Kentaro Torii, Naoshi Uchihira, Tetsuro Chino, Kenji Iwata,

Tomoko Murakami, Toshiaki Tanaka

Toshiba Corporation Research and Development Center
Kawasaki, Kanagawa, Japan
kentaro.torii@toshiba.co.jp

ABSTRACT

Nursing is an important aspect of healthcare services and improvement of the efficiency of healthcare processes is an important issue. Since nursing involves both physical action and information processing, it could be described as "action-oriented intellectual services." There is a striking mismatch between "action-oriented intellectual services" and current information systems, including traditional IT and paper, which are generally developed as tools for deskwork. This mismatch causes inefficiency in healthcare processes and gives nurses a lot of stress. To address this problem, we focus on voice. The voice interface enables hands-free or semi-hands-free message composition. Several voice interfaces, such as telephone, mobile phone (PHS), intercom and voice-mail have been already used for information processing in nursing. But voices are mere audio signals in these systems. Therefore, it is difficult for these systems to handle in smart ways and appropriately deliver them. Smarter voice information processing systems are required in "action-oriented intellectual services." We have been developing a smart voice tweet system for nursing that makes it possible for nurses to input care records, make memos for themselves, and compose voice messages for other staff in a unified way. The system automatically adds to each voice several types of information tags associated with the state at the moment the voice is input. The tags enable automatic delivery to adequate destinations without extra instructions or operations by nurses.

Keywords: tweets, voice, nursing, caring

1   BACKGROUND AND OBJECTIVE OF OUR RESEARCH

More and more countries are experiencing population aging and a rise in the number of patients, and more and more time and money will be devoted to healthcare services. Therefore, promotion of the efficiency of healthcare processes is an important issue. Nursing is an important aspect of health services. We focus on nursing care processes in this paper.

Nurses provide care for patients using their hands and move around hospital wards, staff stations, operating theaters and so on. They also read and write, sending and receiving a lot of information of various types in the course of their work. Since their work involves both physical action and information processing, it could be described as "action-oriented intellectual services."

There is a striking mismatch between "action-oriented intellectual services" and conventional information systems which are generally developed as tools for deskwork. Since nurses' hands are fully occupied during the provision of patient care and nurses move around hospital wards, nursing stations, operating theaters and so on, it is difficult for them to record by keyboard, to compose messages, to telephone other staff and to take telephone calls from other staff at the right moment. Nurses spend much of their time on indirect care (Lemonidou, Plati, Brokalaki, Mantas and Lanara, 1996). The mismatch is one of the significant causes of inefficiency of indirect care. Smartphones and tablet PCs have recently come into use in nursing, but it is difficult to type messages during "action-oriented intellectual services" other than fixed inputs such as multiple-choice and numeric input. Our objective is to promote greater efficiency of indirect care by eliminating this mismatch. The basic concept was described in our previous paper (Uchihira, 2011). In this paper we propose the architecture of a voice tweet system for improvement of the efficiency of "action-oriented services".

2   The architecture of the voice tweet system for automatic delivery to appropriate destinations

2.1   Outline of the voice tweet system

To address this problem, we developed a smartphone application and a smart voice tweet system for nursing that makes it possible for nurses to input care records, make memos for themselves, and compose voice messages for other staff. Figure 1 shows the outline of the architecture of the system, which consists of a smartphone application and a Web server application. The voice interface of the smartphone application enables hands-free or semi-hands-free message composition. Several sensors on the smartphone continuously collect data around the nurse in a background process, and the application periodically sends the data to the server of the system. Several estimators on the server estimate the state (context) of each member of staff who carries the smartphone. We call the states estimated by each estimator "tags". Tags are, for example, specific types of keywords included in recognized voice, location where a member of staff composes a voice tweet, the task a member of staff is executing at the moment a voice tweet is made, the time and the staff ID.

There are two important functions of tags. First, the tags improve comprehensibility of voice tweets for recipients. So, nurses need not say everything about a voice tweet. For example, they can leave out context information (WHO, WHEN, WHERE) in their voice tweets. Secondly, the tags enable the system to decide the type of delivery of voice tweets and to deliver them to an appropriate destination automatically, for example a message for other staff, a reminder of one's task, and inputting to an appropriate database of nursing records.

Figure 2 illustrates typical use cases of the voice tweet system. Nurses are able to compose voice tweets such as patient's pain complaints patient's questions about care processes only by one hand at patient's bed sides. Tags are automatically attached to the voice tweets. For example, a voice tweet about a pain complaint of a patient is delivered to the nurse who cares the patient at a bathroom, and then the nurse is able to easily grasp the patient's pain, and can appropriately assist patient to bath. Since the tweets with tags are categorized automatically, nurses can efficiently hand over information about patients.

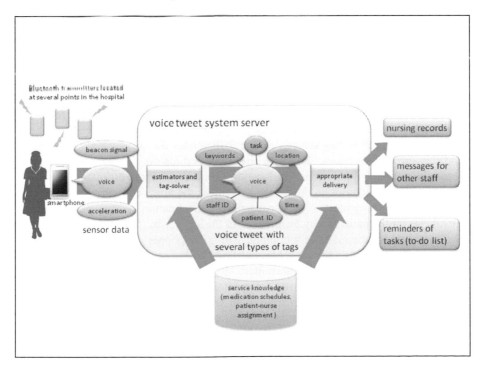

Figure 1 Architecture of the voice tweet system

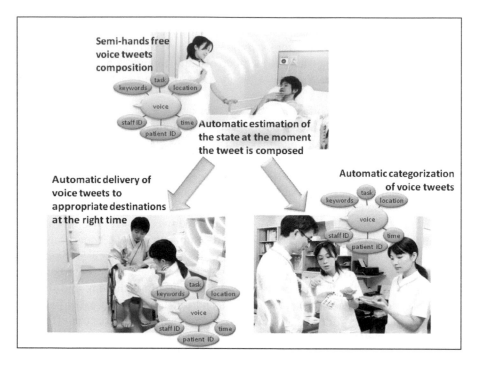

Figure 2 Use cases of the voice tweet system

2.2   Estimators and tag-solver

It is generally impossible for each estimator to estimate exactly the unique true state. The estimated tags of the voice tweet have some ambiguity. So, it is required for some solver which eliminates unrealistic combination of tags and reduces their ambiguity. We call it "tag-solver". Figure 3 illustrates the structure of "estimators and tag-solver" in Figure 1. There are three estimators in the server: voice recognition and keyword extraction, location estimator, and task estimator.

The voice recognition in the server recognizes the voice and extracts keywords included in the recognized voice. For example, patient names and names of medical treatments for the patients may be included in the keywords.

To collect location data in a building with smartphones, we developed a small Bluetooth transmitter that emits a beacon signal including SSID and MAC address of the Bluetooth device on the transmitter. The smartphone receives beacon signals from the transmitters around it, and the application on the smartphone sends MAC address and RSSI (Received Signal Strength Indicator) of each signal to the server. The location estimator estimates the location of each member of staff at each time from the Bluetooth beacon signal.

The task estimator estimates what task was executed at each time from acceleration data and so on. These estimators generate candidates of tags with degrees of confidence. The tag solver in the server searches an optimal set of tags from the candidates using service knowledge such as medication schedules for patients and patient-nurse assignments. An optimal set of tags is a set of tags whose sum of degrees of confidence of tags is maximal among sets which satisfy the constraints derived from service knowledge. The server attaches the optimal set of tags to the voice.

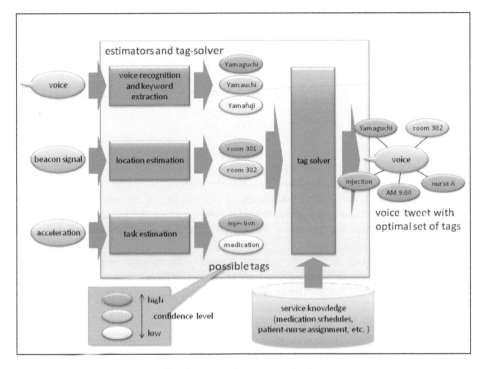

Figure 3 Automatic attachment of optimal set of tags to a voice tweet

2.3   Delivery of voice tweets to appropriate destinations

The system estimates the types of delivery of each tweet according to the attached tags, and decides appropriate destinations of voice tweets (Figure 4). Messages are delivered to the staff who actually need the messages. Reminders by a nurse are sent to her/himself at the right time. Nursing records are reported to an appropriate database. We developed a prototype stochastic estimator for types of voice tweets. Several types of tags such as keywords location, tasks, can be the inputs of the estimator, but here we used only keywords included in voice tweets as described below. When we need to categorize tweets into more complex types than those shown in Figure 4, other types of tags also should be included in the estimator.

First we converted the voice tweets into 3-dimensinal binary vectors according to whether or not the voice tweet includes keyword tags of specific types. The specific types of keyword tags that we selected are 1. patient's name, 2. expressions of vital signs ("body temperature", "blood pressure", "pulse", etc.), 3. expressions of time ("2:00 PM", "after lunch", etc.). We let the binary vector of a tweet $x = (p, v, t)$ where $p, v, t \in \{0, 1\}$, $p = 1$ if the tweet includes a patient's name, otherwise $p = 0$, $v = 1$ if the tweet includes expressions of vital signs, otherwise $v = 0$, $t = 1$ if the tweet includes expressions of time, otherwise $t = 0$. For example, a voice tweet that includes keywords "Mr. Yamaguchi", "body temperature" and no keyword about time is converted into a binary vector $(1, 1, 0)$.

We assume y is a boolean-valued random variable where $y = 1$ when the type of a tweet is nursing record otherwise $y = 0$. Then, the logistic regression model has the form

$$P(y = 1 \mid x) = \frac{e^{\beta_0 + \beta^T x}}{1 + e^{\beta_0 + \beta^T x}},$$

where $P(y = 1 \mid x)$ denotes the probability that $y = 1$ given x. β_0 and β are coefficients which are fitted using training data. As described below, we collected voice tweets in a hospital which are labeled with types by nurses, so we can use a part of these labeled tweets as training set and we can test the fitted model using the remaining labeled data. Models for reminders (to-do list), handover and real-time messages can be built in the same way.

Of course, if staff are able to manually specify the type of each tweet on the application, for example by touching buttons on the display, there is no need for estimation of types of tweets. But in the case that staff are unable to use their hands at all, estimation is indispensable. Voice command is a kind of tag and can be treated in the same way in the architecture shown in Figure 2.

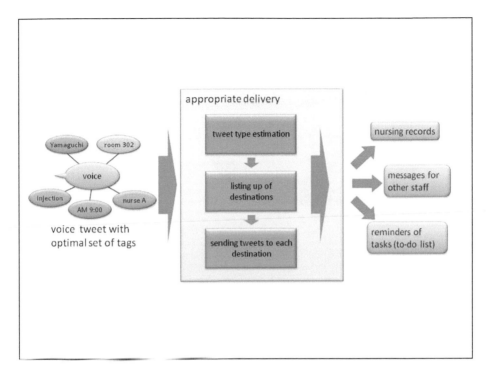

Figure 4 Estimation of delivery types of tweets to appropriate destinations

3   Actual voice tweets collected in a hospital and a prototype stochastic estimator for types of voice tweets

We experimented with this system in a hospital with 199 beds. Twenty nurses used this system for a week. We collected 200 voice tweets by twenty nurses using the system in real nursing care processes. 115 voice tweets were categorized into 4 types by nurses themselves. The categories are 1. nursing records, 2. handovers to the nurses of the next shift, 3. reminders of tasks (to-do list), 4. messages to other staff. As shown in Figure 5, about 50% of voice tweets are about nursing records, 15% of voice tweets are handovers to the nurses of the next shift, 14% of voice tweets are reminders of tasks, and 13% of voice tweets are messages to other staff.

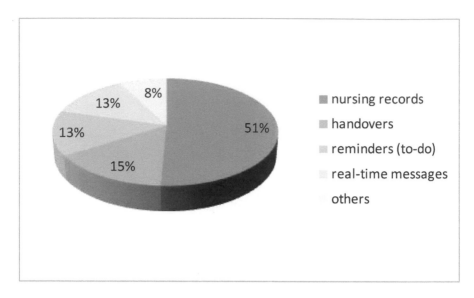

Figure 5 Types of voice tweets collected with the voice tweet system in real nursing care processes in a hospital

We are now implementing and evaluating several estimators and tag-solvers. So, we manually extracted keywords belonging to the types mentioned above and built a prototype stochastic voice tweet estimator based on a logistic regression model that estimates whether a voice tweet is a nursing record or not, using 77 voice tweets randomly selected from 115 tweets. And we ran cross-validation using the remaining 38 tweets. Using the method described above, we executed a total of 10 model constructions and cross-validations using 10 random selections.

The average precision of estimations by 10 models is 89% and the average recall is 71%, and therefore the average F-measure (equals harmonic mean of precision and recall) is 0.79.

We also built models that estimate whether a voice tweet is a reminder (to-do list) or not in the same way. The average precision is 78%, the average recall is 57%, and the average F-measure is 0.64. Models for handover and real-time messages built in the same way do not work well, because those types of keywords do not have significant relations with handover or real-time messages.

Both high precision and high recall are required for higher usability, for example almost 100% precision and 95% recall. If recall were allowed to be lower, precision could be higher, but it would mean more false positives and deterioration of usability of the system, and vice versa. To achieve this objective, we intend to introduce other types of tags and service knowledge into the system.

4   CONCLUSIONS

We proposed the architecture of a voice tweet system for improvement of the efficiency of "action-oriented services" such as nursing and caring. The voice interface of the system enables hands-free or semi-hands-free message composition. And the system automatically attaches several types of tags to the voice tweets, such as keywords included in the tweets, locations where the tweets are composed, the tasks the staff are executing at the moments voice tweets are made. Tags not only improve comprehensibility of voice tweets for recipients, but also enable automatic decision on appropriate types of delivery of the tweets. These two functions will reduce stress in information processing in "action-oriented services".

We built a prototype stochastic estimator for automatic decision on appropriate types based on a logistic regression model using only keyword tags included in voice tweets, and evaluated the accuracy of the estimator using actual voice tweets collected in a hospital. For more complex categorization and delivery, location tags and task tags also should be included in the model.

ACKNOWLEDGMENTS

This research is funded by Japan's national project JST S3FIRE (Service Science, Solutions and Foundation Integrated Research Program).

REFERENCES

Hendrich A., M. Chow, B. Skierczynski and Z. Lu. *A 36-Hospital Time and Motion Study How Do Medical-Surgical Nurses Spend Their Time?*, In: The Permanente Journal The Permanente Journal Summer 2008, Vol. 12(3), pp.25-34.

Hollingsworth J., C. Chisholm, B. Giles, W. Cordell, D. Nelson. *How do physicians and nurses spend their time in the emergency department?*. Annals of Emergency Medicine Vol. 31, Issue 1 , Pages 87-91, January 1998.

Lemonidou C., C. Plati, H. Brokalaki, J. Mantas, V. Lanara. Allocation of nursing time. *Scand J Caring Sci.* 1996;10(3):131-6.

Uchihira N., et.al. *Innovation for Service Space Communication by Voice Tweets in Nursing and Caring: Concept and Approach in Japanese National Project*, 20th Annual Frontiers in Service Conference, 2011.

CHAPTER 45

Service Modeling of Compliments and Complaints and Its Implications for Value Co-Creation

Francisco Villarroel Ordenes, Mohamed Zaki,
Babis Theodoulidis, Jamie Burton

Manchester Business School, University of Manchester
Manchester M13 9SS, United Kingdom

francisco.villarroelordenes@postgrad.mbs.ac.uk
mohamed.zaki@postgrad.mbs.ac.uk
babis.theodoulidis@manchester.ac.uk
Jamie.Burton@mbs.ac.uk

ABSTRACT

The paper demonstrates the impact of using text mining techniques to automate analysis and classification of large amounts of customer compliments and complaints (C&C). The research is using an empirical approach to generate a better understanding of how co-creation processes can be designed based on customer feedback experiences. In order to improve the service propositions, the integration of customer comments as operant resources of the organisation is discussed. A co-creation feedback model is proposed, considering positive and negative comments across three main categories, resources, activities and attributes (positive/negative comments). Finally, the co-creation feedback model enables the mapping of the organisation's service processes from the customer perspective.

Keywords: SD-Logic, Compliments and Complaints, Text Mining, Co-creation

1   INTRODUCTION

The use of the internet and ICT technologies have increased the level of customer participation in the service process and changed the way of communication and interaction with customers (Rust et al. 2006). Recent theories like SD-logic (Vargo and Lusch 2004, 2008) are proposing that customers are value co-creators, active participants of the service process, and the only ones that can assess the value propositions (Payne et al. 2007; Vargo and Lusch 2007). A key part of the customer assessment of value propositions is customer feedback obtained through a variety of mechanisms and recent research into co-creation is looking into the way that customer feedback can be used to enhance business processes (Baron and Warnaby 2011; Grönroos 2008).

Advances in ICT technologies have improved dramatically the way that customer feedback is collected, analysed and disseminated. By customer feedback, in this work, we refer specifically to customer compliments and complaints that are collected via solicited mechanisms e.g., questionnaires (Wirtz 2000). Since the majority of the customer feedback takes the form of unstructured data, the analysis of the feedback could be based on technologies such as text mining which allow unstructured information to be analysed in automated way thus, minimising the errors and delays from human intervention but also giving the possibility to improve the performance and availability of the information across the organisation. There are a number of examples where text mining technologies have been used across different sources such as emails, social networks and call centres (Ur-Rahman and Harding 2011).

This paper proposes a value co-creation feedback model based on analysing customer complaints and compliments and generating a customer-view of the value propositions. An empirical approach is used to generate a better understanding of how co-creation processes can be designed based on customer feedback experiences. In particular, the paper demonstrates the impact of using text mining techniques to automate analysis and classification of large amount of customer feedback data and the identification of operant resources. The proposed co-creation feedback model considers positive and negative comments across three main categories, resources, activities and attributes (positive/negative).

2   BACKGROUND

The two broad research areas of relevance to this work are first, SD-logic and value co-creation and secondly, customer feedback and specifically, complaints and compliments. According to the SD-logic, service is defined as: "the application of specialised competences (skills and knowledge) through deeds, process, and performances for the benefit of another entity" (Vargo and Lusch 2004, p. 5). SD-logic makes an important distinction between two types of resources, operand and operant. Vargo and Lusch (2004) explain that "in a service centred view, firms must focus on operant resources identifying or developing core competences, the fundamental knowledge and skills of an economic entity". Furthermore, according to SD-logic literature, "value is co-created through the combined efforts of firms, employees, customers, stockholders, government agencies, and other entities related to any given exchange, but are always determined by the beneficiary" (Vargo et al. 2008, p. 148). Once the service experience has been performed between the company and the customer, only the customer can assess the real value through their satisfaction level, called *value in use* (Vargo and Lusch 2007).

Previous studies have highlighted the importance of customer's feedback in the achievement of high service quality standards and on the organisational learning process (Bell et al. 2004; Caemmerer and Wilson 2010). Compliments and Complaints are a type of an active form of customer feedback generated by a delight or dissatisfaction episode with the service (Friman and Edvardsson 2003). In particular, complaining customers can contribute to the firm by giving information about what part of the service can be improved and potential service innovations (Blodgett and Anderson 2000; Gruber 2011). From an SD-logic perspective, focusing customer feedback analysis only on specific attributes of a service process is an incomplete understanding of customer experience (Macdonald et al. 2011). Evaluating output attributes such as reliability, assurance, tangibility, empathy, and responsiveness (Parasuraman et al. 1988) represents an oversimplification of the different resources, encounters and activities that interact across the service process (Edvardsson and Tronvoll 2011; Macdonald et al. 2011). Indeed, the classification of attributes into predefined quality dimensions provides only superficial information about customer experience (Caemmerer and A. Wilson 2010), which can possibly distort the feedback analysis, especially when the co-creation process considers many interactions (Payne et al. 2007). Considering that a service centred view of marketing perceives marketing as a continuous learning process directed to improve operant resources (Vargo and Lusch 2004), then customer feedback analysis should provide information about resource performances across the value co-creation process (Tronvoll 2007).

The paper proposes a value co-creation process which considers that the firm has a value proposition and the customer is a value co-creator who interacts with the service proposition and creates value (Payne et al. 2007; Tronvoll 2007). Both, the firm and customer have their own resources and activities that interact across this service process. As a consequence, satisfaction or dissatisfaction episodes are generated by interactions, driven by specific resources that perform over or under expectation. The proposed approach is focuses on extracting the resources and activities that are determinants of a customer positive or negative evaluation. In the proposed model "attributes or quality dimensions" are not seen as the final evaluation of customers of the entire service, but as a positive or negative opinion about a resource or activity in the co-creation process.

3 CASE STUDY

The research approach follows a case study methodology providing empirical evidence of the applicability of SD-logic into real business problems. The research aims to examine co-creation process perspective through identifying the resources and activities that generate customer complaints and complements. Furthermore, the paper examines the ability of text mining techniques to automate the analysis of them adapting a value co-creation process model.

The participant organisation is a car park company in one of the UK airports. The organisation provided a dataset with 1091 customer comments over a week and related to data from a car park survey that sent to customers two days after using the service. Table 1 shows the structure of the dataset. The *Car Parks - Single Improvement Factor* contains the answer of the open question "*What is the single most important factor you feel we can improve upon to enhance your car park experience*" and *Recommendation Rating* represents a net promoter score question that asks how likely is the person to recommend the car park to a colleague or a friend on a scale from 0 to 10.

Table 1 Customer Feedback Data Source

REF	Car Park	Car Parks - Single improvement factor	Recommendation Rating	Car Park Departure Date
1	Car Park E	Barrier did not recognise my pre-booked credit card - had to press buzzer but person very helpful. Bus going out was fine - after waiting 15mins for bus on return we walked - very poor	5	05/06/2011

Currently, comments are analysed manually and coded into different categories and sub-categories based on one annotator's judgements. Normally, the process takes around two weeks to analyse comments and produce the report.

3.1   Text mining process

The objective of the text mining process is to help the organisation to automate the analysis of customer feedback. The process follows a two stage approach as shown in figure 1, namely, *Training* and *Population*. In particular, the process extracts the main concepts to build a domain specific library and map these concepts to four main categories, namely, Company Resources, Customer Resources, Activities (Company, Customer), Attributes (Positives, Negative) (Yu et al. 2011, p. 736). The first three categories are extracted because of their relevance to SD-logic, and to understand the co-creation process (Payne et al. 2007; Vargo and Lusch 2007). Furthermore, resources and activities represent the different interactions between the company and customers (Payne et al., 2008; Kwan and Min, 2008). The last category is extracted to analyse whether the customer experience was positive or negative.

As part of the training stage, a sample of 100 comments were randomly selected and manually annotated by the research team. The manual annotation used the categories and subcategories employed by the company but also extended these with some new or modified subcategories. The revised manually annotated training sample contains a total of 453 concepts annotated and classified to the categories and subcategories. Following this process, an evaluation took place with the customer relation department of the participant company to review and validate the categories and subcategories and the proposed annotations.

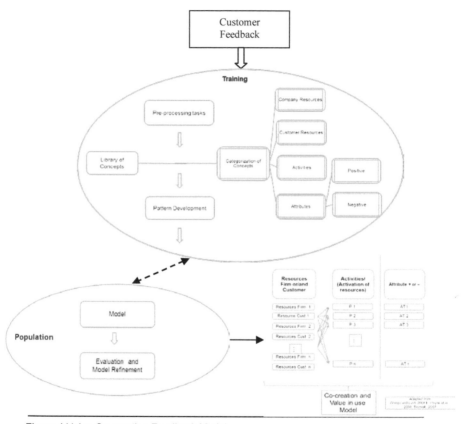

Figure 1 Value Co-creation Feedback Model

In order to map the comments to a designated service process from customer centric view, the approach adapts the co-creation process model (Payne et al. 2007). According to this, compliments and complaints are mapped on the main activities that take part in a service process. Therefore, the manual annotation process helped in the pattern development as the comments has been split into information units to analyze comments using sentence level analysis (Singh et al. 2011). The research employed sentence level analysis for more detailed insightful information about the customer experiences as each comment could express different experiences of various services offered by the participant company.

The patterns used the developed library resources to extract resources and activities from these information units. In the training sample, overall the text model have 72 patterns designed to extract resources and activities of service process where customers were having positive or negative value in use experiences. As result, the analysis showed that there are four main service processes (*Booking, Arriving to Car Park, Parking Car, and Bus Service*) were experienced by customers as shown in figure 2.

Figure 2 Adapted Service Process Model for pattern development

Table 2 shows an example the patterns extracted and the corresponding linguistic structure per information unit. Patterns are grouped based on the highlighted service process (booking, arriving car park, parking car and bus service) which contains the total number of patterns were developed across the different service processes. The pattern structures of the first information unit "*Barrier did not recognise my pre-booked credit card*" is [company resources+customer activity+company activity+customer resources]. The pattern automatically classifies this information unit as a complaint about the entrance and maps it to two service processes booking and arriving to car park. However, the structure of the second information unit "*Had to press buzzer but person very helpful*" is different [customer activity+company resources+company resources+attribute (positive)]. The customer in this information unit compliments the staff. In the same comment, the customer evaluated and experienced the bus service. The pattern of the information unit of "*bus going out was fine*"is structured as [company resources+customer activity+attribute (Positive)]. The customer was neutral in his evaluation about the bus services. In this case, the pattern analysis deals with neutral feedback as a compliment and maps the information unit to a bus service process. However, the customer was not happy about waiting 15 minutes for the bus on his return as he commented "*waiting 15mins for bus on return we walked- very poor*". The pattern structure of this information unit is [customer activity+customer resources+company resources+attributes (negative)].

The second stage, population, applies the model on the complete dataset and evaluates its capabilities through evaluating the accuracy of the annotations generated.

3.2   Results

The prediction capability of the model was evaluated based on the number of comments that were presented in the training sample and the number of accurate annotations. This gives an indication about the missing and inaccurate extraction which will be considered in the model refinement process (Feldman and Sanger 2007).

Regarding the training sample, the model annotated 76 comments from a total of 100. Due to language structure complexity, the model could not extract 24 comments from the training sample. Overall, from the 76 comments, 110 patterns were found. From the 110 patterns, there were three inaccurate annotations, which gives 98% overall accuracy. Finally, 84% of the comments were classified as complaints and 16% of the comments were classified as compliments.

Table 2 Pattern Development and Service Process

Single improvement factor	Pattern Development	Company Resources	Customer Resource	Customer Activity	Company Activity	Attributes	Service Process	C&C
Barrier did not recognise my pre-booked credit card -	Barrier did not recognise my pre-booked credit card	Barrier	credit card	did not recognise	Pre-booked		Arriving to car park & Booking	Complaint
had to press buzzer but person very helpful. Bus	Had to press buzzer but person very helpful	Buzzer/ Person		press		very helpful	Arriving to Car Park	Compliment
going out was fine - after	bus going out was fine	bus		going out		fine	Bus Service	Compliment
waiting 15mins for bus on return we walked - very poor	waiting 15mins for bus on return we walked- very poor	bus	15 mins	waiting		very poor		Complaint

As part of the population stage, the model captured 550 comments from 1091 comments, which represents 50% of the total dataset. From the 550 comments, 694 patterns were found, only 55 were wrong predictions, which gave 92% overall accuracy. The model found 86% complaints and 14% compliments which is consistent with the result of the training sample. Overall, the result was fairly good enough to demonstrate how text mining could be used for automating the analysis of compliments and complaints following a value co-creation process model. Indeed, the model needs improvement and design more patterns to fit the analysis and increase the model accuracy which is considered in future work.

Table 3 and table 4 show details about model accuracy to extract compliments and complaints per service process. The analysis evaluated the customers experience across the different service process from a customer centric perspective. The analysis gives an indication which services are mostly complained or complimented by customers. Therefore, the presented results could contribute to the participant organisation to improve value proposition and enhance the relationships with customers. For example, table 4 shows that highest percentage of customer complaints are inside the car park (298) to use the facilities to park his/her car. In particular, the analysis classified different complaints about company resources such as directions, space, staff, facilities, others parking, and others customer resources that customers addressed in their feedback. Generally from the analysis, customers were complaining frequently about directions, information or signs that direct customers to the car park.

Table 3 Model Accuracy for Compliments

Service Process	Right Predictions	Wrong predictions	Total	Accuracy
Booking	1	0	1	100%
Arriving Car Park	0	0	0	0%
Parking car	83	10	93	89%
Directions	1	1	1	100%
Staff	20	2	22	91%
General	62	7	69	90%
Bus Service	7	1	8	88%
Total	110	13	123	89%

Furthermore, the analysis shows that booking and parking car services represent the main complaints activity. In particular, booking service has complaints reached 54%. In particular, customers were complaining about online booking service on the website, e-mail confirmation, and information about the car park in the website. Price was another subcategory that has frequent complaints by customers, as 71% of the booking comments referred to price resources. Most of the cases price resource is linked to negative activities such as "reduce" and "increased", and also towards negative attributes such as "expensive", "less", "cheaper" and "same".

Table 4 Model Accuracy for Complaints

Service Process	Right Predictions	Wrong predictions	Total	Accuracy
Booking	87	3	90	97%
General	23	1	24	96%
Price	64	2	66	97%
Arriving Car Park	54	1	55	98%
Parking car	298	38	336	89%
Space	71	3	74	96%
Staff	25	4	29	86%
Facilities	9	0	9	100%
Directions	111	3	114	97%
Others car park	65	12	77	84%
Others Customer resources	17	16	33	52%
Bus Service	111	3	114	97%
Total	550	45	595	92%

Figure 4 shows an example of booking analysis that map complaints related to price resources. The words in circle presents the resources of the company and others are mainly the attributes that are classified into complaints.

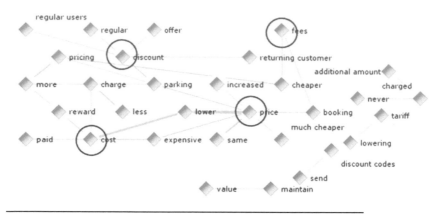

Figure 4 Price resource and associated attributes

4   EVALUATION AND CONCLUSIONS

The evaluation process analyses how the results could be used to improve the organisation service process, value proposition, and enhance the relationship with the customers. It follows a similar approach   discussed in Baron and Warnaby (2011). First of all, the customer feedback for the car parking is a major area of concern because of the large number of complaints received from the customers. Following a demonstration of the approach to the company, the manager of the customer relation department commented "*You definitely have got a good approach around this...*" "*... from my point of view you have developed something that works... you could drag it off to different areas of the business and just be as efficient as possible, ... however no matter how good it is, it always is going to drop off a few data points, but there's a big benefit of using it compared with doing it manually ... Especially as the manual method data is so delayed*".

Getting the analysis done quickly is certainly one of the major advantages but it is also important to consider advantages such as the consistency of the information extracted highlighting areas of concern and allowing the organisation to respond much quicker and in a more informative way. In general, the manager's comments validated the importance of detecting complaints quickly especially if the resource is easy to solve. In terms of value in use of customers, it was highlighted the practicality of identifying some resources of the company which could be improved immediately such as the staff as quoted "*tactically you want to solve the problem straight away... and make sure you improve in a way that affects the majority of people, so it might be looking at improving something... not something like price which can take time, but might be something like staff which you could do straight away*" and "*....and if the customer decides as an example of what they have improved I think the customer is doing the job for you...*".

The paper contributes to the SD-logic development by utilising text mining to analyse large amount of customer feedback and define how the co-creation process can be designed based on customer experience. This could be considered as an alternative approach compared to previous "attribute" or "quality" dimension approaches. Applying resources and activities could help organisations to gain insightful information from analysing customer experiences which could be mapped directly to the service process. Also, sentence level analysis gives a better feedback resolution as the customer is able to comment on different service processes (or parts of) and different resources and their attributes. The customer focus provides however, a very powerful feedback mechanism including the ability to not only facilitate value co-creation but also to integrate customer information into the customer relation management system.

However, the model is still in the early stage of development. Given that the accuracy level of the extraction result is still relatively because of the ultimate need to enrich and train the model with library resources and develop relevant linguistic patterns. The future work will focus in increasing the accuracy of prediction to reach at least 90% for better customer feedback analysis. Furthermore, the model will incorporate structured data from CRM systems to produce rich analysis and profiling about customers. More emphasis on extracting complicated phrases structure such as irony comments.

Furthermore, the research is working to incorporate the customer context into the value co-creation model. This will include an analyser for contextual comments related to personal situational contexts. An example of the customer comments that describe a situational context is: *"A death in my colleagues' family on the morning I was due to travel resulted in the trip being cancelled. Having pre-paid I was informed on your web site that I could not cancel or amend the booking therefore I lost my money. Not happy about that"*. By analysing the context, the organisation could react differently depending not only on the customer information but also the situational context.

REFERENCES

Baron, Steve, and Gary Warnaby (2011), "Individual customers' use and integration of resources: Empirical findings and organizational implications in the context of value co-creation," *Industrial Marketing Management*, 40(2), 211–18.

Bell, Simon J., Bülent Mengüç, and Sara L. Stefani (2004), "When customers disappoint: A model of relational internal marketing and customer complaints," *Journal of the Academy of Marketing Science*, SAGE Publications, 32(2), 112.

Blodgett, Jeffrey G., and Ronald D. Anderson (2000), "A Bayesian Network Model of the Consumer Complaint Process," *Journal of Service Research*, 2(4), 321–38.

Caemmerer, Barbara, and Alan Wilson (2010), "Customer feedback mechanisms and organisational learning in service operations," *International Journal of Operations & Production Management*, 30(3), 288–311.

Edvardsson, Bo, and Bård Tronvoll (2011), "Value Co-Creation and Value-in-Context: Understanding the Influence of Duality of Structures," in *2011 International Joint Conference on Service Sciences*, Ieee, 292–96.

Feldman, Ronen, and James Sanger (2007), *The text mining handbook: advanced approaches in analyzing unstructured data*, Imagine, Cambridge: Cambridge University Press.

Friman, Margareta, and Bo Edvardsson (2003), "A content analysis of complaints and compliments," *Managing Service Quality*, 13(1), 20–26.

Gruber, Thorsten (2011), "I want to believe they really care: How complaining customers want to be treated by frontline employees," *Journal of Service Management*, 22(1), 85–110.

Macdonald, Emma K., Hugh Wilson, Veronica Martinez, and Amir Toossi (2011), "Assessing value-in-use: A conceptual framework and exploratory study," *Industrial Marketing Management*, Elsevier Inc., 40(5), 671–82.

Parasuraman, A., Valarie A. Zeithaml, and Leonard L. Berry (1988), "SERVQUAL: A multiple-item scale for measuring consumer perceptions of service quality.," *Journal of retailing*, Elsevier Science, 64(1), 12–40.

Payne, Adrian F., Kaj Storbacka, and Pennie Frow (2007), "Managing the co-creation of value," *Journal of the Academy of Marketing Science*, 36(1), 83–96.

Rust, Roland T., Ajay K. Kohli, Evert Gummesson, and Eric J. Arnould (2006), "Invited commentaries on the service-dominant logic by participants in The Otago Forum," *Marketing Theory*, Otago, 6(3), 289–98.

Singh, Surenda N., Steve Hillmer, and Ze Wang (2011), "Efficient Methods for Sampling Responses from Large-Scale Qualitative Data," *Marketing Science*, 30(3), 532–49.

Tronvoll, Bård (2007), "Customer complaint behaviour from the perspective of the service-dominant logic of marketing," *Managing Service Quality*, 17(6), 601–20.

Ur-Rahman, N., and J.a. Harding (2011), "Textual data mining for industrial knowledge management and text classification: A business oriented approach," *Expert Systems with Applications*, Elsevier Ltd, (October), 1–11.

Vargo, Stephen L., Paul P. Maglio, and Melissa A. Akaka (2008), "On value and value co-creation: A service systems and service logic perspective," *European Management Journal*, 26, 145–52.

Vargo, Stephen L., and Robert F. Lusch (2004), "Evolving to a new dominant logic for marketing," *Journal of marketing*, JSTOR, 68(January), 1–17.

———— (2008), "Service-dominant logic: continuing the evolution," *Journal of the Academy of Marketing Science*, 36(1), 1–10.

Wirtz, Jochen (2000), "Institutionalising customer-driven learning through fully integrated customer feedback systems," *Managing Service Quality*, 10(4), 205–15.

Yu, Chong Ho, Angel Jannasch-Pennell, and Samuel DiGangi (2011), "Compatibility between Text Mining and Qualitative Research in the Perspectives of Grounded Theory, Content Analysis, and Reliability," *The Qualitative Report*, 16(3), 730–44.

CHAPTER 46

Co-creating Value to Increase Service Productivity – Impacts of Outputs and Inputs for the Product-supporting Service Value Chain

Andrea Rößner, Florian Kicherer

Rainer Nägele

University of Stuttgart IAT
Fraunhofer IAO
Stuttgart, GERMANY
andrea.roessner@iat.uni-stuttgart.de, florian.kicherer@iat.uni-stuttgart.de
rainer.naegele@iao.fraunhofer.de

ABSTRACT

Service delivery is set in a complex environment of interactions, relations, and processes. Service management needs to be able to measure and control all players that are involved within the Service Value Chain in order to ensure a maximized productivity. The paper discloses possible perspectives and approaches how services, which are handled by different actors in the SVC, can provide an additional value for all the actors and the productivity can be managed. The requirements for a KPI driven management tool are discussed: adequate measures for effectiveness, efficiency, and flexibility. From a real world example the issue of managing service productivity via KPIs is being shown.

Keywords: productivity; service management; service value chain; KPI.

1 SERVICE MANAGEMENT AND PRODUCTIVITY: AN INTRODUCTION

In this paper we want to show why and how the subject of value creation in services currently gains such an importance. Because of the broad field of different service offers, it is necessary to distinguish the specific characteristics of services. We focus our exposition on product related industrial services. We stress possible management approaches in order to link the preconditions, the challenges and the approach to the solution of the research problem. Our proprietarily developed dynamic, three phased Service Management Concept is thus presented as a possible solution to this problem. We analyze the main adjusting and operating levers to control the elusive notion service productivity. Since interaction processes are highly relevant, we target the analysis on this section of productivity management. We finally show the practical benefits for service providing enterprises by the example of an industrial project partner of the cooperative research project "ServUp". This gives us the opportunity to introduce possible Key Performance Indicators (KPIs) that give an indication how services can be measured, controlled and assessed in practical application. Describing the experiences of one real world industry partner, we illustrate how the mentioned KPIs might be integrated in an existing service providing system and at the same time be combined with a new coherent IT Software Infrastructure.

The intent of this paper is to set a central theme about the different notions of the broad terms value, co-creation and the interrelations of a "service production system" as defined by a service value chain or a service network.

1.1 Characteristics of the industrial sector and its implications on services

The term "service" is used in very different disciplines like in marketing, in operations management or in computer science. It represents diverse and not easily delimitable denotations–depending on the respective point of view (cp. Alter, 2008). An adequate approach is necessary in order to build a basis for the service value assessment and productivity control in the sense of a co-creation of value. By reason of the service notion being a broad field of research which includes varying properties such as automation, IT- dependency, personalization and customization, repetition, duration, temporal and responsibility aspects, the development of a universal framework for defining them is difficult (cp. Alter, 2008). The complexity that already appears in this short explication of possible influence factors shows the necessity of a restriction for the analysis of services. Considering the special research field of our cooperative research project "ServUp", we focus on industrial, product-related services. According to Spath and Demuß (2003), they are defined in the following way: Mostly immaterial, they are directly or indirectly bound to the existence of the corresponding industrial goods. As a part of the investive services branch, where the demand is mainly characterized by cost-aware enterprises, they differ from the consumptive services where the customers are mostly private consumers. They are frequently employed as a support for marketing or customer's benefit of the related industrial good. (cp. Spath, Demuß 2003). In turn, goods can even be considered (in a service-dominant viewpoint) as the "distribution mechanisms for service provision" (Vargo and Lusch 2004). The circumstance of high cost awareness thus the cost pressure within our focused industry branch affects both the proposal and the cost structure of industrial, product-related services.

As Porter demonstrated with his five forces scheme, many factors have an impact on sales of products and services in general. These branch environmental and internal factors can include the "Threat of New Entrants", the "Bargaining Power of Buyers", the "Bargaining Power of Suppliers", the "Threat of Substitute Products or Services" and the "Rivalry Among Existing Competitors" (cp. Porter, 2008). Referring to these vacillations, the customers and other further parts of the Service Value Chain become a center of attention. Sales in goods and in services of industrial manufacturers thus react very sensible on changes in the economic environment as the direct customers of the service providers do. Speaking about the customer impact on the service sales, the notion and the impact of the "integration of the external factor" needs to be defined with regard to uncertainty and vacillations. It means that the customer as an external factor is more or less integrated in the (pre-) production process of the service. The service provider is thus dependent on this sort of resource. In form of the direct customer or his material, monetary or legal commodities it serves as an input factor for at least one part of the service provision or becomes as a temporally bounded resource even an integral part of the whole service provision process (cp. Gouthier and Schmid, 2003; Bharadwaj, Varadarajan and Fahy, 1993).

1.2   The notions Value and Co-Creation and their interrelations for the SVC

The impact of the service productivity measurement for a clear and systematic industrial service management approach is currently becoming significant. To understand the concept of value co-creation in the Service Value Chain (SVC) a brief discussion about "value" seems appropriate: value-in-use vs. value-in-exchange. This differentiation shows how controversial the outlooks on value creation are. The latter, value-in-exchange, is considered as a traditionally firm centered respectively goods-dominant approach (cp. Vargo, Maglio and Akaka, 2008). Enterprises following this approach for their service offerings focus predominantly on internal factors, thus excluding the customer from the value creation process. They control the specification of the value addition process before the product or service as a result of the production process is being proposed to the customer. Thereby the value added can only be gained at a unique point of exchange (cp. Prahalad and Ramaswamy, 2002).

Contrary, in the customer centered and service-dominant value-in-use approach the customer plays an integral role of the value creation process. Customers are perceived as an influential part concerning the temporal factors and in which part of the service process the mutual (created and achieved) value is being realized. Value creation is hence no more related to industrial boundaries as this perspective even makes the customer a self-dependent competitive factor. The value-in-use concept allows as a result that in several points of exchange value can be created (cp. Prahalad and Ramaswamy, 2002; Vargo, Maglio and Akaka, 2008).

In reference of Salomonson et al. 2011 we determine the term value co-creation as related to the service provider. Service value creation in turn is dedicated to the service beneficiary as this actor of the value creation chain seeks to gain the most advantage in form of value for himself (cp. Salomonson, 2011). The enterprise however seeks to improve the service proposal to enhance the mutual value (cp. Grönroos, 2008). Therefore the service provider is dependent on the information about the service customer. This information exchange process mostly happens directly in the service (pre-) delivery process. That is why the perspective of the two-sided definition of the notion co-creation (of value) is utile for this paper.

The notion co-creation of a service implies the production that consists of the proposal and the acceptance of the service offer, relies on the participation of at least two parties. Seen from the view that the service production is realized while transferring resources by the use of processes and service process participants the whole system can be considered as a work system (cp. Alter, 2008). This leads to the evident finding that the service provider and the service customer jointly form a construct that can be interpreted as a service system (cp. Spohrer et al. 2008, Howahl et al., 2011). The twofold role of the customer as client and deliverer of production factors that enable the service delivery leads to a "supplier-customer" relation. Seen from the co-creation point of view, we therefore focus our analysis on this important bridge of the SVC.

We introduce a concept that helps to analyze the main influence factors on SVC-included entities and allows the measuring of the results of a service process. According to each of the three service specific phases Service Development, Service Management and Service Delivery, the main aspects for analyzing the existing status of the Service (Work) System are presented and demonstrated via examples from "ServUp". The following figure 1 shows the composition of the three phased Service Management Concept and its main objects of investigation (technology, processes and the service-supplier-customer interaction).

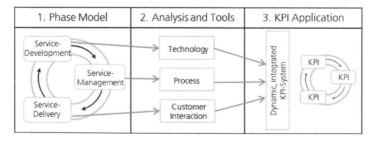

Figure 1 Composition of the three-phase Service Management Concept. (Own source.)

2 A DYNAMIC THREE PHASED SERVICE MANAGEMENT CONCEPT

As a basis for the assessment of services we present a framework that enables an overview of a dynamic, three-phase Service Management and Assessment Concept. We also keep including the service value chain concept while using this analysis instrument. Thus, we show possible approaches for the service analysis and assessment within a service lifecycle. The framework seen below allows different views on management aspects for the SVC. Each phase of the service lifecycle is analyzed in a SVC-specific way. The background substantiates from the practical research experience within the joint research project "ServUp". Figure 2 shows the model discussed in the following paragraphs.

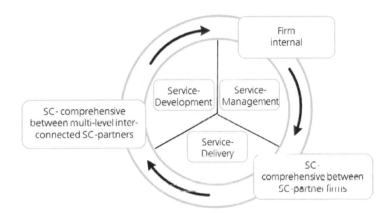

Figure 2 The Integrated Service Management Framework. (Own source.)

From the viewpoint of a physical product Supply Chain, a component manufacturer delivers its products to the next part of the supply chain, the "converter". This enterprise will buy the components and use them to build a more complex product. This step might be repeated several times before the physical product reaches the end-customer. The latter in turn is the source of the value creation process seen from the Value Chain perspective. Industrial, product-related services play an important role for sales of the concerning physical product. That is why we focus on the service characteristics that create a value for the (end-) customer and how this value can be increased by the co-creation of value.

Changes of a service offering do not necessarily affect only the next part of the service value chain, i.e. the customer that pays for the delivered service, but the end-consumer. To make this effect clear, we focus in this paper on one single service offer that already exists: the customer service hotline. The possible KPIs for assessing this service proposal will be discussed in the next chapter and how they affect the SVC.

2.1 Service Development and technologies for SVC-improvements

During the development phase the basis is provided to improve the service offer all over the SVC for internal and external customers. In this context, not only the existing service offers are being tested for improvements, but also the possibilities of new services development are included in the reflections. Particularly the advancements in information and communication technology (ICT) are significant for the (further) development of service proposals. In our example case, the service provider has reacted on the needs of the service beneficiaries and technological improvements. The component manufacturer thus is about to establish an IT-based Hotline Supporting System to facilitate the information transfer and capture. Productivity gains are realized in (automated) information gathering and provision in the background. The hotline agent has fast access to the customer data and the internal information needed to serve best the request of the customer via a standardized interface.. For the caller the advantage relies on a fast solution of his service request, leading to less standstill and non-productive periods for the service customer. Automation might even provide room for personalization, freeing up the agent from information retrieval tasks. Nevertheless, such a technology-driven innovation needs to compete with complexity that increases by the active integration of the external factor, meaning that service skills are still all the more necessary. Technological innovations need to have acceptance, otherwise work, the advantages for the SVC are lost.

2.2 Service Management and Processes for the Visualizing of SVC Elements

According to Vargo and Lusch (2004), service can be interpreted in the form of an "application of specialized competences (skills and knowledge), through deeds, processes, and performances for the benefit of another entity or the entity itself (self-service)". Well-defined and thus well visualized processes can be appropriate to enhance the service productivity. By using a so-called process snapshot, the whole process might be well-arranged. Information about process-participating customers, the process activities and their interaction stages as well as the required information and the used technology is necessary (cp. Alter, 2008). Again the integration of the external factor causes complexity, a structured analysis and visualization of the concerned processes is necessary. The analyzed process has to be improved in the sense of efficient and transparent information- and knowledge flows, particularly when the service beneficiary is directly involved in the service provision process. Successful service process management depends on the transparency of responsibilities and of the scopes of action and instructions for all parties concerned. To realize the correct and clear visualization of such a complex and multi-entity service process, we propose the instrument of service blueprinting. Blueprinting was developed in the 1980s by Shostack and since then developed for service specific contexts by research scholars as Kingman-Brundage (cp. Fließ and Kleinaltenkamp, 2004; Shostack, 1982). A service blueprint creates transparency about the interrelations between the concerned activities, the process, the service interfaces, and the various responsibilities of service employees and the (co-creating) customers. In the following Figure 3 a part of the Hotline Service Blueprint of our project partner is illustrated.

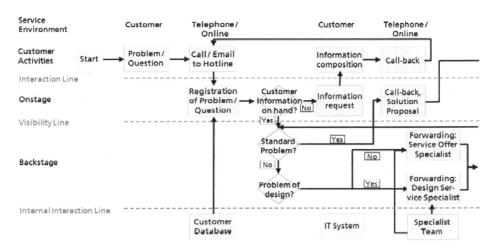

Figure 3 Snapshot from the hotline specific Service Blueprint. (Own source.)

The Interaction Line separates the customer co-creation-based activities from the Onstage activities that are visible for the customer but not in his sphere of action. Seen from a co-creation viewpoint, some potential for improvements in form of an active integration of the service customer can be identified in this sphere (cp. Schwengels, 2003). Backstage activities in contrast are invisible for the customer and therefore arranged behind the Visibility Line where service relevant operational activities take place (cp. Eversheim et al., 2003). The Internal Interactivity Line separates the service supporting activities such as an internal IT-support. Efficiency considerations of processes within the Service Blueprint might even encompass the consideration of individual workloads. As an example, the job of the hotline agent requires a high attention level. As this means a high workload, he needs some unproductive time periods to regenerate.

2.3 Service Delivery and specific skills for supporting Interaction in the SVC

As a co-produced service „requires attention to the actions and responsibilities of both the service provider and the customer" (Alter, 2008) service relevant skills are necessary between these elements of the SVC. The interaction process between the service provider and the customer is important. In our example the necessary skills of the hotline agent at the first tier co-creation contact point in the SVC are explained. Following Prahalad and Ramaswamy (2002), a successful co-creation of value in service needs to take into account the experience of the customer and his ability and willingness to cooperate with the service provider. Special skills are required form the service staff to fulfill the customers' needs and expectations as well as to gain the relevant information of the contact with the customer. This information again can mean a competitive advantage for the service providing enterprise. Generally speaking, attentiveness, perceptiveness and responsiveness are the skills necessary to extract customer information (cp. Salomonson 2011).

3.   MEASURING BENEFITS OF SERVICE PROVIDING USING SPECIFIC KPIS

Via co-creation value for the enterprise as well as for the service beneficiary is being created. The question about the measurement of product related services arises. For services the input-output ratio needs to be defined. The assessment mechanisms established in production environments are not globally applicable for the service sector. Since generally no physical products result from the service provision, only "soft factors" like the customer satisfaction as well as intervals for service production steps or partial results are assessable

While working with the real world industry partner on the management aspects of services, the concept shown in figure 4 below allowed us to jointly develop KPIs for assessing and thus controlling services proposed from an industrial centered perspective. A crucial fact for service enterprises is to establish efficient portfolios: Which service proposals are highly relevant to the customer? Which services should better be adapted to the customers' needs or eliminated from the portfolio? The advantage of a better adapted service portfolio might lead to the identification of valuable pro-actively proposed concepts for service providers and beneficiaries raising productivity by an even more efficient resource allocation.

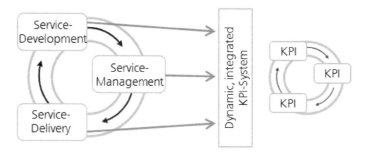

Figure 4 The aggregated service lifecycle productivity assessment needs to propose at the same time phase-specific KPIs. (Own source.)

In the logic of the three phased service management concept and again within the practical experience framework of a customer hotline we present possible KPIs for improving the measurement of service proposals.

The mostly resource-based KPI agent capacity utilization per day (as a fixed interval of time) provides important information of why and to what extent vacillations in customer demand (can) have an influence on the internal, organizational structure that is needed to handle this service customer demand at a time. It gives ideas for operating levers like the IT-based support that our project partner realized in order to improve the handling of these customer-driven external fluctuations. Another resource-based KPI is the calculation of support costs, based on the time needed for the completion of the entire service process from the customer demand until the delivery of the solution to the service beneficiary. It allows filtering the relevant from the non-profitable service proposals that are to be improved or to be eliminated from the portfolio.

The group of process oriented KPIs that have been developed in the course of "ServUp" includes a service ticket based and a time focused KPI. The service proposal-focused KPI is about the assessment of the mean duration, i.e. the mean passed time from the hotline ticket opening until the ticket in each case concerning the time needed for working on one specific ticket. The second process KPI implies the ratio between the number of closed hotline tickets to the number of incoming new tickets at a specific, fixed period. The "throughput time" of the temporally organized service demands makes the efficiency of the handling process transparent. These KPIs in combination with a process analysis give an approach for possible levers in the process and thus service management improvement.

The last group of KPIs focusses on an assessment of the interaction between service hotline staff and service beneficiary. The ratio between the operating expense per customer and the number of tickets per customer in a fixed time interval gives transparency about the productivity of the hotline communication. If for example the operating expense is very high in relation to the number of customer-correlated hotline tickets, the question arises whether the staff-customer interaction is inefficient or the problem might stem from a quality aspect of the corresponding product. Another KPI that allows a measurement of interaction processes is the First Pass Yield Ratio (FPYR). This KPI illustrates the number of hotline tickets which could be closed by the first level support. It thus provides an interpretation of the knowledge- and interaction skill-level of the corresponding hotline agent. If the FPYR-level shows a need of improvement, in turn possible management measures for the hotline staff might be trainings, technological support or a combination for improving the interaction within the service provider-customer-interface.

4   CONCLUSION

As seen in the specific example of the industrial manufacturer, some KPIs can be derived from already existing production contexts. Other service specific KPIs may have to be specifically designed. A transfer of the KPIs given in the specific example into other service business areas and their particular circumstances, such as maintenance or proactive service offers, needs to be further investigated. The way to implement this in practice will require some effort–both intellectually as well as organizational. A pitfall might be too quick adaptions of the main contents of generic KPIs with little variations into different service domains. On the other hand it is in question how the contribution of the customer and thus the data necessary for the efficient use of the KPIs in the concept presented in this paper can be achieved. A KPI centered approach inherently suggests a quantitative nature, yet customer interaction might have large areas that need to be assessed in a qualitative manner. The factors that show a considerable time lag before they fully manifest (i.e. measure something quantitative now and see qualitative results later) will be of particular interest. This opens another research field including the necessity of further research in the service interaction and productivity domain.

ACKNOWLEDGMENTS

The authors would like to acknowledge that the underlying project "ServUp" is kindly co-funded by the German Federal Ministry of Education and Research (BMBF) with the reference number 01FL10083.

REFERENCES

Alter, S. 2008. Service system fundamentals: Work system, value chain, and life cycle. *IBM SYSTEMS JOURNAL*, 47 (1): 71–85.

Bharadwaj, S.G., P.R. Varadarajan and J. Fahy. 1993. Sustainable Competitive Advantage in Service Industries: A Conceptual Model and Research Propositions. *Journal of Marketing*, 57 (4): 83–99.

Eversheim, W., J. Kuster and V. Liestmann. 2003. Anwendungspotenziale ingenieurwissenschaftlicher Methoden für das Service Engineering. In: Bullinger, H.-J., A.-W. Scheer (eds.). 2003. Service Engineering. Entwicklung und Gestaltung innovativer Dienstleistungen. Berlin: Springer, 2003: 417–442.

Fließ, S. and M. Kleinaltenkamp. 2004. Blueprinting the service company. Managing service processes efficiently. *Journal of Business Research*, 57: 392–404.

Gouthier, M. and S. Schmid. 2003. Customers and Customer Relationships in Service Firms: The Perspective of the Resource-Based View. *Marketing Theory* 2003 (3): 119–143.

Grönroos, C. 2008. Service logic revisited: Who creates value? And who co-creates? *European Business Review*, 20 (4): 298–314.

Howahl, F., P. Hottum, H. Fromm and G, Satzger. 2011. Fundamentals of a productivity model for service systems. Academic Conference "Understanding Complex Service Systems Through Different Lenses", Cambridge Service Alliance, September 2011.

M. E. Porter. 2008. The Five Competitive Forces That Shape Strategy. *Harvard Business Review* January 2008: 23–41.

Prahalad, C. K. and V. Ramaswamy. 2002. The Co-Creation Connection: Companies spent the 20th century managing efficiencies. They must spend the 21st century managing experiences. *strategy+business*, 27: 1–12.

Salomonson, N. et al. 2011. Communicative skills that support value creation: A study of B2B interactions between customers and customer service representatives. *Industrial Marketing Management*, 41 (2012): 145–155.

Schwengels, C. 2003. Kostenorientierte Entwicklung von Dienstleistungen. In: Bullinger, H.-J. and A.-W. Scheer, (eds.). 2003. Service Engineering. Entwicklung und Gestaltung innovativer Dienstleistungen. Berlin: Springer, 2003: 507–530.

Shostack, G. L. 1982. How to Design a Service. Designing a Service. *European Journal of Marketing*, 16 (1): 49 – 63.

Spath, D. and L. Demuß. 2003. Entwicklung hybrider Produkte. Gestaltung materieller und immaterieller Leistungsbündel. In: Bullinger, H.-J. and A.-W. Scheer, (eds.). 2003. Service Engineering. Entwicklung und Gestaltung innovativer Dienstleistungen. Berlin: Springer, 2003: 467–506.

Spohrer, J., S. Vargo, N. Caswell and P. Maglio. 2008. The Service System Is the Basic Abstraction of Service Science. In: Proceedings of the 41st Annual Hawaii International Conference on System Sciences, p. 104.

Vargo, S. L. and R. F. Lusch. 2004. The Four Service Marketing Myths: Remnants of a Goods-Based, Manufacturing Model. *Journal of Service Research*, 6 (4): 324–335.

Vargo, S. L., P. Maglio and M. A. Akaka. 2008) On value and value co-creation: A service systems and service logic perspective. *European Management Journal*, 25: 145–152.

Author Index

T - #0323 - 071024 - C492 - 234/156/22 - PB - 9780367381110 - Gloss Lamination